D1017492

the match

COMPLETE STRANGERS,
A MIRACLE FACE TRANSPLANT,
TWO LIVES TRANSFORMED

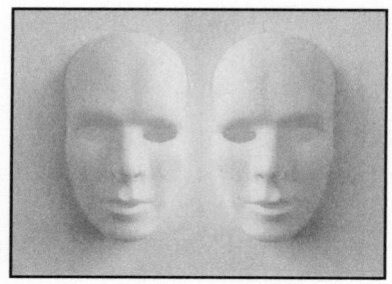

susan whitman helfgot
with william novak

Simon & Schuster
New York   London   Toronto   Sydney

Simon & Schuster
1230 Avenue of the Americas
New York, NY 10020

First Simon & Schuster hardcover edition October 2010

SIMON & SCHUSTER and colophon are registered trademarks of Simon & Schuster, Inc.

For information about special discounts for bulk purchases, please contact Simon & Schuster Special Sales at 1-866-506-1949 or business@simonandschuster.com.

The Simon & Schuster Speakers Bureau can bring authors to your live event. For more information or to book an event, contact the Simon & Schuster Speakers Bureau at 1-866-248-3049 or visit our website at www.simonspeakers.com.

Designed by Nancy Singer

Manufactured in the United States of America

10   9   8   7   6   5   4   3   2   1

Library of Congress Cataloging-in-Publication Data
Helfgot, Susan Whitman.
    The match : complete strangers, a miracle face transplant, two lives transformed / Susan Whitman Helfgot, with William Novak.
        p. cm.
    1. Maki, James—Health. 2. Helfgot, Joseph—Health. 3. Face—Transplantation—Biography. I. Novak, William. II. Title.
        RD523.M35 2010
        617.5'205920922—dc22

2010022476

ISBN 978-1-4391-9548-2
ISBN 978-1-4391-9551-2 (ebook)

for joseph

He had a face like a blessing.
—*Cervantes*

the match

chapter one

For a dying man, it is not a difficult decision because he knows he is at the end. If a lion chases you to the bank of a river filled with crocodiles, you will leap into the water, convinced you have a chance to swim to the other side.

—Dr. Christaan Barnard

Monday, April 6, 2009, late morning.
Brigham and Women's Hospital, Boston.

a calculated quiet has fallen over the normally frenetic and noisy cardiac intensive care unit. Nurses stand in groups of two or three, speaking quietly. Others attend to their patients' life-or-death concerns in slow and measured movements out of respect for what has just happened.

Death has come to the floor. Not just any death, although all deaths are tragic in the cardiac surgical ICU, but the death of a patient who has become a fixture here, and for many of the staff, a friend. Joseph Helfgot had battled heart disease for more than a decade, and for the past two years he was cared for by the people who now stand around in stunned silence.

He came so damn close, inching up on the list of patients waiting for a new heart. Only a third of them ever get there, and he was one of the lucky ones. Just two nights ago the New England Organ Bank finally matched the heart from a man who had just died. Word spread quickly. Helfgot was watching late-night TV while his wife slept. The phone woke her up. "Mrs. Helfgot, it's Dr. Lewis. I think we found a heart. Don't rush, take your time, but start putting things together."

"Joseph, we have a heart!"

Ignoring the doctor's advice, they rushed to the hospital, barely taking time to say goodbye to their two boys, who were camped out in the family room, half asleep, ready for bed.

"Bye, Dad. We'll see you after the operation."

At the hospital e-mails flew around the floor. Off-duty staff were copied: Joe's getting a heart! That was Saturday night. Two days later elation has turned to grief.

Earlier that morning.

Dr. Jim Rawn, the surgeon who has orchestrated Helfgot's day-to-day care during his frequent stays in the ICU, steps into the surgical unit. It is barely 6 a.m., but the place is jumping. Two new hearts came in over the weekend, Helfgot's and another one.

Dr. Rawn likes Helfgot, a market research executive who works in the movie business. He knows he has broken a cardinal rule: Don't get too close. But sometimes a heart patient pierces through the cloak of aloofness that intensive care physicians wear like armor. Hollywood Joe, as the nurses call him, is one of them. Rawn has learned the hard way that he shouldn't become too attached. Although the Brigham's cardiac unit is one of the finest in the country, not every transplant patient who comes in here will walk out the door. Better to check your emotions before you come to work, because it hurts too much when you get close.

But sometimes you can't resist, and Joseph Helfgot can shatter the toughest of façades. When he's not knocking on death's door, it's hard to believe he's a patient at all. In the ICU his bed is usually littered with half a dozen movie scripts. Piles of yellow legal pads filled with notes cover the top of his hospital tray, and a second tray on the other side of his bed holds his laptop and his BlackBerry. He has set up shop here. The nurses call it his bed-quarters, a term his employees have been using for years. Even before he got sick Helfgot loved to work from bed.

In the hospital he regales anyone who cares to listen, as well as a few who don't, with behind-the-scenes stories about the "real" Hollywood. It isn't uncommon to find a heart specialist with a gaggle of medical students crowding around Helfgot, who is propped up in bed with a movie booming loudly on his laptop. "So I'm watching *Public Enemy*. You know, Jimmy Cagney plays a gangster? They're doing a kind of remake, with Johnny Depp as John Dillinger. So which do you like better? The scene over here, where Cagney shoots the guy?" The students lean in. "Or this one, where he kisses the girl?" He fast-forwards the movie as they stare blankly at the screen. "Christ, how old are you guys anyway? Do you even know who James Cagney *is*? How about Jean Harlow? You know, the blonde? Crap, never mind."

Some of the nurses adore him—mostly those he hasn't driven half-crazy with his perpetual list of demands. With a few of them he has the kind of relationship they have with their hairdresser. He can also be exasperating.

"I'm sorry, Judy, but I need this ice-cold, please."

"Mr. Helfgot, we're talking about liquid potassium. It's medicine, not a cocktail. We don't serve it on the rocks."

"Just bring me some ice, please," he says, flashing a petulant smile.

Judy shakes her head. The first time she met him, he asked if she was married.

"No," she said, busily attending to an occluded IV. "Put your arm out and try to hold still, will you?"

"Why not?" he asked, inches from her face as she checked his line, looking for the chink in the tubing.

She stared back at him blankly, thinking, Boy, *that's* personal. But then she heard herself saying, "Good question. Why the hell am I not married?"

"You'll find somebody," he told her. "You're pretty."

Helfgot's wife was there, and he turned to her and said, "Susan, do we know anyone for her?" To the nurse's embarrassment, which conveniently masked how much she was enjoying this conversation, Joseph and Susan began ticking off names of the single men they knew, and why this one or that one would be suitable or not. A year and a few dates later, some with Helfgot's bachelor friends, Judy was still single.

"I'll get your ice, but you have to drink it all at once, and not over the course of the next two weeks. Your K is so low you're going to crash."

"Okay. I promise."

Dr. Rawn steps quietly into Helfgot's room. This morning, right after the transplant, there aren't any scripts on the bed. Just Helfgot. He hasn't woken up, but it was a long surgery.

"He's not waking up," the nurse says. The nurse is worried, but he's trying not to show it.

Rawn stands over the bed and does his own quick check. He lifts Helfgot's eyelids and examines his pupils, which are dilated. He takes the flashlight from the wall and shines it right into Helfgot's eyes. Nothing happens, no contraction at all. *Shit.* He calls for a CT scan and the nurse picks up the phone.

An hour later, after the scan, Rawn hovers around the computer screen at the physician's desk outside Helfgot's room, anxiously waiting for the results to upload. He nods at Susan, who has just arrived.

"Not awake yet?" she asks the nurse.

"You know how long it takes for Mr. Helfgot to wake up from surgery."

She does indeed. This is his fourth surgery in a year and a half. Although this is the big one, the one they have long hoped was coming, she is numb from the constant fear of his death. She hasn't had any real sleep in more than a year. This whole medical adventure has been a prolonged road trip through hell.

"Wake up, Joseph!" she shouts in his ear. Sometimes hearing a familiar voice does the trick. She lifts his eyelids and then shakes him a few times. An angry vitals monitor picks up the disturbance and sends out an alarm. Susan reconnects an EKG lead and instinctively pushes the reset button on the monitor high above her head. After a year of bringing her husband in and out of the ICU she knows her way around the machinery, although she also knows that she shouldn't be touching the equipment. She turns to the nurse and says, "You didn't see that."

Dr. Rawn, waiting at the computer, is watching through the glass wall of the room and thinking that Susan would make a great nurse. But why is it taking so damn long for these results? Finally a lateral view of Helfgot's head appears on the screen before him. He stares at it, not wanting to believe what he is seeing. A third of the right side of the brain is in darkened shadow, and some of the left side as well. God damn it! Massive ischemia. The brain architecture is gone. Clots must have traveled up during surgery, closing off the blood supply to Helfgot's head.

Rawn spots Greg Couper, Helfgot's heart surgeon, and motions for him to come over. "Greg, take a look."

Couper pulls off his glasses and peers at the screen. He tries not to show any expression as his stomach sinks down to his toes. During the surgery he had found a clot in Helfgot's aorta, which was not a good sign. He mentioned it to Susan when they spoke after the operation. He wasn't sure she grasped the significance of the information, but she had been with Helfgot's son, and it wasn't the time or the place to raise fears. He had found clots before; they don't always mean a bad ending, but often they do.

"Have you spoken to his wife?" he asks Rawn.

"No."

A few nurses are looking over in their direction with *We know something's up* expressions.

"The heart is doing well?" Couper asks.

"Banging away." With a respirator still bringing air into Helfgot's lungs, the heart can continue beating.

Stupid! Couper thinks. This whole thing is just too stupid. After all that work to keep him alive. After the artificial heart pump he'd implanted a year and a half ago to keep him going while he waited for a heart, which almost killed the poor guy from so many complications. But Joseph Helfgot was a fighter. He was determined to make it to his son's bar mitzvah, which he talked about all the time. Well, at least he got there. And now, a few months later, he dies getting the transplant? Damn it.

Couper is tired. He performed a second heart transplant only hours after scrubbing out of Helfgot's surgery, and he hasn't been to bed in over twenty-four hours. Transplants are a strange and unpredictable business. Sometimes it's quiet for days on end, and at other times it feels like sheer insanity, with hearts flying in like planes over LaGuardia. And you never know when one of your patients is going to die.

Like right now.

A former college wrestler, Couper had an instinct to hang on, to cling to something to keep his patient off the mat. He stares at the composite X-ray of Helfgot's brain, and the image makes him angry. He wants to pick up the monitor and hurl it through the nearest window.

"Okay," he says, "let's get the heart checked out. Maybe someone else can use it."

"We're doing one more scan just to be sure," Rawn tells him, but he knows the test is pointless. This movie is over. Couper goes off to check on the other transplant patient who has come up from surgery.

Dr. Rawn prepares himself to walk into the room and inform the widow, who doesn't yet know she's a widow. For a long time he was never sure whether to call her Mrs. Helfgot or Mrs. Whitman. Finally it just became Susan. Now it's Susan the widow.

"So, what's going on?" she asks.

He has always been candid with her. She has a knack for medicine and has been a tireless advocate for her husband.

"He should be awake by now. We did a scan. It doesn't look good."

She suddenly knows that this is the worst moment of her life.

"Dr. Couper found a clot in the aorta during surgery," she says. "He's stroked out, hasn't he?"

"We're doing another scan."

"Why?"

"We just are. There might be something . . ."

They stand there for a moment. "It's okay," she says. "I get it."

He nods, unable to speak.

"So that's that, isn't it?"

He nods again.

"I'm really sorry, Jim," she says as she reaches out and touches his sleeve.

I had thought again and again about how not to fall apart when it finally happened, because I knew it would. The odds were always against us. It soothed me to go off in my mind and practice his death. It made me feel safer, more prepared.

And now Jim Rawn is looking at me, his eyes telling me Joseph is dead. Now there is no inside of me—only outside. All my rehearsing has paid off. The inside will come back. Later.

Just like that, it's over. No more ideas to try, no meds to tweak, no specialists to consult, no waiting for a damn heart. Just like that.

Now there will be crying children, phone calls to make, a casket to buy, plane tickets and food to arrange—all the unrelenting busy-ness that is the sole blessing attached to death.

On another floor of the hospital Esther Charves, a family coordinator with the New England Organ Bank, has just finished a case. On the elevator she recognizes one of the chaplains, a southern woman who has just left the ICU and who knows Helfgot well. Among other topics the chaplain and Helfgot liked to argue about how to make good barbecue. "Joseph," she would admonish him in her thick drawl, "you shouldn't be eating that stuff at all. Too much salt." But most of the time Joseph ate what he wanted.

The chaplain tried to go easy on Helfgot, for although his dietary habits could be abominable, she admired the way he never complained about his bad luck—and he had a lot of it. In her line of work the standard question is "Why has God done this to me?" Helfgot never asked. Maybe he had an answer, or maybe he knew that it didn't help to wonder.

She knew patients who just lay there, like inmates on death row, cursing their bad luck. Helfgot wasn't one of them. Sometimes he would get up and stroll around the ICU, trying to cheer everybody up. One day he handed out sushi that Susan had smuggled in.

"Try this urchin. It's *amazing*."

"Urchin? Um, no thanks."

"No, really, you *have* to try this. It's unbelievable."

"I'd love to," one patient told him, "but the doctors have put me on an urchin-free diet."

"Esther," the chaplain says, "you may have a case up on six. Heart transplant gone bad."

"How old?" Young people dying are the worst.

"Late fifties, maybe sixty."

Esther gets off on the sixth floor and enters the unit, where a few nurses are crying. She has witnessed this before, but not often. Jim Rawn greets her and fills her in, adding, "You may want the heart too."

A retransplant of a newly transplanted heart? "I've never seen that," she says.

"I haven't either, and I've been here ten years. But we think it's viable, and Dr. Couper wants to take a look."

She glances into Helfgot's room. Close to a dozen people, mostly hospital staffers in scrubs and white coats, surround a small woman Esther can barely see. She must be the widow.

Esther is optimistic. A family that has just received an organ, even if it went badly, will likely reciprocate if they have the chance. She'll come back in a little while, when things are more settled.

Her cell phone rings. It's Chris Curran, the head coordinator at the organ bank. "Are you on the cardiac unit?" he asks her.

"I just got here."

"They say the heart's working fine."

"That's what Jim Rawn just told me."

"We may have a match in the Midwest," says Curran. "Their team is checking it out."

On Saturday, when the organ bank matched Helfgot with a heart, his name was moved from Waiting to Transplanted in the national database. Curran has never met Helfgot, but for years he has watched his name moving up and down the list. It's always satisfying when a patient finally makes it to Transplanted. Now they will have to move him over to Deceased. He looks at the information next to Helfgot's name: type O, age sixty.

"Esther? He could be the one we've been waiting for. We'll check it out on our end."

"Are you sure? The heart's a big enough deal. And I don't remember any of us ever asking for a retransplanted heart. I haven't even met Mrs. Helfgot yet. It's way too soon to go there."

"But do you think it's possible?"

"Chris, I just got here. I'll call you in a little while. I need to walk."

Esther drops the cell phone into her pocket. It's too soon, but her heart is beating faster anyway. They've been on the lookout for quite a while. She peers back into the room. The widow isn't there; she must have left during the phone call. A few doctors are at the patient's bedside. Esther recognizes one of them. He looks upset.

A few days ago the organ bank team thought they had their special donor, but it didn't pan out. This one could work. Helfgot's family might be open to it. Esther walks around the unit, past the room of the other man who was transplanted over the weekend. She pulls out her phone.

"Chris? I'll speak with her about the heart when she comes back—but only if you're sure there is someone who will take it. And if that works out, I'll talk with her about the other thing."

Esther wonders when Mrs. Helfgot will be back. Maybe she'll be the one, the one who will say "Yes, you can have my husband's face."

chapter two

Monday, April 6, 2009.
A veterans' home in central Massachusetts.

James Maki stares at his computer screen. A few doors away an old church at the foot of the hill peals off the familiar Westminster chimes. As the hour strikes, Maki stops to count. Eleven o'clock. The toolbar in front of him gives the time, but it's too tiny for him to see. His right eyelid is stitched shut. The nerve that opens and closes it was destroyed four years ago in the accident.

One of his housemates is cooking something downstairs. It smells like grilled cheese, although it's a little early for lunch. He'll go down in an hour or so. On the screen three cards show up: a ten, an eight, and a three. He's holding a queen and a seven. He folds.

This hasn't been one of his better days, but he keeps playing. Texas Hold'em eats up a lot of time, and time is all he has. The screen clears and he pulls an unsuited deuce and seven—the worst pair you can draw.

A car will be coming at five to take the residents to Applebee's for dinner, but Maki won't be joining them. Maybe he can talk the woman who runs the house into bringing something back for him, like one of those sundae shooters in a cup. She knows about his sweet tooth.

He'd love to go with them, but it's not a good idea. People don't want to look at a big crater where a person's face ought to be, especially when they're out to dinner. And because the top of his mouth is missing, half of what he tries to eat ends up on his lap. His housemates, who are also middle-aged Vietnam vets, have their own problems. Some are missing arms or legs, or are using wheelchairs, or are on oxygen, and they draw enough stares without him tagging along. Jim Maki draws more than stares. People have been known to scream when they see him. No thank you. Maybe he'll have macaroni and cheese tonight. A pair of jacks? Now that's more like it.

He keeps clicking away. Up three hands, down the next two. He plays so much that he barely pays attention and shifts over to the game show on his flat-screen TV. He's got a pretty nice setup, all things considered. True, the room is small and the floor is linoleum. Wall-to-wall carpeting would be nice. But the TV and the computer are top of the line. His brother, John, who lives in Seattle, bought the computer when Jim arrived here a year ago from the rehab place in New Bedford.

John hasn't been back to visit for a while. He hates to fly because he gets claustrophobic on planes. But he came several times when Jim was in rehab to make sure his younger brother was well cared for. Because Jim is a sharp dresser, John bought him some beautiful clothes. Back in the day Jim used to shop at Louis, Boston's premium men's clothier.

Jim's wife, Cynthia, from whom he is separated, advised John to return the clothes to wherever he bought them because they soon would be stolen. She was right. Most of them disappeared within a few days.

He looks out the window, past the cheap white polyester curtain that hangs down the middle in a loose knot. It's a miserable day, but even with the rain it's good to look outside. Someone said this morning that Opening Day might be rained out. Downstairs the phone is ringing, and a few seconds later he hears someone's name called out from the foot of the stairs. So the call isn't for him. But someday it could be. Someday Dr. Pomahac could be calling him.

He can't remember much about the day he ended up at the Ruggles subway station in Boston, where he fell onto the tracks and lost his face.

He was living in a halfway house in Malden and sticking to his methadone treatment. But he was still using drugs whenever he could, trying not to get caught and be thrown out. He was strung out that night and somehow fell onto the tracks. He lost the use of his right hand, which dangles from his arm. He can live with that, but his face hit the third rail dead on. Thank God he can't remember any of it.

Dr. Pomahac operated on him ten times, until there was nothing more he could do. Then Jim saw him on TV, talking about face transplants. He sat transfixed as Pomahac explained how this procedure, which was still mostly theoretical, might eventually help horribly disfigured patients.

"I saw you on TV," he said at his next appointment. "Do you think I could have a face transplant?"

"We're working on it, Mr. Maki." The young Czech surgeon knew that if anybody needed such an extreme intervention, it was Maki. "But we're not quite there yet."

Then one day, seemingly out of nowhere, Dr. Pomahac said, "If we could give you a new face, which carries some grave risks, including the possibility of getting cancer from the rejection drugs you would have to take for the rest of your life, would you still want to do it?"

Maki didn't hesitate. "Absolutely."

He'd had so many tests, and with each one he worried that they'd find something wrong, that something would change Dr. Pomahac's mind and rule him out. So far that hasn't happened. He's still hanging on. The doctor has told him he will get a face as soon as they find a donor. But who knows when that will happen. Or if it ever will.

It's hard not to think about it every time the phone rings. Jim knows this could take a long time, but the waiting has been hard. He feels awful knowing that someone has to die for him to get his face. At the hospital he has talked about it with the psychiatrist. Dr. Pomahac says that he won't look anything like the donor, which is a relief. But the whole thing is still pretty weird.

In the distance he can hear what sounds like one of his housemates talking on the phone to one of his children. He instinctively looks over at a picture of Jessica. Beautiful Jessie, with her huge almond eyes and exotic cheeks, a perfect blend of him and Cindy. She stares back at him with a big smile under a straw hat. Eighteen, with creamy white skin and not a wrinkle on her face, she is a perfect flower. Soon she hopes to go to Korea to teach English. Her grandmother would be so proud.

His thoughts turn to his mother, and he wishes she were here to comfort him. Mary Maki was a sweet and gentle woman who never turned her back on him, even after all the hurt he put her through. Lately he's been thinking about her a lot. He wasn't sober when she died more than ten years ago, and it is only recently, now that his mind is finally clear, that he feels the crushing grief.

Fifty-two years earlier. Winter 1957. Seattle.
Mary Maki looks out her kitchen window at the swirling clumps of sticky white snowflakes. They have been falling steadily for over an hour, and she can barely see across her yard. Seattle receives its share of snow each winter, but the mild sea breezes blowing in from Puget Sound usually melt it pretty fast. This morning looks like one of those times that the snow will stick. Maybe they will have to cancel school.

She runs her spatula across the bottom of a skillet, gently prodding the eggs into a fluffy mass. Her husband left for the university at the crack of dawn, off to another early breakfast meeting of some committee or another. It's a wonder he has any time to teach. If this snow keeps up, he may come home early for a change.

"Boys," she calls out, "your breakfast is ready."

John Jr. and Jim, who are nine and seven, take their seats at the table for toast and jam, eggs, and fruit. There are no boxes of Cheerios or, heaven forbid, that Kellogg's Special K everyone is talking about. She pours milk into plastic cups and sits down next to her children.

"Mom, you think there's gonna be school today?"

"I don't know, Jim. Eat your breakfast."

The phone rings. It's the wife of one of her husband's colleagues, who tells Mary that school is closed.

"Yay!" the boys shout.

She smiles at her two adopted sons. "Sit down, boys. Finish your breakfast."

"Mom, can we go sledding?" Jim asks as he shovels eggs into his mouth.

She considers Jim's face as he swallows the eggs with a large gulp of milk, his high, wide cheekbones peeking from the rim of the cup. Where did he get those cheeks? The older he gets, the more certain she is that the half of him that isn't Japanese is something other than white American. There's no doubt about the half-Japanese part; his smoky black eyes under almond eyelids attest to that. But for the past year or two she has wondered about the other half of his ancestry. Maybe he's part Native American?

"I think you should stay in the yard today."

"*Please?*" he begs.

His brother joins in the appeal: "*Please, Mom?*"

They are the only Japanese family in the neighborhood. The people on their street had a meeting when they learned that John and Mary Maki had put an offer on the house, which is slightly grand by the modest standards of this blue-collar area. A few residents tried to think up ways to keep the Japs from moving in. Pearl Harbor wasn't that long ago, but this is about more than the war. John and Mary find it hard to imagine that German Americans would be treated this way.

Professor John Maki, who is known as Jack, is a noted Asian scholar at the University of Washington who worked for the U.S. government during the war. He may be Japanese, but he is educated and well regarded by Seattle's academic community. On 125th Street, just a block from the lake, the bigotry is mostly tucked away behind closed doors. But it's definitely there.

It was easier when the children were little. Mary kept them in the yard and invited children from families they knew to come over and play. But now the boys are older, and a wider world is out there, waiting for them. She watched sadly one day as they stood at the edge of the yard while a group of boys ran down the street in cowboy hats and holsters with toy guns. Jack was adamant that their boys should stay in their yard.

Just after they were married Jack and Mary had experienced a far more shocking form of prejudice. Soon after Pearl Harbor, like other Japanese Americans on the West Coast, they were forced into an internment camp. The Makis were fortunate: their stay ended after a month, when Jack was summoned to Washington, D.C. One day you're a security risk; the next day you're working for a government intelligence agency. Most of their relatives paid a higher price for their ancestry.

"I'd rather you boys played in the yard," Mary tells them. "Maybe you and your friends can build a snowman."

"Mom, that's *boring*. We want to go sledding. We've got the best hill in the whole neighborhood." Jim is already out of his seat and wiping his milky mustache with his sleeve. Before she can object, he grabs his boots and coat and dashes outside, sending a flurry of white powder into the room. It was true: kids came from several blocks away to sled down their street.

"John, don't let him out of your sight. And don't go past the corner. Do you understand me?"

"Yes, Mom. You're the best mother there ever was!"

As Mary scrapes the half-eaten breakfast plates into the sink, she is already planning lunch. Maybe Jack will come home early. A nourishing soup? She opens the door of the fridge and starts pulling out vegetables. While other housewives may wonder whether to use light cream or half-and-half when they prepare Campbell's condensed tomato soup, Mary makes everything from scratch. There are no frozen TV dinners in Mary Maki's shopping cart.

Her father, Ichi, was Isei, a Japanese-born immigrant. A Christian farmer, he stuck to traditional ways, even insisting that Japanese be spoken in his American farmhouse. He did well and eventually installed a train track on the farm to ship potatoes and rhubarb to the railroad station, where his produce was added to the freight traveling east. He and his wife, Rin, raised five children on the farm.

Often, when Mary bends down to smell a flower or pull a weed, she is reminded of her childhood. In the spring she and her siblings would run barefoot through the fields. The watery sun would warm the soil, and early life would erupt from the ground. As the shoots became stronger the kids would rake off the protective straw so the little plants could drink in the sun. She knows that John and Jim are not that different from those tiny slips. They too need room to grow.

She met her husband during college. Their decision to marry was hastened when he was invited to Japan to study. Jack had been raised by his adoptive parents, the McGilvreys, a Scottish family. They took him in as a baby after answering an ad placed by his natural mother, who couldn't afford to raise him herself. Mary never met her adoptive father-in-law, who died while Jack was still in high school. Shortly before the wedding her father came to the young man with a suggestion. "McGilvrey is not a Japanese name.

Maybe you should change it to something more appropriate now that you are getting married." He didn't have to add *to my daughter.*

Jack wanted to please his new father-in-law. The *Mc* in Mc-Gilvrey sounded a bit like Maki, a common Japanese surname pronounced Mah-kee. "What do you think, Mary?"

And so they became Mary and John Maki. Officially he was John M. Maki. His middle name was now McGilvrey, a tribute to his adoptive parents.

Mary opens the front door and blinks through the blinding snow. Children in bright winter coats fly down the steep sidewalk. The white scene looks like an Impressionist painting. Shouts and laughter ring through the moist, cold air. The boys are fine. She goes back in the kitchen to put up her soup.

Winter 1957. New York City.
Night has fallen on the East Coast. Eight-year-old Joseph Helfgot is eating dinner at a small table in the back room of a tiny corner grocery store on Manhattan's Lower East Side. The room is unheated and freezing cold so nothing can spoil the merchandise, says his mother. In the summer she stops selling cold cuts because it's too expensive to keep the refrigerator going.

The boy pushes his fork around a plate of matzoh brei, a horrible dish his mother makes at least every other day by shredding stale Streit's matzoh into a heavily oiled pan on a small electric burner next to the table.

The only time Rachel Helfgot cooks at their apartment is on Saturday morning, when the store is closed, and she hopes God will forgive her for lighting the stove on Shabbos. On Friday nights Joseph's father hurries off to shul down Avenue B for the evening prayers, taking the boy with him. Rachel stays at the store with her oldest child, Pauline, where they light the Sabbath candles and say the Hebrew blessings in the back room.

There's a Catholic church around the corner, and this once-Jewish neighborhood is now home to Italians, Poles, and a growing number of Puerto Ricans. The store is busy on Friday nights because workers get paid on Fridays, and although it's the Sabbath Rachel can't afford to close. Friday night is when the goyim buy, and her family needs the money.

Joseph picks through the matzoh brei, looking for a piece that isn't burned. The leftover matzoh spits and burns in the pan while Rachel runs out to the front of the store to wait on a customer. Although her husband is there, Rachel always adds up the order on the back of a paper bag, jotting down the numbers with a tiny pencil and adding them up at lightning speed. She can't read, but she learned how to add. She says Naftali, her husband, always makes mistakes. He lets customers buy on credit, but Rachel never does. Invariably she smells the matzoh brei burning, and she runs back to the pan, whips up an egg, and tosses it into the mess.

"Yosel, *zine* dinner. Eat a *bissel* more."

His friend Iggy's mother makes really *good* matzoh brei, sweet and light with cinnamon and sugar mixed in. Iggy's mother has her hairpiece done every other week. Rachel has hers done just once a month and lacquers her blond wig with large quantities of hairspray to hold her over. Iggy's family has their own bathroom, but the Helfgots have to share theirs with another tenant. It's cheaper.

At the table Joseph traces words into Hebrew letters in his blue notebook. He wants to finish his homework tonight so his father will let him play stickball tomorrow. After school today, before it got dark, he and his friends cleaned up a spot at the bottom of the park and tomorrow they'll play. Joseph writes the letters, but his fingers are cold, even with gloves on. *Alef, bet, gimmel* . . .

There's a litter box under the table and the room reeks of cat urine. It's nine o'clock and he longs to go back to the apartment, but his mother has decided it is safer for him at the store. She has set up a cot in the room that Joseph used to share with his sister. Now Joseph shares the room with a man who is thin and smells bad. Pauline's bed is in the parlor. The thin man smokes and snores, and his breath smells funny, like Naftali's when he drinks wine. Joseph is afraid of him, but the boarder pays Rachel five dollars a week to sleep there. After school Pauline stays upstairs at a neighbor's until her parents come home from the store. Rachel doesn't want her kids to be alone in the apartment with a strange man.

At night, while the thin man snores, Joseph tries to dream up ways he can make money so his mother won't need to have a stranger in the house. But he is too young to work.

His father pushes the curtain aside and enters the cold room. "You study?"

Joseph holds up his notebook. "See? I'm almost finished."

"*Gut zin.*" Good son. "I go to du bank now." He jingles a canvas pouch holding the night receipts. He leaves from the back door, going through the alley rather than calling attention to his departure out on the street. He heads for the night depository up on Avenue C.

The moment he's gone Joseph jumps up and races through the curtain. Enough studying. He has been sitting for three hours. The front of the store is deliciously warm and smells like ground coffee. Rachel is putting things away for the night. Less than five feet tall, she stands on a stepladder to make room on the shelf. Joseph starts to munch on leftover pieces of broken cookies that she always puts out on a plate for the customers.

"Yosel, give me dus cans." He hands her tuna fish tins from a display next to the cash register. Rachel reaches high up into a cupboard, the number tattooed on her right forearm at Auschwitz visible as she shoves the cans to the back of the shelf.

Two young men have come in. They begin to collect items from around the store, expensive things like toothpaste and soap. They pile them on the counter and then go back for more. It's a large order, more than thirty dollars, Joseph is thinking. He is happy for his mother. Nobody ever buys this much.

Rachel has been watching from her ladder and thinking the same thing. No one in this neighborhood has that kind of money. Three bottles of shampoo? "Another can, Yosel." He knows from her voice that something isn't right.

"Done!" she says too brightly, coming down off the ladder. She begins to add up the order, writing on the back of a paper bag, placing each item neatly inside another bag as she goes along.

"Yosel, put dus into the other bag," she says, reaching down under the counter. His eyes widen as she slips a butcher knife under her apron, gripping the handle tightly through the cloth with her right hand.

One of the men is perusing the coffee selection, picking up cans and setting them down again while his friend goes outside. Through the window Rachel sees the man look up and down the street while he lights a cigarette. His friend continues to study the back of a coffee can. The man outside looks up and down the street as he takes a drag. She moves quickly from behind the counter to the door, where she snaps the bolt shut, locking it.

"Put dus can back on the shelf or I kill you," she says in a low voice. She draws the butcher knife from under her apron and takes a step toward the customer.

Joseph's hands are sweating as he clings to the edge of the counter. His mouth is dry. He can only partially see what is happening, just his mother standing there with the knife. Shelves block his view of the man.

"Lady, let me out!" Joseph hears him say as something heavy hits the floor. Joseph watches a coffee can roll into the center of the aisle.

Rachel unlocks the door and pushes it open. "I see you again, I kill you," she says as he slips past her.

The two men run off down the street.

"Yosel, pick up dus can."

chapter three

Monday, April 6, 2009, afternoon. Intensive Care Unit,
Brigham and Women's Hospital.

a n ICU nurse calls his off-duty colleague. He wants to be the one to tell her. In September they had gone together to Jacob Helfgot's bar mitzvah. Had there been enough room in the temple, Joseph would have invited the entire hospital.

"Lisa, it's Kevin." He hears a car honk. "Are you driving?"

"I'm on the expressway."

"I have to tell you something, some bad news. Don't wreck the car."

"What?"

"Joe won't wake up."

She is silent while she tries to absorb this.

"Who's on?" she finally asks.

"Jim's in there, talking to Susan. The organ bank is here too."

"What happened?"

"He threw a clot. Maybe a few of them."

"I'm on tomorrow."

"You don't have to take him, Lisa."

"Yes, I do."

Why is this room so dark? They've spent millions on floor-to-ceiling glass to keep us from feeling claustrophobic and they've got the damned blinds closed. I distinctly remember opening them. Then somebody dragged me downstairs for coffee. I didn't want coffee. I pull on the beaded chains and light streams into the room, spilling over Joseph's bed. The nurse looks up from his chart.

"Do the blinds have to be closed?"

"No, Mrs. Helfgot. It's just respect for your husband."

If it's dark right now, I will die. And I have never been able to get this nurse to call me Susan. He's always so shy, and today especially.

More people are coming into the room, some of the cardiac doctors.

One says, "I'm so sorry, Susan."

"I know. It's okay."

"Your husband was a wonderful man"—*wait, I've already heard it, please don't say it again*—"funny, brilliant, noncompliant, intense, original, outrageous"—*or another phrase*—"a boost to morale, one of a kind, in love with you, passionate about his kids"—*or something. I'm not listening.*

It's not that these words are inaccurate. It's that nobody has spoken the deeper, more important truth: "Your husband was so alive. And now he is so dead."

Jonathan, my stepson, is sitting on the long couch along the far wall. He took the red-eye from Los Angeles on Saturday night to be with his dad during the transplant. He slept late this morning because of the time change and arrived at the hospital expecting to find his father awake. I didn't call him. I waited until he walked into the room to tell him. I wonder if he's angry that I didn't call and make him race over here. But for what?

Now we're waiting for Ben to arrive from school. He'll graduate in a few weeks and start New York University in the fall. Joseph was over the moon about that. When I walked into the kitchen last December,

Joseph was sitting at the table with his metal heart pump clacking away a mile a minute—whoosh, click, whoosh, click. It could speed up or slow down just like a real heart. From the sound alone I knew he was excited about something.

"I got in! I got in!" Ben was jumping up and down like a little boy.

Joseph was crying, but then, Joseph cried at everything—including, I swear, a Charmin toilet paper commercial.

"How's Dad doing?" Ben has called his mother from school.

"Come over to the hospital when you get out of class, okay? Jon's here. There's a lot going on."

"Okay. I'll be there around three." Ben isn't worried. There was always a lot going on with his father.

"Love you, sweetie. Be careful."

As Jon and I wait for Ben to get here, I shake hands and hug people and thank them for stopping by. Dr. Lynne Stevenson, Joseph's cardiologist, walks in, regal and aloof. Her eyes are swollen and red. Joseph was angry that she didn't come to Jacob's bar mitzvah, but he got over it.

"You really won't be there?"

"No, Joseph. I don't mix with patients outside the hospital. I just don't."

"You really mean that? After all this, everything that's happened, you still won't come?" She has saved his life more than once.

"No, I'm sorry. I can't."

Lately, though, they've grown a little closer. She is holding something in her hands, and I recognize the videotape of Joseph's mother describing her life in Auschwitz, although I'm not sure "life" is the right word. Joseph lent it to her when he learned that her daughter was writing a paper for school on the Holocaust. She tells me they watched it together. She hands it back to me, tying up loose ends because she knows I won't be coming back. Experience has taught her to see through this moment and out the other side. It's true, I guess. I won't be coming back. I never imagined that I'd miss this place, but as long as I had a reason to be here, it meant Joseph was alive.

She hands me a card and says, "Don't open it now."

The room is crowded and getting too loud. Jon is on the phone, trying to reach his older sister, but Emily is in Paris on vacation, where her cell phone doesn't seem to be working. Now he's talking to Joan, his mother, informing her that her ex-husband is dead. I wish I had thought to call her so she could brace herself to comfort her children. I'll call later.

Joseph is on the bed, his chest falling and rising, down and up, the respirator pushing air in and out until Emily gets here and we work everything out. Maybe we can donate his kidneys. The doctors seem to think so. They told me someone would be here to talk about it, probably tomorrow.

Suddenly Ben is standing in front of me. He sees all the people in the room, and he knows.

"Are they sure, Mom?"

"I'm so sorry, Starshine."

He tries to be brave, but his chin and chest are quivering. Of all the Helfgot children, Ben is the one with the coolest demeanor, at least externally. It must be the one-quarter British phlegm from my side of the family.

The doctors have been waiting for Ben. I'm not sure why, but they are leading us into a conference room.

Susan and the two boys pass within a foot of Esther Charves as they are led down a long hallway. She tries not to meet the family until the doctors have explained that there is nothing left to be done for their loved one, still pink and warm to the touch, but dead nonetheless. Sometimes families cannot grasp it. They ask for one more test, one more procedure, anything that might alter or at least defer this terrible truth. Esther will sit down with Mrs. Helfgot only after the meeting where any lingering questions the family has will be answered. Meanwhile she will continue to gather information, trying to learn everything she can about the potential donor. Sometimes that's impossible. Sometimes a driver takes a slippery curve on a dangerous road and dies in a strange hospital, far from home.

This case is very different. She knew it would be the minute she arrived and saw the expressions on the faces of the doctors and nurses. Esther has learned the basics—Helfgot's line of work, the names of his children, and so on—but never has she put together such an intimate portrait of a potential donor in just a few hours. She has been collecting comments all day about a man who must have been easy to know, who had insisted on being known.

"Mr. Helfgot drank only fresh-squeezed juice. His wife brought it every day. He loved fresh juice of all kinds, and she wanted to make him happy."

"He was an intellectual who also loved Judge Judy and Costco. He called it applied sociology."

"One day he had Chinese food delivered to the whole floor. Sometimes pizza on Friday, or big baskets of candy or cookies."

One of the nurses told Esther about the day Helfgot noticed that she was feeling down and asked her if something was wrong. She confided that her fiancé had broken off their engagement. "What, is he gay? He must be gay. In your case there's no other explanation. Don't worry. There's someone else around the corner."

Esther's colleague Meredith Pitzi has been plowing through Helfgot's medical records. She wonders whether the years of heart-failure medication may have compromised his kidneys and liver, making them unsuitable for donation. This will be one of those times when they won't know for sure until they inspect the organs at the time of recovery. The organ bank has found a potential recipient in another region who is in desperate need of a heart and who is a match. Couper is on the phone with his doctors now. When the call is over there's a quick huddle. "They want the heart," he says. "They want to know when."

Good question. Meredith and Esther know it's too early to talk about *when*. Wheels are already in motion in case this is the face they've been waiting for, but they don't want to mention it yet, not even to the doctors. They want to be certain.

"We'll know more in a few hours," Esther tells Dr. Couper. "You're sure about the heart? I don't want to say anything to his wife until you're sure."

Families dealing with sudden death often find solace in organ donation, knowing that part of their loved one will live on. Esther doesn't want to raise this hope and then have it shattered. This family has been through enough.

Because Helfgot is a registered organ donor, Esther doesn't need his wife's permission for his kidneys or liver, or even the new heart. But a face is something else entirely. She will ask Mrs. Helfgot only when she is pretty sure the answer will be yes. Slow down, Esther, she tells herself. One step at a time. Even Dr. Pomahac, the surgeon waiting to perform the nation's second face transplant, doesn't yet know that they may have a face.

All day long she has watched Mrs. Helfgot through the glass, greeting visitors. They come in, approach the bed, and touch Joseph's hand or his shoulder. Or they grasp the bedrail and gaze at him in silence. Some of them have broken down, but not Mrs. Helfgot. She seems strong. Given what may soon be asked of her, she had better be.

Why are we in a crowded conference room? I want to go back and sit with Joseph. Who is that bald man in the blue shirt with his sleeves pushed up? He looks about my age, but in this pressured place, I wonder. A week in the tropics might take ten years off his face. He's a neurosurgeon, and he's going on about the swelling of the brain and how they could go in there and relieve the pressure, but it may not . . .

"Excuse me," I butt in. "Go into his head? Why would you do that?"

He clears his throat. "Well, as I said, to try to relieve the pressure."

"But he's dead," I hear myself say. Ben starts to cry.

"We just want you to feel comfortable. If you would like us to try something . . ."

The room becomes unbearably silent. No one wants to be the first to make a noise, even to scrape back a chair and stand up. That would mean moving on, and no one here is up to it. There has always been something else to try, but that's not true anymore.

When the room clears out I tell Jonathan and Ben that we have to go home and tell Jacob, the youngest Helfgot. Ben will ride with Jon. They need to be together.

"If you guys get home before me, don't say anything until I get there."

I work my way through the hospital complex to reach my car. Each building is an achievement, a dramatic statement underscoring the power of the institutions and contributors that have created this huge hospital: the Shapiro Center, the Tower, the old Women's Lying In, the Amory, the bridges connecting Children's Hospital, Harvard Medical School, and Dana-Farber Cancer Center. There's a billion dollars of real estate here and the best technology money can buy. But none of it could save my husband.

The walk soothes me. Finally I am alone and anonymous. All day people have been staring at me. How is she holding up? I know they mean well. I'm on the Pike now, the hospital's long pedestrian spine that connects all the buildings. Connors Health Center for Women? Take Exit 4. Amory Building? You can go up the elevators this way, ma'am, or cut through and go that way.

Right now I'm going that way, toward home, to tell Jacob. I don't know what I'll say, exactly, but it will probably be sealed in his memory for the rest of his life.

"Mrs. Helfgot," someone calls as I clear Exit 7. It's one of the thoracic surgeons. "I heard Mr. Helfgot got his heart!"

I guess the news hasn't made it to Exit 7. I paste on a smile and walk swiftly past him.

At the end of the Pike I enter the original Peter Bent Brigham Hospital. It smells of old wood, its stately Roman columns visible through the revolving door. Under a dimly lit alcove sits a case holding the 1990 Nobel Prize in Medicine, which was awarded to Dr. Joseph Murray, who performed the world's first successful organ transplant right here, at this hospital, in 1954. It was a kidney, and he transplanted it not in an operating theater, but in a quiet room away from probing eyes that would surely widen in horror at the idea of taking a warm organ out of one body and putting it into another. But what is unthinkable in one decade can be commonplace in another.

I have come here often when it was all just too much. There are pictures and display cases holding old instruments, and it brings me peace to be reminded that Joseph and I are part of a much bigger picture, that the struggle to keep him alive, so enormous to us, is a soft breeze blowing a tiny ripple across a vast ocean. I glance quickly at the alcove as I head out the door.

Luis, the valet, hands me my keys. He has been so good to me, standing in snow and rain and blistering sun through the seasons of my years at the Brigham.

"Buenas noches, señora. ¿Señora?"

"¿Si?"

"¿Estás bien?"

"No, no estoy bien. Perdona."

To this day I don't remember driving home, or what I said to Jacob when I got there. He sat on Jon's big lap and cried while his brother held him for a long time until both their tears were spent. Jon told me I was brilliant, and for this I will always love him. He now admits that he can't recall a single word I said, just that it sounded right.

chapter four

Tuesday, April 7, 2009, morning. Intensive Care Unit.

Lisa Kelley, Joseph Helfgot's intensive care nurse, stands over her patient, wiping down his arms with a cool moist towel before moving on to his legs. She's keeping him clean as she helps him through this one last ordeal. Last year on a Tuesday morning, they would have been playing cards. They played a lot of Crazy Eights after a bad reaction to heparin required the amputation of Helfgot's toes; card games helped him exercise his mind after an unusually long postoperative delirium. It took him a week to tell a diamond from a heart, and another week to remember that eights were wild. After that Lisa stopped letting him win.

She was here two days ago, on Sunday morning, as the Helfgots waited for the transplant operation to start. Like many of the doctors and nurses who knew the couple, she had popped in to help them pass the time during the interminable delay. It wasn't a party exactly, but the mood was anticipatory, relaxed, and happy. Given what Joseph Helfgot had endured, and how strong he looked, no one imagined he might not make it through the transplant.

Was it really only two days ago?

"Lisa, lovely Lisa," Helfgot sang. "What brings you here?"

"I work here, remember? So, are you nervous?"

"I am, but excited too. I feel like I'm going down a tunnel. I'll be really happy when it's all over."

Her eyes well up as she recalls their final conversation.

"Lisa? Do you know when Mrs. Helfgot is planning to get here?"

Esther Charves has spent the night in the hospital. She had hoped to grab a few hours of sleep on a couch, but she and Meredith were up all night, drinking coffee and consulting with doctors on the organ bank's list about the potential suitability of Helfgot's organs. Organ bank policy is to wait up to an hour for each doctor to return the call before they move on to the next name. It makes for slow going.

"If you're ready to talk, I'll call her."

"I'm ready." Esther will start by asking about the heart. If that goes well, she will ask Mrs. Helfgot how she feels about donating her husband's skin, which can be used in reconstructive surgery. And if *that* goes well, she will move on to the big question—but not right away. Her immediate priority is meeting Susan Helfgot and supporting her, no matter what she decides. For now, Esther will put the face transplant out of her mind.

Dr. Greg Couper spots Susan entering the ICU. They haven't spoken since right after the heart transplant on Sunday night, when he plopped down in a vinyl easy chair in the visitors' lounge, his hands dangling tired and free over the armrests, and told her the surgery had gone well, but there had been a clot in the aorta. Her stepson, Jon, was with her, and Susan hadn't asked about the clot. Then Couper went off to start a second heart transplant. She couldn't imagine how he was able to do two of these in succession, but there was no choice. Life and death don't happen on a schedule.

Couper had first met the Helfgots eighteen months earlier, when he installed a VAD, a ventricular assist device, in his patient's chest. Joseph came into the hospital extremely ill and was put at the top of the region's heart transplant list. But when a heart never came, and Helfgot's own heart finally gave out, Couper implanted the VAD to buy them some additional time. It turned out to be all the time they had.

For the first few months after VAD surgery Joseph dropped off the heart waiting list while he regained his strength. When he was healthy enough for a transplant operation, he went back on the list.

A simple machine, really, considering its value, a VAD diverts blood out of the heart and into a metal container about the size of a canteen. When the container is full, a metal plate inside is pushed down by a motor, forcing blood through a tube and directly into the aorta, the main blood highway. From there the patient's blood branches out to the rest of the body.

VADs are lifesavers, but they usually don't last more than eighteen to twenty-four months. And they come with major restrictions. Patients must stay away from electrical outlets, and even blankets or clothing that could cause static electricity and short out the wiring. Forget about taking a shower or a bath.

If the machine malfunctions, an alarm screams out and often the patient must be hand-pumped manually, usually by a family member with a cool head, until they can get to the hospital. Helfgot's VAD had malfunctioned twice. Susan will never forget her frantic middle-of-the-night phone call to Dr. Couper when the alarm sounded. The diagnostic testing she tried showed nothing was wrong, but something clearly was. Couper talked her through it, and the problem turned out to be a faulty electrical cable that needed repair. The boys had awakened from the alarm, and rather than scaring them further with yet another sudden dash to the hospital, they switched to the external battery pack and waited for morning to go into the hospital to have it fixed.

For the next three nights Susan couldn't sleep. Even when everything is working perfectly, VAD life is a special kind of hell. But it's better than dying.

This morning Susan greets Dr. Couper and waits for him to say he is sorry. Everyone has been repeating this morbid little mantra, and she has come to expect it.

Instead he rubs his white goatee and says, "I gave my action figure to my little nephew."

Susan has forgotten all about that. Around the time Couper installed the VAD, Joseph was working on the marketing of *Iron Man*, starring Robert Downey Jr. as a brilliant scientist who almost dies in an explosion. To survive he builds himself an iron heart. Joseph used to call Dr. Couper the real Iron Man because he put metal VADs in people.

Just after the movie came out, a deliveryman showed up at the ICU reception desk with a large box on a dolly. It sat there for a while until the charge nurse got tired of tripping over it and finally ripped open the top with surgical scissors. A typed note lay on top: "For Joseph Helfgot's hospital friends, from Paramount Pictures." The box was filled with T-shirts, action figures, coffee mugs, visors, and mouse pads, all bearing the *Iron Man* logo. The action figures went first. Some of the doctors fought over them like little boys.

Couper puts his hands on Susan's arms and squeezes tight— his way of giving her a hug. He is too formal for anything more. He finally says the words she has been expecting and even waiting for.

"I'm so sorry. The clots. He was probably gone eight minutes in. We just didn't know."

Dr. Rawn has joined them. She looks at them both and says, "Please, don't feel bad. It's nobody's fault."

She wonders what it must feel like to work so hard and fail. They must go through this type of frustration fairly often.

"Mrs. Helfgot?" Susan has been looking out the window at the sheets of rain running down the glass. She turns around and sees a short, middle-aged woman in a business suit standing at the door.

"I'm Esther Charves, with the New England Organ Bank." As Susan leads her over to see Joseph, she says, "I feel I know your husband. So many people have told me stories about him."

They stand by the bed watching his chest rise and fall. Susan begins telling Esther about her husband.

As they make their way to a conference room where they can talk in private, Esther learns that Susan and Joseph were together for almost thirty years.

"Your husband was a registered organ donor?" Esther begins. She is still waiting to receive a copy of his organ donation card from the organ bank.

"Oh, yes, he was really into it, especially after waiting so long for a heart. A few weeks ago a documentary film crew came to our house and interviewed us. Joseph was so happy that he could talk about organ donation."

Esther knows all about the film crew. The organ bank's in-house lawyer has been drafting language for a special consent document that Susan will need to sign if the face transplant happens. Face transplants are big news, and protecting the identities of both the donor and the recipient will not be easy. A television production team for a series called *Boston Med* has been at the Brigham since January. They have been filming patients at home who are waiting for organs, including Helfgot, and hoping to follow them through their transplant surgeries, if and when they occur. But the day Helfgot got his new heart they were off filming another transplant.

The producer has the hospital's permission to film the country's second face transplant, if and when it occurs at the Brigham.

With three weeks left in their five-month shoot, it looks like he may get lucky. But nobody could have predicted that the face would belong to Joseph Helfgot, a man they have already interviewed.

For now Esther tells Susan the most urgent news, that a potential recipient in another city is desperate for the transplanted heart.

"It can be used again?" Susan says. "That's incredible." She feels no connection with this newly transplanted heart. It is not her husband's heart, the one that beat in his chest throughout their days and nights together, the one she could feel and almost hear during their most intimate moments. This new heart never really became a part of him. But she knows that the family who donated it will feel a loss if they learn the recipient has died. The news that their loved one's heart might save someone else will be huge for them. Had it been Joseph's heart, she would have wanted to know that the gift had not been in vain.

"It's amazing, but apparently it can go to someone else. We're not sure about the kidneys and the liver. We'll know more later. Your husband was sick for such a long time."

"Whatever you can use, please take it."

"Mrs. Helfgot . . ."

"Please, call me Susan."

"Susan, sometimes we are able to recover tissue. How do you feel about that?"

"What kind of tissue?"

"Well, muscle, bone, veins, and sometimes even heart valves. Sometimes tissue is used right away, but it can also be preserved. The tissue on your arm, for example"—she holds out her forearm and runs a manicured fingernail along the inside. "And the back and the buttocks are often used for breast reconstruction after a mastectomy."

"Really? Well, whatever you need." Susan laughs. "I'm sure Joseph would enjoy the idea of winding up on a woman's breast."

Esther smiles. She knows it's too soon to ask for the face.

They've only just met. But she already feels that the answer will be yes.

"People really loved your husband."

Susan nods.

"One of the doctors told me he never saw anyone fight so hard to live."

"Yes, it was in his blood. His mother survived Auschwitz. She was such a fighter."

"Did you get to meet her?"

"I did. She was the toughest person I ever knew. And after what she went through to survive, there was no way Joseph wasn't going to fight."

"I think he fought for you and the children."

Esther leaves the room to call Chris Curran at the organ bank. "I just talked to Mrs. Helfgot. She said yes to organ and tissue donation. She even made a joke about tissue. I'm going to ask her."

Chris Curran swallows, and his mind shifts into high gear. If Mrs. Helfgot says yes to Esther's next question, there will be a lot to do, and quickly. And there's a heart going out of the region. Tricky.

"Are you sure?"

"I'll be really sure before I ask."

A little later Esther and Meredith are in the nurses' lunchroom, catching a bite and planning their next step.

"He matches up," Meredith says. "His HLA, antigens, age, everything. But Dr. Pomahac still hasn't seen the face."

Esther knows that, but she worries that if Pomahac visits the unit, the whole hospital will be buzzing. He's been on television talking about face transplants, and everyone knows that if the Brigham does one, he'll be the lead surgeon. There is no reason for a plastic surgeon to be on a cardiac floor. What if someone sees him and mentions his visit to Susan before Esther has asked her? What if somebody calls a reporter?

A few nurses enter the lunchroom, talking about a conference they attended yesterday. Helfgot was supposed to have been the guest speaker. The topic was "life on a VAD."

"They announced it from the podium," one of them says, "that Joe finally got his heart. But not what happened after that."

"Maybe there's something about the conference on the hospital's website," Esther says after the women leave. "It might be posted. If there's a picture, Pomahac can look at it."

A quick search shows that Helfgot's picture pops up in several places. They find a nice profile shot of him. Perfect.

A woman with creamy white skin and auburn hair walks into Joseph Helfgot's room. Pam Levine is his surrogate daughter. Her father, Sol Levine, hired Joseph to teach sociology at Boston University in the 1970s. Pam was a little girl when she first met Joseph. She was sitting in pajamas on her father's knee during a poker game in her parents' dining room.

"Daddy," she whispered, staring at Joseph's enormous Jewfro. "Why does that man have so much hair?"

Years later Joseph hired Pam fresh out of college to work for his fledgling movie research company. She was a gifted marketer, but her real specialty was managing the boss. One of the major studios eventually hired her away, but she and Helfgot stayed close, becoming even closer when her father died suddenly in 1996. Although Pam is a glamorous Hollywood businesswoman, today, in a baggy sweater and jeans, she could almost pass for a graduate student. She has obviously been crying.

She and Susan embrace. "When I landed, I went to the house."

"Have they found Emily yet?"

"She's on her way." Pam looks down at Joseph and rubs his shoulder.

"He looks so good, Sue. How can he look so good?"

"I know. It's weird, isn't it?"

"Have you thought about what you're going to do?"

"We're just waiting for everybody to show up. The rabbi is coming tomorrow to help us say goodbye."

"I mean with Joseph gone. After."

"Oh."

What am I going to do? Finally finish my teaching degree? Sell the house? Take a long trip somewhere? God, I have no idea. I can start by tossing out all the Splenda and the salt-free sauces. But what am I going to do?

Pam's question terrifies me. I close my eyes to hide and see a color-less wall. It's so very, very close. Too close. I take a step back to see it better. Why am I holding my breath? I suck in air and the wall is gone. Everything is white.

5 p.m. Helfgot's bedside.

The widow has left to attend to matters at home. A neurological exam required to determine brain death was performed earlier in the day, and the time has come to pronounce Helfgot officially dead. An intimate group gathers. Ideally such a group includes someone who is unknown to the family. Today Dr. Gentian Kristo, a surgical resident, will follow to the letter the protocol established by the hospital in keeping with the Uniform Determination of Brain Death Act. Dr. Couper will sign off as the attending physician.

This is all much more complicated than it used to be. Advances in modern science have made it possible to keep a body alive after brain death: ventilation machines deliver rich air; IV lines pump fluids and medicines keep hearts beating and maintain blood pressure; feeding tubes through the stomach deliver sustaining nutrients. Even when the mind is gone machines controlled by other minds can keep a body stable, suspended as if in a cocoon, ghostlike, hovering between this world and the next in a permanent vegetative state.

The heart keeps beating after brain death. Unlike other nerves in the body that are attached to the brain stem, it has its own rhythmic nerve center and can keep going without instructions from the brain. All it needs for its cells to keep working is oxygen. As long as a machine can supply the lungs with oxygen, a body can be kept "alive" for some time. In the famous (but unusual) 2005 case, Terri Schiavo's body was kept "alive" for fifteen years until a court demanded that the ventilator be turned off.

A strong light is placed within millimeters of Joseph's eyeball. His large dilated black pupil, ringed by a hint of blue, remains fixed as he appears to stare straight ahead. Pupillary reflex—gone. A doctor turns Joseph's head and checks for doll's reflex. But his eyes do not move.

Next Dr. Kristo lightly touches a cotton swab to Helfgot's closed eyelid. There is no twitch, nothing. Corneal reflex—gone.

Then he lightly stabs the skin in sensitive spots with the sharp end of the cotton swab. No response. They continue checking reflexes until they come to the final test for apnea. The ventilation equipment is turned off. The room becomes silent. A minute goes by, then another. Nothing. No gasp for air, no primordial struggle for survival. Joseph's lungs are screaming for oxygen, but the cries go unheard. His brain is gone and no longer commands his body.

They quickly reconnect the ventilator because there are still goodbyes to be said and lifesaving organs to protect for others. For a little longer, Joseph Helfgot will float between worlds.

At 5:13 p.m., on Tuesday, April 7, 2009, Dr. Gentian Kristo pronounces Joseph Helfgot legally dead. Cause of death: cerebrovascular accident. The two doctors sign the death certificate and go off to care for the living.

"Chris," says Esther, "it's me again. They've declared him. His wife went home for a while. She'll be back around seven. I told her we would do the oral history with Meredith when she returns. I'll ask her then."

"We should tell Dr. Pomahac."

"If you do, let him know it would be better not to come up here until I talk to the family."

"I'm all over it."

Dr. Bohdan Pomahac, the director of Brigham and Women's Burn Center, is on his way out the door. It's early for him, but his wife is going out with her friends, a rare event. With two small children, she barely gets a moment to herself. His phone starts to buzz. The words "I'm sorry, but whatever it is, I have to go home" are already half-formed on his lips. But when he sees the number on the screen, his mouth goes dry.

"Dr. Pomahac? It's Chris Curran. We may have the donor."

"Where?"

"Right in the hospital. He's a heart transplant that didn't work out. Esther Charves is there. The family has already said yes to tissue. Esther hasn't asked for the face yet. His wife is coming back around seven to do the history. We'll know then. I'll call you the minute we know for sure."

Pomahac hits a name on his speed dial list. Julian Pribaz, director of the Harvard Plastic Surgery Resident Program, was his teacher and is still his mentor. A year ago the two men flew to Brussels and trained together to perform this almost unprecedented operation.

"Julian," Pomahac tells him, "we may have a face."

chapter five

Tuesday, April 7, 2009, early evening.
Brigham and Women's Hospital.

esther needs to clear her head. She glides down the escalator and is swept along with the crowd of people spilling into the street. Daylight Savings Time has kept the twilight at bay, but in the misty drizzle it feels like night. A strong cold wind smelling of the sea rips down Francis Street, but she doesn't mind; after thirty-six straight hours indoors she finds the fresh air exhilarating. She walks toward Huntington Avenue, collecting her thoughts and preparing herself for the conversation she is about to have with Susan Helfgot. She has had to do many difficult things in her years at the organ bank, but this one is in a category all its own.

Nothing in her life has been easy. She grew up near the sea in Bristol, Rhode Island, a quiet town that juts out into the cold north edges of Narragansett Bay. Bristol isn't far from the mansions and yachts of Newport, but only if you're measuring in miles. When Esther was little the town was solid working class.

Her father was a construction worker, and she remembers a time when they didn't even have a telephone.

As a teenager she would babysit for the children of an older cousin in a house filled with books and magazines, so Esther became a big reader. She won a partial scholarship to a good private high school, but she barely had time for homework. Because her mother was very sick Esther took a part-time job and helped care for her siblings. She has a powerful memory from early childhood: she is waving to her mother, who stands at the window of a Boston hospital. At the time children were not allowed to visit, and to Esther the hospital looked a lot like a prison. It was the Brigham in an earlier incarnation.

Esther went on to Boston College, and later to Boston University. She married early and had three children, but stayed in school and became a nuclear medicine cardiac technologist, performing isotopic scans at a big heart clinic.

A warm and compassionate woman, she was friendly with some of the patients who came in for imaging, often with their children. Some were waiting for new hearts, and although they knew the odds were against them, they remained hopeful. Over time a number of her patients died waiting.

It was a sad progression, and Esther grew tired of watching it. Eventually these patients were unable to walk, and they started arriving at the clinic in wheelchairs. A few months later they might be on oxygen. Finally there would be a missed appointment that was never rescheduled. Sometimes word would trickle into the clinic that one of their patients was given a new heart. Far more often, though, the story ended badly.

One patient, Jerry, was forty-seven, with five sons. Esther had grown attached to the boys, and when their father died in 1999 she attended his funeral. Jerry's five sons stood next to the casket, lined up as progressively shorter versions of their dad, right down to the little five-year-old, who stood there bravely, shaking hands. She heard later that the family was evicted from their home and living in a tiny apartment.

For Esther, Jerry's death was the last straw. By now her marriage was over, and her youngest child was already in college. She knew all too well what it was like to wait for a heart. Now she wanted to work on the other side of the equation, the supply side. The New England Organ Bank had an opening, and Esther was hired.

As a family services coordinator she sits with grieving mothers, weeping wives, and devastated husbands on the worst day of their lives. And she asks them for permission to open up their loved ones and remove organs that will help other patients, complete strangers, many of whom will die without these extraordinary measures. It is a very demanding job. Often, at such a moment, the person she's speaking to would say no to anything, even a gift of a million dollars, because her husband or his wife went out this morning to pick up a carton of milk and never came home. Sometimes all the survivors need is a little time to absorb the news that their loved one is really gone and to consider the request. She can offer them at least that.

She still remembers the first request she ever made. She was trembling. This was the job she wanted, but until then she hadn't fully understood how bold she would have to be. Like anything else, you get used to it—not completely, but mostly. Now she never flinches. She believes a family has the right to hear the question she is asking on behalf of the person who has just died. "I am gone," she imagines them saying. "Please let someone else have a chance to live. There's been enough tragedy for one day."

Esther is not a saleswoman. She embraces their noes and yeses with equanimity. She knows that some people want to have all the available information and will still say no, and she feels she owes them the same respect and dignity she would give to someone who says yes. She hopes families will say yes because lives are at stake, but also because she knows from experience that most families find solace in donating. The decision to help another family—complete strangers—may be the one thing they can hang on to in the brutal days ahead.

These conversations don't always go well. Sometimes family members don't support the wishes of the deceased who has signed an organ donor card. Sometimes they become angry at Esther. Back when these requests were made by phone, one woman shouted, "Don't ever call me again!" As if she would ever call again.

Once, a mother whose child had just died spat out, "Go away. Leave her be. If I can't have her, neither can you!"

Esther was stunned, because a nurse had told her that the eighteen-year-old had been obsessed with organ donation. She had raised money for the organ bank and hoped to become a transplant surgeon. The girl died waiting for a liver that never came.

Esther quietly gave the woman her card and left. She always gives her card, even when the answer is a defiant no. Maybe in the dead of night they will call with second thoughts. It's rare, but it happens. Six months later the mother called her. "You were right," she told Esther, crying. "Now I have nothing. And neither does anyone else. I'm so sorry."

One night the sister of a young accident victim slammed her fist down on the table, shouting over and over, "Show me the money! Show me the money!" as her fist kept time with her words. Then her brother joined in. They continued pounding and chanting as Esther backed out of the room, blinking away tears and humiliated by their behavior. She could still hear them on the other side of the ICU door as she waited for the elevator.

But more often than not the families say yes. And with that yes somebody, somewhere will receive a precious and anonymous gift.

"What do you think?" a colleague asked Esther as they drove to the meeting with Dr. Pomahac at the organ bank.

"I'm not sure I could do it. I can't even think about how I would ask somebody for his wife's face. What about a child's face? No, I don't think so," Esther said.

But then she met Pomahac, a quiet young man with a receding hairline and sharp brown eyes. Esther reads people for a living. She knew he was nervous, this unassuming surgeon with his soft European accent. He seemed worried about their reactions when he asked for help in finding facial donors for his new program at Brigham and Women's Hospital.

She couldn't miss his anguished expression as he described how some of his patients were unable to swallow, to breathe, or to speak. They endured surgery after surgery, with only modest improvements. Some of them were unable to go out in public because of their injuries. Some were young men who lost parts of their head on the battlefields of Iraq or Afghanistan. Some were marred by cancerous tumors. He showed pictures to demonstrate how muscle and bone from a face donor can rebuild a life in a way nothing else can.

He explained that it wasn't someone's entire face he needed, just parts of a face. But everyone in the room that day knew that to a grieving family, this distinction would mean little. Dr. Pomahac knew it too.

After seeing the pain in his eyes as he described his wounded patients, Esther knew that, if she had to, she could ask the question. But she sorely hoped she would never have to.

The organ bank drew up a script that she and her colleagues could use. But they all knew that any script they tried to follow would soon be abandoned. This was uncharted territory. They would have to follow the conversation, wherever it went.

Susan Helfgot's quick consent to Esther's first question, and her composure, her smile, and her willingness to donate her husband's skin tissue—all these things make Esther believe that Susan will say yes. But how is she going to ask? How is she going to phrase the question?

She hears Dr. Pomahac's words and sees a picture in her mind of the man he has been trying to help, a man who is horribly disfigured from a fall. How it happened, she doesn't know. She knows only that he is married and has a child. And that he can hardly breathe, or swallow, or smile. He can't even step outside without ridicule.

Susan, someone is suffering horribly. Can you help him?

She utters the simple prayer she always recites before she meets with a donor family. She asks God to help give her the right words, and to allow her to be helpful to these grieving people, even if they say no to her request. She takes one last breath of fresh air and heads back inside the building.

At the door to Joseph Helfgot's room she says, "Susan? May I speak with you again? There's something else I need to ask."

chapter six

Tuesday, April 7, 2009, 7 p.m.
Intensive Care Unit.

Susan has returned to the hospital with Pam, who doesn't want her friend to have to meet with the organ bank lady by herself. The three women make their way to the conference room, where Susan and Esther spoke earlier. The darkness out-side is almost complete, but no one moves to turn on the lights. Diffuse light from the hallway filters through the glass panels. Pam curls up in a chair while Susan sits on a couch, facing Esther.

Esther reminds herself that she is here to ask a question that Joseph cannot ask of his family. Given everything she has learned about him, she is certain that if he could, he would grant her request. This belief calms her as she begins, speaking softly: "Susan, there is something I have to ask you, something Joseph didn't sign up for. When he signed the organ donor card, it was for organs and tissue. What I'm going to ask you is a big deal."

With Esther scattering caveats and throwing up warning signs, Susan wonders what's coming. She can't imagine what it might be, but feeling a need to help Esther along, she offers a joke. "Don't tell me you want his Jewish nose?"

Esther's head snaps back a fraction of an inch. Then, resisting the temptation to respond to Susan's surprising guess, she sticks with the script in her mind. "If I told you there was a man who was horribly hurt in a fall, who can't eat normal food or talk on the phone, who sometimes has to breathe through a tracheostomy tube, would you want to help him?"

Well of course she would. For more than a month last year Joseph had both a trachea tube and a feeding tube. It was awful. He wasn't even allowed to have a sip of water. She can still hear him saying, "Please, just a piece of ice. My throat is so sore and dry."

"Esther, what exactly are you saying?"

"We have a man who needs a face."

Pam's arms fly up and she gasps.

"Oh my God," Susan says. "You really do want Joseph's nose."

For a moment nobody speaks.

"I think I have to do it," Susan finally says. "It wouldn't be right not to. But I have to call my children."

Pam thinks her friend is too exhausted to make such a big decision so quickly. "Sue? Are you sure?"

"If Emily and Jon agree, and Ben too."

She calls home. "But not Jacob." He's so young, she is thinking. Let's not burden him with this.

Jonathan answers the phone, and Susan explains that she and Pam are with the organ bank lady. "Yes, they still want the heart." She asks him to get Emily and Ben on the line. Jacob is playing video games with a friend.

When Emily and Ben have joined the call, Susan begins: "Guys, listen to me. There's a man in real trouble who needs help. He can't eat or talk or breathe very well. He fell very badly and lost his face. Yes, his whole face. I don't know how. That's not important, is it? They want to know if we will let them have your dad's face to help fix him."

There is no immediate response. "Guys? What are you thinking?"

Emily asks, "What do *you* want to do, Sue?"

"I want to do what Daddy would want us to do."

"He would want to do it," says Jon.

Susan listens as Emily explains it to Ben. "Do you guys want to think about it for a while?"

"Thinking about it won't change anything," Emily says. "It's what Dad would want to do. Jonathan, what do you think?"

"I think so too."

Sue hears Emily talking to Ben, who says, "Yeah. We should do it."

"Okay, then that's what I'm going to tell her. We'll be home soon. I love you guys."

"We're going to do it," she tells Esther.

Esther steps out of the room and calls Chris Curran. "She said yes."

"Oh my God. Esther, are you sure?"

"No."

"*What?*"

"Kidding."

"I gotta call everybody. Bye."

Curran starts punching in numbers. He calls the head of public relations for the organ bank. He calls the head of the organ bank, and somebody alerts the chief medical director. He tells the tissue service manager to start organizing her recovery team. Next comes the organ bank's lawyer, who has been sitting at her desk reviewing the consent documents. "Chris," she tells him, "you can't give Dr. Pomahac the go-ahead until Mrs. Helfgot signs the consent. And Meredith still has to take the social and medical history. Something could come up, or she might change her mind. Maybe she won't sign. I'm faxing it over to Esther right now."

Chris knows it's not a done deal, but he is dying to call Pomahac. He knows he's waiting at home and probably climbing the walls.

8 p.m. Suburban Boston.

Bo Pomahac isn't quite climbing the walls, but he is watching the clock and wondering when his phone is going to ring. Hanka, his wife, has left for her dinner with friends.

"Daddy?" His little girl is looking up at him. "Daddy, are you listening?"

"What, sweetheart?"

"Daddy, I need a newspaper for school. We need a picture cut out of a newspaper."

"Okay, honey."

"*Daddy?*"

"What?"

"We don't *have* a newspaper. That's what I'm trying to tell you."

"Okay. Let's look for something we can use. Get your scissors and we'll find some magazines, and then it's time for bed, okay?"

He keeps looking at his cell phone, the time passing slowly, as he reads his son and daughter a bedtime story. He wishes Hanka were home. He wishes his phone would ring and that the kids were already asleep. Please call, he keeps thinking. Please call.

"Daddy?"

"What?"

"Daddy, *read*." The kids giggle.

"Sorry."

• • •

8:30 p.m. Intensive Care Unit.

Meredith has joined them in the conference room, where the air is charged with an energy that's impossible to measure. Esther and Meredith, Susan and Pam—four conspirators bound by the still secret knowledge that something big is unfolding here.

Meredith spends an hour taking Joseph's social and medical history from his wife. Susan likes her. This nurse from the organ bank is a little unconventional, with a funky tattoo and rings in her ears—quite a change from the staid crowd that normally roams the ICU. She delves deeply into Joseph's history with very personal questions. "You knew him for how long?"

Susan winces to hear her husband referred to in the past tense, but she knows she'll have to get used to it. "Almost thirty years."

After some questions about foreign travel, she asks, "Did your husband use drugs recreationally?"

Pam bursts out laughing.

Esther says, "Susan, you can only answer what you know about."

"That's what I'm afraid of!" They all laugh.

"Well, he taught undergraduate courses at Boston University. One class was Drugs in Society, and the other was Sex in Society."

Meredith grins widely. "For real?"

For real. Years ago, as part of his research, Joseph had shadowed a cocaine dealer, and Susan knows that her husband—with the very nose she has been joking about—had been known to sample the goods. He sometimes smoked pot, but Meredith and Esther are concerned about the kind of drugs that require needles, which Joseph never even considered using. They also know that to be eligible for his heart transplant, he was examined very carefully.

After the history is taken, Esther brings in the release for Susan to sign. She scribbles her signature. She is spent, completely exhausted. It is done.

As they leave the hospital everything seems muffled, as it does when you're leaving a very noisy sports event. After dropping off Pam, Susan cries on the way home. She tries not to, but she can't help it. She doesn't want to look like a wreck for the kids, but after holding the tears at bay for all this time, she can no longer suppress them.

Twenty-four years earlier. 1985. Boston University:
An energetic professor in his mid-thirties stands outside Morse Auditorium in Kenmore Square. Inside, the audience of four hundred undergraduates is stirred up. Today, Helfgot has told them, they will meet a special guest, a woman named Michelle. She used to be known as Michael, and in a few minutes she will be telling them about her sex-change operation at Johns Hopkins in Baltimore.

Earlier this morning, as they left the house together, Helfgot had asked Susan, his girlfriend, if she would come to his class today. "Michelle will be speaking," he said. "You know, the woman I had on the show last week?"

When he isn't teaching, Helfgot is on the radio as a local version of Dr. Ruth Westheimer, the hugely popular sex therapist during the 1980s. His Sunday night show on WHDH has a big audience. It breaks up the monotony for drivers returning home from Cape Cod at the end of a summer weekend, or from ski trips in Vermont and New Hampshire in the brittle winter cold. The professor is a bit of a local celebrity, and a political hot potato for both Boston University and the Catholic Archdiocese of Boston, who aren't exactly charter members of the Joseph Helfgot Fan Club.

"I don't know," Susan told him. "I've got a lot going on today at work."

"Please? We can have lunch after class, okay?"

• • •

"Michelle, you go ahead. Sit where you like, but somewhere in the middle. I'll introduce you in a few minutes. We'll see how everyone reacts when you stand up."

Susan arrives a moment later. "I can't believe you talked me into coming," she says. But that's not really true. He's very persuasive. They head inside, Joseph leading Susan down the center aisle toward the stage.

"Professor!" a student shouts. "I can't believe what they can do!"

"Christ, they think I'm *her*!" Susan whispers.

"Exactly!" He grins impishly as he jumps onstage and grabs the microphone. "And doesn't she look great?" He beams down at Susan in the front row, who shoots back an angry look just as someone whistles from the back of the room.

"There! Who whistled?" Helfgot whips around and begins pacing back and forth, brightly lit in a spotlight, the mike loose in his hand. Phil Donahue meets Mick Jagger, and both sides of him are delighted to be up there.

"That's perfect. That says it all! How to get a girl. Amazing. You don't even need language. Just whistle." He pauses and thinks for a moment. "Okay, so it's birdcalls. That's what it is, human birdcalls." The students are transfixed. "Do you guys remember the famous scene from *To Have and Have Not*?" Apparently not. "C'mon, you know, where Lauren Bacall tells Bogart to 'just whistle' if he needs her?"

There are nods of recognition, but not many. "So if you were a linguist, what would you say?" And off he goes on a rant about sounds and sex and beauty, rolling them into one big blur. Everyone is laughing and students are shouting back. Susan is trying to be invisible.

"Today," the professor announces, "we're gonna talk about how beauty is in the eye of the beholder. It's time to meet our special guest." He looks at Susan and gestures for her to come up. But Susan remains seated, as he knows she will, and Michelle stands up and starts walking toward the stage. There are some audible gasps.

"What is the lesson?" he shouts out. In a recent class he mentioned that men are sometimes attracted to transvestites, but rarely to women who used to be men.

"Don't you ever pull anything like that ever again!" Susan says when the class ends, punching Joseph on the arm. It's not a playful punch, and she's pleased when he recoils in pain.

"You're a jerk," adds Michelle.

April 7, 2009, 10 p.m. New England Organ Bank.
Chris Curran is on a late-night conference call with seven of his fellow employees as they try to cover every base: travel arrangements, legal considerations, the needs of the tissue recovery team, publicity control, family support, and much more. There is a plane to charter for the out-of-town cardiac team, who will have to get to and from the airport. It's not clear yet whether Helfgot's liver and kidneys can be salvaged, but the potential recipients for those organs are local, thank goodness.

The organ bank's hospital liaison will have to work with the Brigham to reschedule surgeries and soothe ruffled feathers. Pomahac and his team will need two operating rooms that are near each other for a minimum of twenty-four hours. This is a big request, as operating room space is hard to come by. Tying up two of them for so long with virtually no advance notice will be tough on everyone in surgical services.

"When do we think this thing will actually start?" the liaison asks.

"The family is having a bedside service tomorrow afternoon," says Esther. "Any time after that, I imagine."

After a few more kinks are worked out, they ring off, each with a list of tasks that will keep them working a while longer.

At 10:40 Chris Curran calls Pomahac. "We're set," he says, and Pomahac sighs loudly on the other end. "Esther will call in a few minutes. We'd rather you didn't go in tonight. Maybe in the morning, early, when nobody is around. Everybody knows you. Is that all right?"

Pomahac's mind is racing. He needs to call his team, get them together first thing tomorrow.

"Fine." He hears the car in the driveway. Hanka is home.

Twenty minutes later. Intensive Care Unit.
Esther stands over Joseph's bed. "We did good," she whispers. "We got it done." Although she had never met him when he was alive, Esther feels she has been speaking for him, that they have a kind of partnership.

Then she calls Pomahac. "I'm with the donor," she says. "Meredith says everything checks out beautifully."

"Thank you, that's great. How does he look to you?"

"He's an attractive Jewish man. Dark hair. He was supposed to speak at a nurses' meeting yesterday, and there's a good picture of him on the hospital website. A nice-looking man with blue eyes."

"Where's the picture?"

"Search his name. It'll come up." She spells it.

"I'll take a look," he says. This is it. He feels it. This is the one, the face Jim Maki has been waiting for.

Pomahac looks at his watch. It's late. He'll call Jim tomorrow. Let the man sleep. At least somebody will get some rest tonight.

chapter seven

Fifty years earlier. September 1, 1959.
Pacific Ocean.

a huge wave slams against the hull of the ship, stirring the boy from his deep sleep. "John," he whispers from the top bunk. "You awake?"

Another wave crashes as his older brother sleeps on. Jim hops down, instinctively landing with his feet wide apart for balance. He shimmies over to the porthole and stands on tiptoe, peering outside. As the ship pitches wildly toward the sea he is pushed against the curved steel wall. Dark foamy water rushes up to meet his gaze. He is thrilled by the thought that the ship's window might dip deeply enough to actually touch the ocean. But just when it seems that they are about to kiss the sea, the swirling water disappears and up they go again, butterflies rising in his belly as they ascend to the top of the next swell. Now he sees only gray sky through the glass. Then, suddenly, they lurch, his skinny knees turning to jelly as he braces himself on the aquatic roller coaster.

"John, wake up! It's my birthday again!"

"What time is it? Go back to bed."

But Jim is pulling up his pants as he holds the metal bunk for balance. "I'm going up."

"It's too rough. Dad will be mad at you."

"Not on my birthday!"

During the night the ship crossed the International Date Line, so it's Jim's eleventh birthday all over again. How lucky can you get? His mother promised him a second cake. Yesterday they gave him white with chocolate frosting; maybe today it will be lemon, his favorite.

He gets his sea legs running down the hallway and bursts through the metal door. Skipping up the mesh stairs, he skids out onto the deck, which is slick from salt spray. It is windy out here, and barely dawn. An endless sea of blue stretches out before him and meets a strip of crimson that fades to blackness. Any minute now it will turn into sunrise.

During the war this ship was converted into a floating hospital. Now it is tired and showing its age. It's not really a cruise ship, because it also carries freight. But to Jim it's impressive, a huge rocking mountain.

When they sailed to Tokyo almost a year ago, it took twenty-one days. That's when he learned how to walk on a ship. The boys' father was being sent to Japan to study something important, their mother explained, a legal document that had to do with the war. Jim heard that his father had a Fulbright, a great honor for professors. The youngster, who loved words, thought it was a good name for a man who was so full of knowledge and so clever.

Now they are heading home to Seattle. Butterflies fill his stomach again—not the kind he gets when something exciting is about to happen, but the ones that make him sick to his stomach. Hoping the feeling will go away, as it sometimes does, he makes his way to the cracked Ping-Pong table. He's dying to play, but John is still in bed. A man stands against the railing. Jim remembers seeing him before. The ship is not that big.

He picks out a paddle from the basket under the table. He never played before this trip, but now that's all he wants to do. Even with the boat listing back and forth, he can return pretty much anything that comes his way. John isn't as good, though, and he gets mad when Jim smashes the ball straight at him. Sometimes he misses, and they fight over who has to retrieve it.

"I *told* you not to hit it like that. I'm not going after it."

"Yeah, well, you're the one who missed it."

"Then I quit," John ends up saying, throwing down his paddle. It's a ploy. He knows Jim will go after the ball so they can keep playing.

"Mister, you want to play?"

"No thanks, son. Where's your brother?"

"He's still in bed."

"Maybe you should go back down. It's pretty choppy. Does your mother know you're up here?"

"Yeah." Embarrassed by the lie, he edges away from the man.

He isn't sure whether he is happy or sad that they are finally going home. Tokyo was a strange place, with strange smells and even stranger people. A lady came in every day to help his mother clean. Everyone lives like that in Tokyo, or so he thinks. He and John had expected that the boys in Tokyo would look just like them. But they didn't. Neither do the boys back in Seattle.

In Japan there was a big baseball diamond at the end of their street. Jim would run down there every day after school to see if there was a pickup game. At home he plays shortstop and third base, but in Japan he was often the pitcher. In Seattle he plays on a Little League team, and in the spring he intends to pitch.

In Japan they had a television and the boys watched all kinds of programs. They don't have one in their Seattle house because they live at the bottom of a steep hill and there's no reception.

Once their dad spoke back to the man on the television set in grammatically correct Japanese. The boys laughed, because next to the real thing, Jack Maki's accent sounded pretty bad.

"What's so funny?"

"Dad, you sound silly!" they said, erupting into fresh peals of laughter.

They went to the American School, where the classes were in English. But Jim picked up a lot of Japanese. He would talk to anybody, including the man at the noodle stand who took the grubby coins Jim picked up from the train station floor. Their lives in Japan were so different! Back home, they mostly had to stay in the yard. But in Tokyo they took three different trains to get to school, and their father thought that was fine. Japan, he always said, was a safe place to live. Even so, Jim and John weren't supposed to stop anywhere on their way to and from school, though they often did.

Their mother would nod her approval when Jim recited Japanese words and phrases. Mary spoke the language perfectly. She grew up speaking it on her parents' farm.

"John, I guess you and I have tin ears," Jack Maki would say as Jim and Mary exchanged a few words. John never did pick up much Japanese. He worked a lot harder in school too. Jim got straight A's in fourth grade in Tokyo. John got mostly Bs, but everyone knows that sixth grade is a lot harder.

"I can't find Jim." Mary is holding an ice bucket and trying to stay calm. "John says he went up on deck."

Her husband purses his lips in a frown. "I'll take a look."

"Jack, it's very rough."

"Don't worry, he hasn't fallen over. How's John?"

"Seasick."

Jack Maki walks toward the stairs. It's always something with that boy. Why can't he stay put and just do what he's told?

Jack is happy to be heading back to his job at the University

of Washington. He is even happier to be returning to the ordinary luxuries of home. Tokyo in 1959 has still not completely recovered from the brutal bombings during the war that left much
of the city in rubble.

He will soon be fifty, and for the first time in his life creature
comforts are important. That certainly wasn't the case the first
time he stepped on Japanese soil more than twenty years ago.
He and Mary had married quickly so she could go with him on
his teaching fellowship. They had met the year before, and he
couldn't bear to leave her behind.

He can't quite believe how they managed in one tiny room,
with a cold-water faucet in the corner and a gas ring heater that demanded a steady diet of coins to provide what little heat there was.
He bought a small iron grill called a hibachi, and between that
and the heater they managed to get their room up to about fifty
degrees in the winter. The house had one bath and no hot water.

War broke out between China and Japan during their stay,
but they weren't aware that the winds of a much greater war were
gathering around them. When Jack's mentor in Seattle suffered
what turned out to be a fatal heart attack, they rushed home and
Jack took over some of his classes. From then on, his future at the
university was assured. Years later Jack learned that intelligence
officers from the Japanese Imperial Navy had been following him
during his stay in their country to see if he might make a good spy.

When war broke out everywhere, Jack came to understand the
world in a new way. Although he had been picked on as a boy for
being Japanese, being adopted by a Scottish family had partly protected him from prejudice. But not always. Once a farmer told Jack's
Boy Scout troop not to hike through his field. The scoutmaster argued with the man, who didn't want a "Jap boy" on his property. In
the end the group retreated and had to walk all the way around. It
was worse in college. When Jack signed up for ROTC, the Reserve
Officers' Training Corps, he was rejected for being "not qualified."

He had planned to major in journalism until the dean cautioned him that no American newspaper would hire an ethnic Japanese. He switched to English literature, where he excelled. For his foreign language requirement he picked Japanese "for no reason other than idle curiosity," he wrote later. When the secretary of the English Department advised him that with his Japanese face he would never get an appointment teaching English literature, he switched to Japanese literature, and eventually to Japanese history.

After Pearl Harbor he and Mary were forced to pack up their Seattle apartment and move into army barracks in the town of Puyallup, squeezing into a tiny space with five other families. Seven thousand Japanese Americans were evacuated to Camp Harmony, with new arrivals being assigned to specific areas of the camp, depending on their education and background. "We weren't fearful or angry," Jack wrote later, "since it was accepted as a wartime necessity. Even though we were Americans, we were inescapably identified with the enemy, and we accepted it." He pointed out that it didn't even occur to them to demand fair treatment. This was well before the civil rights movement of the 1960s empowered many minority groups. It was a very different time, and the world was at war.

After a month at the camp, in the spring of 1942, Jack and Mary were sent to Washington, D.C., where they worked for the government, reading and analyzing enemy propaganda and Japanese radio broadcasts. All around them other analysts were doing the same for communications out of Germany and Italy.

After the war the State Department sent Jack to Japan, where he was stunned at the level of destruction. His job was to write a report on all the fractionalized government ministries. He helped monitor the first election, and he wrote to Mary every single day. Back home he earned his doctorate at Harvard and then returned to Seattle and settled into the life of an academician. He and Mary tried but were unable to have children. They adopted John, who was born in 1947, and then Jim, in 1949, when he was seven months old. Both boys were half-Japanese.

Up on deck Jack finds his younger son leaning against a wall.

"Jim, where have you been? Your mother is worried."

"Nowhere. I don't feel too good, Daddy."

"Go lie down," Jack tells him. He is angry, but also relieved. He too was a little concerned but was able to hide his apprehension behind his wife's.

Down below, Jim says, "John, do you think we can play Ping-Pong a little later?"

"Jim," his mother says, "please stop talking. You boys should lie still and try to rest."

"But it's my birthday!"

"Okay," says John. "We'll play later. But no smashes, okay?"

"Okay."

Wednesday, April 8, 2009, 9 a.m.
A veterans' home in central Massachusetts.
Before leaving his room Jim Maki pats down his dark brown, arrow-straight hair with his good hand. Although he's fifty-nine, there is no gray. I may have no face, he thinks, but my hair's still good. Down in the dining room he sits in one of the clunky captain's chairs, ready for the weekly house meeting. The residence is run by a private charity that helps Vietnam veterans with medical problems. These are men who have nowhere else to go. They are all disabled, and they have all struggled with drugs.

The room is covered with washable vinyl wallpaper with a loud striped pattern in white and Kelly green. A cheap print of a man with long hair in a brown robe hangs from the center of one wall. His hands are tied together with rope and he looks to be in pain as he gazes up to the heavens. All that's missing is the crown of thorns, or perhaps a halo. As the meeting begins, one of the residents is complaining that his room is too hot at night.

When it's Jim's turn to speak, he addresses a man across the table. "I gave you razors," he tries to say, "and other stuff too. But then you went and took my ——." It's hard to make out what he's saying. He tries to say each word slowly and deliberately, like someone with Parkinson's or cerebral palsy. He keeps wiping his chin to sop up the drool that escapes as he tries speak.

The woman who runs the house repeats what she thinks she has heard. "Dave, what do you think about what Jim said?" Dave, in a wheelchair, is missing a leg. Like Jim and the other four residents, he has severe medical problems. One of the men is terminally ill. Jim tries not to get close to anyone; the place is too transitional.

"You don't know what you're talking about," Dave mumbles.

Jim shrugs. It's easier than talking, and he has made his point. There's nothing to gain by arguing with this guy. These meetings are pretty much a waste of time. The phone rings in the kitchen and the director steps out to answer it. She calls Jim to the phone, which doesn't happen very often.

"Mr. Maki, it's Dr. Pomahac. I have some great news. I'd like you to come in right away. We found a donor."

"Really?"

"Yes, we did."

Pomahac hears Jim say something that sounds like "Then let's do it."

The meeting dissolves into a round of "Good luck" and warm wishes. The men are not close, but good news is hard to come by in this house. And not one of them would trade anything he's been through—and they have all been through a great deal—for a single day without a face.

The car service is called, and Jim gets in alone. At the start of the long drive into Boston they pass the donut shop where the guys go to buy pastries. If Jim could smile now, he would. After the operation he'll be able to eat whatever he likes. He can't wait to walk into that shop and buy something for everyone in the house. They all seem to like the lemon and raspberry donuts. He's always loved lemon, anything with the flavor of lemon, ever since he was a kid. He feels like a kid right now, on Christmas morning. A present is waiting at the hospital just for him, the biggest present anyone could ever receive. He can't believe it's really happening.

"This is Cynthia. Leave a message."

"Cindy, they found a face," he mumbles into her voice mail.

They've been married for thirty-one years, but they haven't lived together for almost twenty-five of them. Still, she is his wife. And there's Jessica, who is graduating from college next month. From college! She'll be excited that her father is getting a face. Maybe they will visit him in the hospital. He sure hopes so.

Butterflies are starting up in his stomach. He can't decide whether it's the motion of the car or the excitement he's feeling as they approach the hospital. Dr. Pomahac said to come right away, so they must be ready. Good. He hates waiting.

It's tough to be this excited and have nobody to talk with. He thinks about the things he might do after the surgery. Maybe he'll coach Little League. Maybe there's a kids' football league where he could help out. He'll probably need to get his bad eye fixed first. And he still can't use his right hand, so maybe coaching is out.

Enough of that. He won't think about all the things that are still wrong. Today is a day to focus on what's going right.

He swallows and leans his head back, trying to relax. What did he do to deserve this good fortune? Is it possible that his mother is somehow helping him from afar? He hopes she is watching out for him. Would she even recognize him with a new face?

chapter eight

Forty years earlier. April 20, 1962, midnight. Seattle.

Well-heeled citizens in black tie sip champagne from flutes with crystal stems in the shape of the new Space Needle, which is being christened tonight. They stand in the wind and the dark, straining to hear the opening remarks at the 1962 World's Fair.

A few miles away Jack Maki is just getting home after another long day at the university. The Phi Beta Kappa meeting dragged on, and as the presiding officer he stayed until the bitter end. Then he returned to his office to review his notes for tomorrow's senate meeting; next year he will become its president. He graded some of the student papers on his desk before finally heading for home.

After checking on John, he pokes his head into Jim's room. Fast asleep, both of them. Jim, who is twelve, is beginning to shoot up, his baby fat gone, his limbs lean and sinewy from many hours of basketball. Jack is proud of his talent, but he worries that sports are taking over the boy's life. It's time for more homework and fewer hoops.

He looks at Jim's peaceful face, his unbridled energy bottled up in sleep. He has been patient with his son, and Mary has been a saint. But Jim can't seem to behave. Jack cringes every time he walks through the front door and Mary tells him of yet another mishap. Yesterday it was a fight down the street.

"He called me a Jap."

"So? That boy is nothing to you."

"Dad, I had to do something."

"Yes. What you had to do was turn around, walk home, and get started on your homework."

"You don't understand."

"Don't I? Here's what I understand: if it weren't for education, we wouldn't be in this house. Education is what got your mother and me out of the camp during the war. Grandpa Ichi lost everything. They gave him a check for the farm, but it wasn't enough to buy a flower stand."

"They said my eyes were slanted. They said I squint."

"Grandpa is an old man, Jim. When he dies, he will have nothing to leave to his family. Can you imagine what that feels like?"

But as he stands above his sleeping child, he realizes that a twelve-year-old can't possibly know what that feels like. Even in the dark he can see a bruise on Jim's cheek from the fight. It's a cheek that looks a little exotic. Mary thinks Jim is part Native American, and that sounds right to Jack.

Jim apologized for fighting, but Jack held firm. "No baseball tomorrow. You're staying home. I don't want you outside. Do you understand?"

"Dad, no! The team needs me. They're gonna lose if I don't play. They'll hate me!"

"Go and get ready for bed."

• • •

It's always something with this kid. Last week it was five dollars that Mary left on the kitchen counter. "Mom said she was going to give me some money," he told his father. He turned to his mother and said, "I didn't think you would mind."

"Jim, you have to ask me first."

"If I catch you doing that again," his father said, "you'll be grounded. I don't care if you miss a game. Do you understand, son?"

"Yes, sir."

In Japan Jim once bought lunch for the entire fourth grade. The teacher, concerned about the amount of money in the boy's pocket, had called Mary. Then there was the street vendor who chased after him when Jim grabbed something off his cart on the way home from school.

His grades have been slipping. Junior high can be challenging, but Jim is smart. He picked up a lot of Japanese, and he knows everything about sports. Last week, in the back of the car, he gave an impromptu lecture on the migratory habits of polar bears. He's got a mind, that boy, if only he would use it. He needs to start high school on a sound footing. Maybe he should skip basketball next year to focus on his schoolwork.

Jim turns in his sleep, his long legs shifting under the covers. He will soon be taller than John, who is two years older. Things don't come as easily for John, but he tries hard in school. He is so quiet. Jack can't believe how two boys raised under the same roof can be so different.

He gently closes Jim's bedroom door. It's peaceful at night, the way a home should be.

"Dr. and Mrs. Maki," the principal begins, "your boy is what we call hyperactive. He has trouble sitting still."

"He's a very good athlete," Mary says. "He's also very smart."

"Fighting can't be tolerated in school, Mrs. Maki."

"But the other boy provoked him."

"We know what a good boy he is," says the principal. "But he needs to channel that energy. He can't go around getting into fights. I'm sending him home with you now. He can return on Monday."

Jack and Mary stand to leave. Shaking hands with the principal, Jack assures him it won't happen again.

"Dr. Maki, this really needs to be the last time."

"Yes, we know. Thank you."

Jim has been sitting outside the office. He has heard every word. At least he won't have to go to school tomorrow.

A few months earlier. Seattle.

"Jim, I just received a telephone call from a man who says he's a coach for a Gil Dobie League. Do you know anything about this?"

Jim has been sitting on his bed, pretending to do his homework as he listens to his father on the downstairs phone.

"He's putting a team together."

"Well, he's coming over to speak with me."

"Oh." Jim knew the coach was going to call. He tried out the other day, although he knew his dad would probably get angry. He wanted to see if he could make the team.

After tryouts he walked up to the chain-link fence where they posted the roster. Almost everyone made the team, but it was still a thrill to see his name up there: JAMES P. MAKI.

"Um, coach? I'm Jim Maki."

"I know who you are. What's the matter, son?"

"I'm not sure I can play. I wanted to see if I could make the team, but I don't think my dad will let me."

"Is he worried about injuries?"

"No, he's worried about my grades."

"Well, school's the most important thing, right? Are you having trouble passing your classes?"

"I can do okay when I want to. I get mostly Bs and Cs. My dad wants all A's and Bs."

"Do you want to play, Jim?"

Jim shrugged. "I guess so."

"Would it help if I talked to your dad?"

"Maybe."

It was worth a try. The kid was fast, he had great hands, and he could throw. Maybe a wide receiver, conceivably a quarterback. He could be college scholarship material. This wasn't the first time the coach had to make a house call.

"Hi, Mr. Maki, the name's Dick."

Jim and his father are stacking logs on the side of the house. Jack Maki looks up. "*Dr.* Maki," he says, smiling with a nod. "Jim, hand me another log."

The coach puts his hand back in his pocket. It is brisk, the air clear with a hint of fall. Fireplace weather. Football weather.

"You two need a hand?"

"No, we're fine." Jack stacks another log neatly on the pile.

"Dr. Maki, you're familiar with the Gil Dobie League?"

"No, not really. Football, isn't it?"

"Players are selected from around the city to be on teams. If they're good, they can join a senior league, the best from all over Seattle. Gil Dobie coached at U. Wash many years ago."

"I'm aware of that." Dr. Maki smiles. "I teach there."

"Yes, Jim told me. So you know he coached the Huskies to thirty-nine straight victories."

"That's interesting."

Interesting? The coach can see where this is heading. "Your son wants to play, and he's pretty good."

"Jim already plays Little League. That's sufficient. And he will not be permitted to play this spring if his grades don't improve."

"Dr. Maki, he's got it in him to be very good at football. He might be scholarship material."

"I'm a college professor. We have other avenues for scholarships."

"I can see your son playing quarterback if he works at it, Dr. Maki."

Quarterback! *Please, Dad.* Jim crosses his fingers behind his back, his palms sweaty.

"Jim needs to focus on his studies."

"Excuse me for saying so, sir, but it has been my observation that when a young man finds something that is very important to him, he will usually try to protect it. Jim says he can handle schoolwork when he tries."

"That's exactly right. When he tries."

The coach stands there quietly. Antagonizing the boy's father will get him nowhere.

"I'll hold Jim's spot for a week. Let me know if you change your mind."

He gets back in his car. Through his rearview mirror, he watches the boy run into the house. He can't quite make out if he is crying.

April 8, 2009, early morning. Hospital conference room.
Late last night Dr. Pomahac called his team members to tell them the good news. He swore them to secrecy, and most of them didn't sleep much after the call. Pomahac didn't sleep at all.

Now they have gathered in a very crowded conference room. Thirty-eight doctors, nurses, and technicians will attend the surgery. Pomahac reviews each step, checking everything one last time with each of them. The donor family has planned a bedside service with their rabbi at four o'clock this afternoon, and then they can begin. They expect the surgery to take more than sixteen hours.

Dr. Christine Kim, Jim Maki's hospital psychiatrist, is also there. Pomahac asked her to work with Jim almost as soon as he had him in mind for a face transplant. Even now the surgeon is haunted by what he saw the night Maki was carried into the Emergency Room after his face-first fall onto the electrified rail. He has shown Dr. Kim pictures of Maki that were taken after several rounds of plastic surgery. He didn't share the pictures he took that first night, when Maki came through the door. They were too horrible. So he asked his wife to sketch some black-and-white pencil drawings based on those photographs.

"I haven't mentioned a transplant to Mr. Maki yet," he told Dr. Kim the day she viewed the disfigured profile on the screen. "Would you meet with him and see what you think?"

Maki was guarded in their initial encounters. He hadn't requested a psychiatrist, and he wasn't sure why Dr. Kim was taking an interest in him. She seemed to be assessing him for something, but what? Years of living dangerously had taught him to be wary. Only later, when Dr. Pomahac asked if he would consider face transplant surgery, did he figure it out.

Her job was to determine whether Maki would be able to cope with what lay ahead, although nobody knew exactly what that might entail. He might be the first American to have a face transplant, with all the public scrutiny that would follow. For the past four years he has had virtually no contact with the outside world, let alone with members of the press, who will surely come calling if and when his identity becomes known after the surgery. He has been living in virtual seclusion, slowly recovering from the subway accident and the ten surgeries he has already endured. He is also a recovering drug addict. And he carries the emotional scars of a war gone wrong and, ever since the accident, physical scars as well.

Dr. Kim is well aware that war veterans sometimes have special issues. A couple of years ago she was urgently paged to attend to a patient who was having a major panic attack. He had fought in Korea and became hysterical when he heard a helicopter landing on the hospital roof. The loud thumping of the blades near his window brought back terrifying memories of the war. Now, as her colleagues drone on in their medical jargon about Maki's impending operation, she wonders again about his mental health. From all indications the medical team is ready. But what about Mr. Maki? She thinks he'll do well, but her psychiatric assessment is based as much on art as on science, and there is always room for an unexpected outcome.

In another part of the hospital, almost a city block away, two men arrive at Joseph Helfgot's room. Dr. Marcelo Suzuki is a professor of dentistry at Tufts University School of Dental Medicine, and with him is a young resident.

Suzuki, one of Maki's doctors, is a maxillofacial prosthodontist, a dentist who specializes in building parts of the face that stand in for a person's eyes, nose, or ears. Most of his patients are cancer survivors and burn victims. Dr. Pomahac has asked him to do something about the gaping hole he will leave when they remove Helfgot's face. Suzuki is here to design a mask, a mask made of silicone. He is a master craftsman, and many of his prosthetic facial devices can pass for the real thing. He's had complicated cases before, but this is completely unprecedented.

"Can I help you?" asks Lisa Kelley, the ICU nurse, as she adjusts a catheter.

"We're here from Plastics, to do some modeling."

"Of what?" she asks.

He is careful not to disclose too much. If she doesn't know about the transplant, it's not up to him to tell her.

He is barely listening as he stands in shock, gazing at Joseph Helfgot's face. The man has a beard! Pomahac didn't mention *that*.

He can't set a mold over a beard. The family will be coming in this afternoon for a final goodbye, so shaving it off is not an option.

He mumbles something about Vaseline as he and the resident start removing items from a black bag.

Lisa has been a nurse for a long time, and she knows when to back off and mind her own business. She leaves the men to their work, but she continues to watch them. She feels protective of her patient. She is also wondering what on earth these guys are up to.

Dr. Suzuki takes a large glob of Vaseline and begins to spread it over Helfgot's beard. Then he takes some kind of white casting material that looks like clay, puts it on top of the Vaseline, and works it to the sides before adding more to the cheeks and neck. When he is done, he and the resident stand around for a few minutes making small talk. He taps the white stuff a few times to check its consistency and then starts pulling the hardened material off Helfgot's face. It resists, sticking to the beard, and Lisa hears him mutter something under his breath. The resident is looking at her, so she turns to the computer screen, trying to appear engrossed in something else.

When they leave a few minutes later she sees that Helfgot's face is bright red where the clay has dried. But it doesn't seem to be clay. It's too shiny and too thin. Little white specks of it, together with Vaseline, are embedded in his beard. She puts some cream on her fingertip and massages it onto his cheekbone. She hopes Susan will get here before she has to go.

Lisa is leaving for Africa tonight, along with others from the hospital, including Kevin McWha, who has often nursed Joseph, and Dr. Rawn. It's their second year volunteering as a Brigham-based team at a heart clinic in Rwanda. Last year, when Joseph learned about the project, he helped the nurses pay for their travel expenses. She hopes she'll have a chance to say goodbye to Susan. She is sorry this is happening as they are leaving the country. She is sorry this is happening at all.

A doctor enters the room whom she recognizes from a surgical rotation a few years ago. She remembers him having a foreign accent. Dr. Pomahac nods to her and takes a quick look at Joseph. "We're good," he says to himself before he vanishes. What was *that* about? When she looks through the glass at a young physician, he gives her a *Don't ask me, I just work here* shrug.

April 8, 2009, 1 p.m. Emergency Room.
Jim Maki has been downstairs for a couple of hours in a special area of the Emergency Room. Everyone seems happy to see him. One of the nurses who took care of him during his most recent surgery stopped by with another nurse to help him pass the time. Since then it's been quiet. He'd like some lunch, but they won't let him have his usual thick shake, which seems unfair. Everyone around him looks busy, but he has nothing to do but wait.

Dr. Kim arrives. She is looking for signs of distress, but it's difficult with this patient, who has almost no face to observe.

"How are you doing, Mr. Maki?"

"Fine."

She marvels at his apparent calm. "Are you nervous?"

"Nope."

"Why not?"

"I just never am. I've been under the knife before. I'm ready to go." She believes him. This will be his eleventh operation under Pomahac's steady hand. "Do you know about the donor? Or how he died?" So he has been thinking about that. She wondered when that would happen. Recipients are almost invariably curious about the person who is giving them an organ, and their curiosity is typically accompanied by feelings of guilt. Somebody had to die to make this donation possible.

Kim has been reading up on transplant recipients ever since Pomahac asked her to work with Mr. Maki. She knows that after some time has elapsed, organ banks help facilitate an anonymous exchange of letters between recipients and the donor families. It is left up to the recipients to decide if they would like to become personally known to the family—assuming, of course, that the donor family wishes to be in touch. Kim can't imagine that happening here. Given the nature of the transplant, she is fairly certain the donor family will want to maintain their privacy.

All she knows is that the donor died in this hospital and that he was roughly the same age as Maki. Ever since the new HIPAA rules about privacy it is difficult to know anything about someone else's patient. And even if she did know something about the donor, she would not be allowed to tell Jim Maki.

"I don't know how he died," she tells him. "All I know is that he was a patient here and that he died a couple of days ago."

"Do you know how old he was?"

"Yes. He just turned sixty."

"I'll be sixty in September. But you don't know how he died?"

"No, I really don't."

"Was he married?"

"I don't know that either."

Maybe he too was in Vietnam. He wonders if their lives had been at all similar. He thinks this guy probably grew up in Boston.

"Do you know what he did for a living?"

"No. You know as much as I do. I realize how frustrating it must be not to know more, but the hospital is very strict about patient confidentiality. I'll stop by a little later. Try to stay relaxed, okay?"

Jim has one other question on his mind, but he doesn't ask it. Although Dr. Pomahac has repeatedly assured him that he will not resemble the donor, he can't help but wonder. This man who died, whose face he will be getting—what did he look like?

chapter nine

Vesalius stands at a table, his hands covered in sticky blood. He is all of twenty-four, and the students who surround him are not much younger. Arriving from Leuven, Belgium, after a heated argument with his adviser over the best method of blood-letting to use on feverish patients, Vesalius so impressed the Padua professors that they soon awarded him a doctorate. Young men flock to his workroom, and on this late winter afternoon they strain to watch his surgical technique. As the winter twilight descends, he slices open the abdomen of a young female cadaver. Working quickly, before nightfall, is part of the challenge, part of what makes this exciting.

His most promising student, a young man named Gabriele Fallopio, stands opposite his mentor, assisting him and recovering one of the ovaries while Vesalius removes the other. Now, quickly, to the other room to preserve them in brine until tomorrow's light. Fallopio, who is obsessed with human anatomy, is fascinated by the mysterious origins of human life, and these walnut-shaped spheres, connected to thin, anonymous tubes, are part of the puzzle.

The students are excited. Tonight there will be a dinner to celebrate the private publication of the professor's anatomical drawings, and some of theirs too, in a small collection. Vesalius has arranged for a local bookbinder to put them together, and in a few hours they will finally see the finished volume.

The class has become much more lively since their teacher arrived. Vesalius likes it here because the Venetian Republic permits him to conduct his research on the human body without interference from the Church; almost anywhere else he wouldn't be allowed to work on cadavers. The students realize that his dissections represent a significant advance in the field, but they have no idea how much his work on human anatomy, and some of theirs as well, will influence the future of medicine.

Their drawings will become the basis for a groundbreaking work, *De humani corporis fabrica—The Fabric of the Human Body*. What will evolve into a seven-volume atlas will still be in print in the twenty-first century and will form the scientific basis for the field of developmental morphology. Centuries after its publication Charles Darwin will be heavily influenced by the work of Vesalius when he formulates his theory of natural selection.

April 2008. Experimental Morphology Lab at
Université Catholique, Leuven, Belgium.
Julian Pribaz, the director of Harvard's Plastic Surgery Training Program and a reconstructive microsurgeon at the Brigham, landed in Belgium early this morning with Bo Pomahac, his colleague and former student. They are jetlagged, but their adrenaline is high as they walk the narrow streets where Vesalius hurried along almost half a millennium ago. The visitors from Boston are here as guests of the head of the anatomy department, Dr. Benoît Lengelé. Lengelé travels frequently to France, where he collabo-

rates with Dr. Bernard Devauchelle and his colleague Dr. Jean-Michel "Max" Dubernard from Lyon, who stunned the world with the first successful hand transplant in 1998.

Several years later, in 2005, Dubernard shocked the world again with the news that he had transplanted a human face. This announcement was not well received. Many doctors denounced the transplant as nothing more than a stunt. They believed the medical risks did not justify an operation that was not lifesaving.

But not everybody felt that way, especially Dr. Joseph E. Murray, who performed the world's first successful organ transplant in 1954, a kidney donated by a living donor to his identical twin brother. The double surgery that made Joseph Murray famous was performed in Boston at what was then known as the Peter Bent Brigham Hospital. It took a while for the judges in Stockholm to get around to it, but in 1990 they awarded Murray and his colleague, E. Donnall Thomas, the Nobel Prize for Medicine.

As a young physician during World War II, Joseph Murray had worked at Valley Forge Hospital in Pennsylvania, the country's leading burn center. Some soldiers arrived with such extensive burns that skin grafts normally taken from other parts of their body were not an option; there was simply no skin left to take. Instead Dr. Murray and his colleagues would graft skin from fresh cadavers. But sooner or later the soldier's body would reject the foreign skin. Still, these grafts bought precious time for the soldiers' wounds to heal from within, and for other parts of the body to recover enough to allow for an autograft, where rejection was not an issue.

Murray had trained at Harvard Medical School, and after the war he returned to one of its teaching hospitals. His desire to solve the fundamental problem of rejection led to an association with Peter Bent Brigham Hospital's Surgical Research Laboratory, which was established in 1912 by Dr. Harvey Cushing, the father of modern neurosurgery. As the Brigham's new surgeon in chief, Cushing insisted that his lab be housed in the Harvard

Medical School's quadrangle around the corner, where he was able to use animal specimens and introduce students to the latest surgical techniques.

Years later it was there, under the guidance of more experienced physicians, that Murray and his colleagues successfully transplanted kidneys into dogs. Eventually they were able to carry out the first successful human kidney transplant between identical human twins, born from a single egg that split in the womb. They shared duplicate DNA, making rejection virtually impossible, as each brother's immune system recognized the other as itself. In 1954 the kind of DNA testing that would demonstrate that the twins shared the same genetic material did not yet exist; DNA's chemical structure had been discovered only a few months earlier. Instead, the young men's fingerprints were analyzed at a nearby police station. They were similar in every way, much more so than even fraternal twins.

Half a century later Dr. Murray was thrilled when he heard about the first face transplant. A few days later he attended a meeting with Elof Eriksson, his successor as chief plastic surgeon at the Brigham. Referring to the naysayers, Murray said, "Elof, this controversy is exactly the same one I went through after the first kidney transplant." Back then a research fellow who had wanted to study with Murray was warned off by his chief, who dismissed Murray and his colleagues as "a bunch of fools." And a Harvard professor who was one of Murray's closest friends took him aside and said, "Joe, please don't get involved in this. It will never work, and it could ruin your career."

Joseph Murray was frustrated and disappointed by the negative reaction that Max Dubernard was getting. He knew that face transplants could repair broken lives, but there was a more personal reason for his dismay. Forty years earlier Dubernard had been Murray's student. Murray liked the young, chain-smoking French doctor who had come to Boston in the mid-1960s to study under his guidance. They became close, and by the time Dubernard returned to France he had learned a great deal about human organ transplantation. It was said that he had also left behind a few broken hearts.

As he reminisced with Eriksson about his newly famous protégé, Murray asked, "So when do you think *we'll* jump in?"

"We'll get there," Eriksson replied. He ran through a mental list of the young surgeons rising up at the Brigham. They all had talent, but talent wasn't the issue. In 2005 the United States wasn't ready for face transplants. They needed a young doctor with the necessary fire to bring this advanced surgery to the Brigham.

The doctor who best fit Eriksson's description was Bohdan Pomahac, but Eriksson didn't know that yet. Neither did Pomahac—not until 2008, when he sat in a lecture hall and listened to a presentation by Max Dubernard. Eriksson had just established a visiting lecturer series in honor of Joseph Murray, and Dubernard was given the honor of inaugurating the event.

Dubernard told the audience that he had been thinking about facial transplantation ever since he had performed the first hand transplant. He imagined a single surgery that would encompass the transplantation of part or all of a human face: muscles, nerves, blood vessels, bone, and skin. He conceded that enormous advances had been made in plastic reconstructive surgery, but even after multiple operations the most severely disfigured patients were left with woefully inadequate results.

For some of these people war injuries had shattered their jaws and cheeks. Others survived blazing infernos with their skin and nerve endings burned away, which required weeks of debridement to remove slowly decaying flesh. Still others had been badly mauled by animals or attacked by savage tumors.

Lifting even part of a donor's face with millimetric precision and fitting it perfectly onto another person was close to miraculous. And as with any transplant, the recipients would have to take immunosuppressant medications for the rest of their lives. These drugs have serious side effects, which can be deadly. Although skin is an organ, along with the heart and the kidneys, some questioned whether a face is really necessary for survival. Was it worth the risks that such a transplant would entail?

Dubernard told the audience that his own doubts were erased when, at Dr. Devauchelle's request, he visited a woman named Isabelle Dinoire, who had survived a brutal attack by her dog after she overdosed on sleeping pills. She had fallen and hit her head, and some people believed the dog had been trying to wake her up. Whatever the reason, the savage encounter left the poor woman without a nose, lips, right cheek, and chin.

After one look at Dinoire, Max Dubernard jumped into action, taking Devauchelle with him. He thought about how he would feel if Isabelle were his daughter. This woman needed help, and these two doctors were going to do everything they could to provide it.

They began to treat her wounds in a new way, forgoing the usual repair regimen of one surgery after another. Instead they focused on intensive physical therapy to keep her face muscles strong and scar tissue at a minimum as they waited for approval from the French authorities. While the committee considered the question of a face transplant, the French surgeons performed facial dissections on cadavers, practicing over and over again. Practice and more practice, they believed, was the key to success in the operating room.

Pomahac sat there transfixed as he listened to Dubernard. He had first discussed facial allograft with Dr. Eriksson soon after the French team performed the groundbreaking surgery in 2005. He even began to formulate a protocol for face transplants at the Brigham, with the idea that they would be offered to patients who were organ recipients and already on lifelong immunosuppressant drug therapy, so a face transplant wouldn't subject them to any additional risk. Pomahac received preliminary approval in 2007, but nothing had come of it because he didn't have any patients who had already received an organ transplant and who also needed a face. How many such people could there be?

Now, as he sat in the lecture hall listening to Dubernard, he thought about several of his patients, including Jim Maki. They needed faces, desperately. By the time the lecture was over Pomahac knew this was something he was going to push hard for. There was simply nothing, *nothing* that could help these patients short of a face transplant. He had to try.

After the lecture Eriksson introduced him to Dubernard, who told Pomahac and Pribaz, "You guys have to do this," adding that his door was open whenever they were ready. Dubernard understood, just as Joseph Murray had half a century earlier, that only two things would quiet the scathing criticism he had received for the Dinoire surgery: time and more facial transplants. Who better to make that happen than Murray's successors?

Back in 2005 American medicine was not ready for face transplants. After Isabelle Dinoire's surgery, the American Society for Reconstructive Microsurgery released a position paper opposing them. In their opinion the potential for rejection, coupled with increased risks of infection, cancer, and diabetes from immunosuppressant drugs, far outweighed any physical improvement that a face transplant might provide.

And anyway, how important were those improvements? "Most patients with facial deformity adapt quite well and accept their physical appearance as 'self,'" the authors of the paper argued. "The psychologic repercussions of a facial transplant on family and friends of both donor and recipient cannot be underestimated. The ethics of inflicting an untried, and potentially fatal or deforming remedy for the purposes of advancing science must be carefully weighed against the Hippocratic credo of doing no harm."

Bo Pomahac, himself a member of the society, understood these reservations, and for the most part he shared them. But his focus was on extreme cases, patients whose problems went well beyond disfigurement, who often needed a feeding tube to eat and a tracheostomy tube to breathe. How they looked was the least of their problems.

It took a while, but when Pomahac finally received approval from the Brigham's review board, he took Dubernard at his word and made travel plans.

He and Pribaz flew to Brussels last night and now sit across a table from Devauchelle and Dubernard. The review board and their colleagues at home have agreed that James Maki is the person most in need of a face transplant, and the Boston doctors have brought his entire medical history with them.

They discuss Maki's case, and the French team agrees that he is the logical choice. For the next three days, under the watchful eyes of the men who first performed this groundbreaking surgery, Pomahac and Pribaz practice on cadavers in Dr. Lengelé's lab. Pomahac is impressed by Devauchelle's surgical technique. He is a skilled guide, and Dubernard is a brilliant communicator. By the fourth cadaver the Americans have reduced the time it takes them to recover a face from four hours to an hour and a half.

Before they fly back to Boston, Pribaz and Pomahac are taken to dinner by their hosts. Incredibly this is the first time the entire French team has been back together since the face transplant they performed in 2005. They have brought along a surprise guest: their patient, Isabelle Dinoire.

She looks fantastic, Pomahac thinks. He watches her eat, savoring each bite without difficulty. Although she speaks softly, her voice is clear and strong. She dines easily, talking and chewing and sipping water without any apparent effort, and laughs at jokes. She even slips outside with one of her surgeons to enjoy a quick cigarette. Now *that* would make a hell of a picture, Pomahac thinks. But of course the French do things differently.

"I have a lot to do when we get back," he tells Pribaz on the plane back to Boston. As it turns out, he has exactly one year.

chapter ten

June 30, 2005. Malden, Massachusetts.

for the past few months James Maki has been living in a half-way house on Boston's North Shore. As a former heroin user he's been on methadone for years, but he still scores other drugs, mostly pills, which are easy to find on the streets of the city. He pretends to be sober and is very good at fooling people, which is one reason he has survived for so long. As long as he doesn't bring drugs into the house, they can't throw him out. And he can't afford to get thrown out because he has no place left to go.

Cynthia has long since washed her hands of him. When their daughter, Jessica, was born in 1985, he made all kinds of promises that the past was the past. He was in jail the day Jessie was born, after a policeman found a large amount of marijuana in the trunk of his car. Cindy was with him, pregnant with Jessie. When he was sent away, his lawyer took Cynthia under her wing. They were friends from college, and the lawyer persuaded her to take the LSATs, the law school admission test, and perhaps go on to law school. Cynthia didn't do well on the test, but she went back to school and became a paralegal.

On the day Jessica was born they let Jim out of jail to see his new daughter. The prison warden had coached basketball at a western Massachusetts high school and remembered Jim as a gifted player. He knew he wasn't dangerous, merely troubled. A police officer dropped him at the hospital, promising to return in a few hours. When Cynthia gave him the baby to hold, he kissed her tiny forehead and took in her baby smell. He couldn't believe they had created such a beautiful child.

Cynthia got her own apartment, and Jim showed up the day he got out. He promised there would be no more drugs and that he would stay on his methadone treatment and clean up his act. He said he would find a real job rather than picking up the odd painting or housecleaning gig like so many of the users he knew.

He was so glad to be out of that hellhole. The guards treated him as if he were stupid, as if he were nothing more than dirt. If you took more sugar packets than you were supposed to, they called it stealing.

On his first night back with Cindy he went out to celebrate his new freedom. He didn't return until the following night, high as a kite. Although she was furious, Cynthia couldn't bring herself to tell him to leave. He would come and go, and she let him hang around when he wanted to.

One evening when Jessie was in preschool a social worker showed up at their door. A neighbor had called, concerned about some of the activities at Cynthia's apartment. Jim's mother was very sick, and Cynthia's mother was seriously depressed, so Jessie stayed with a friend while Cynthia got some counseling. The social worker gave her an ultimatum: If Jim didn't move out, they would take away her daughter.

Jim or Jessica. That was a tough year for Cynthia. Her mother-in-law, the one person who had always been there for her, was completely bedridden, slowly dying of rheumatoid arthritis that was attacking all of her bones. Cynthia felt alone, vulnerable, in a way she never had before.

Mary Maki was a love, kind and sweet and gentle. On weekends Cynthia would bring Jessica to spend the day with her grandparents. When Jessie heard "We're going to Grandma's today" the little girl's eyes would light up in delight. They would drive to Amherst, in the western part of the state. Dr. Maki had been teaching at the University of Massachusetts since Jim was a senior in high school and had built a Japanese-style home.

Jessica would walk around the property with her grandmother, exploring its delights. They would pick flowers together, and Mary would sit with the little girl and arrange them, along with the twigs and moss that Jessie found, into small masterpieces of design. Jessie would present the arrangement to her grandpa, and she and Jack would play cards together while Cynthia and Mary sat in the kitchen as the rice steamed. But then Mary began to fade.

Jim came to see her shortly before her death. He was not allowed to visit his wife and daughter, and he missed them terribly, especially now that his mother was dying. He had never felt so low in his life.

His father and a hospice nurse were there when Jim arrived. "Don't say a word about Jessie having to stay with a friend," Jack told Jim. "Your mother is already worried sick about what will become of you after she's gone, and I don't want her to start worrying about Cynthia and Jessie."

But Jim couldn't help himself. His mother had always comforted him when things were bad. She knew what to say to make things better. Jim told her everything, and after he left, Jack found her in tears. She died soon after that visit, and Jack blamed his son for hastening her death. He never forgave him, but he did write Jim a check after her funeral. Jim was entering a new halfway house and needed some spending money. "But if any of this goes to drugs," Jack said, "there won't be any more, ever. This is the very last time I'm going to bail you out."

A few days later the woman who ran the halfway house called Dr. Maki. He had asked her to let him know if Jim was looking for money, and apparently he had asked for a loan. Jack had given him two hundred dollars, more than enough to cover incidentals like toothpaste and shampoo. The professor was true to his word; he never gave his son any more money. When Jim called he would grunt a few words and hang up.

After his mother's death Jim moved around from place to place, dropping in on Cindy intermittently when he needed food or a little money, or a place to take a shower and sleep for a day. She was never able to turn him away. He was, after all, her husband and Jessie's father. And he didn't have a mean bone in his body.

She blamed Vietnam for his troubles. That was where he first tried heroin and where he saw some terrible things. She sometimes thought there was a demon inside him.

Once he was so desperate for a fix that he threw a television out of his parents' bedroom window. He was sick with need, but Jack refused to give him money. The television landed on a solar panel below, shattering the glass. Jim was so apologetic afterward. "You don't know the grip heroin has on me," he said. "I'm sorry it happened, but there's no rectifying it." Not much later he entered rehab yet again. No matter how many times he cleaned up, though, Jim and trouble always managed to find each other.

In 1999, while visiting Cindy, he came out of the shower and put on clean clothes. He walked a few feet and collapsed, straight to the ground. Jessica was thirteen, and she watched in horror as her father lay there, convulsing.

"Jessica!" her mother screamed. "Go outside and stay there!" An ambulance rushed Jim to the hospital. He had suffered a brain aneurysm that should have killed him then and there. Thank God he'd been at Cynthia's. It was touch and go for several days, and he struggled for a long time to get back on his feet. But even after another lease on life, he soon turned back to drugs. That was the last straw for Cynthia. She was done with him, completely done.

Now, on this fine summer morning in 2005, Jim has jumped on the train from Malden heading into Boston. It's easy to score in the city, and he has a little money from working for a guy who pays him to clean houses. This guy's no dummy. Everyone he hires uses drugs, and he knows it. He pays them first thing in the morning, rather than after they've finished working. That way they can go off and score, and come back and work like crazy men, not even stopping for lunch. If they don't return to work, they won't get paid the next morning, which means they won't get their fix. He's got it all figured out, this guy.

It's one of those days you hope will never end—not too hot, but nicely warm with a gentle breeze. New Englanders wait months for a day like this. Night has fallen, and Jim and a few buddies are hanging out in Southwest Corridor Park, a five-mile strip of grass and trees laid down when the old Orange Line tracks were sunk underground a decade earlier. There's a pickup basketball game under the lights, and Jim watches the young men running up and down the court. He's jealous of their youth, but he enjoys watching them play. It's late, and he should probably be getting on a train back to Malden. It could be after midnight, but he's not sure. He is feeling no pain. Time is moving slowly, like it always does when he is high. He is dizzy and a bit shaky, but he'll be all right. He'll sit here until the dizziness fades, and then he'll get on a train home. The Ruggles stop is nearby. A lot of people are still out, walking around, sitting on park benches, trying to extend this beautiful evening. There's no rush. He'll just sit here a while longer.

• • •

A few hours later. Trauma and Burn Unit.

Bo Pomahac is on call, dozing in a chair. He's exhausted. Having completed his surgery program just last year he gets the tough shifts. Hanka and Bo had their first child, a girl, barely two years ago, and now they have a baby boy. The little guy is one now and is trying to pull himself up to a standing position. Their daughter is a moving ball of energy. How Hanka keeps up with them all day he'll never know.

He is startled out of sleep by his beeper calling: a burn victim, somehow electrocuted, is coming in. The paramedic in the ambulance is briefing the Emergency Room staff on the speakerphone as Pomahac rushes in. A man has fallen face-first onto an electrified subway rail at the Ruggles station. Miraculously he still has a pulse.

They crowd around the speakerphone. The paramedic is saying he's never seen anything like this.

"How bad is it?" asks one of the doctors.

"Burns. The middle of his face, or what's left of it, it's actually kind of gone. His right arm is really burned up, especially his hand. Jeez, he's a mess."

"Is he breathing?"

"Intubated."

They hear the wail of the siren approaching outside. The paramedic keeps talking as he and his partner bring the man in on a gurney. The ER team brace themselves, preparing for the worst as they try to picture what they have just heard. But it's impossible. What does a man without a face look like?

The gurney bursts through the door. Pomahac stands back, behind the triage team. He takes one look at the man and he's stunned. He has to do something, but what? He has never seen anything like this. The man has no nose, no upper lip, and no facial skin. Pomahac clamps his teeth together, the adrenaline rushing through him as he fights off waves of revulsion.

"There's no ID on him."

"I don't think he's going to make it," says one of the docs.

"Get Social Services down here. We've got to find his family."

"We're losing him!" Without a family member to stop them, they resuscitate. But nobody believes he will last the night.

chapter eleven

Wednesday, April 8, 2009, 2 p.m.
Helfgot residence, Brookline, Massachusetts.

S usan is making final preparations for Joseph's funeral, which
will take place Friday morning. It's a little tense as both the
temple and the funeral home try to accommodate the family.
She called them on Monday, and Jewish law encourages a timely
burial. But relatives are still arriving from abroad; Joseph's daughter, Emily, has just come from Paris, and a cousin is flying in from
Israel. The Helfgots are not a big family. Joseph's mother, father,
and sister are all dead, and almost all their other relatives perished
in the Holocaust. Joseph's adult nephew, Bobby, is playing cards
with Jacob at the kitchen table. Bobby looks lost. Uncle Joseph
was his stand-in father, and his mother, Joseph's sister, is dead.

Passover starts tonight, and funerals are not normally held
on major Jewish holidays. The director of the funeral home has
been in touch with the secular Independent Workmen's Circle
Cemetery, a fitting resting place for this hardworking child of
immigrants. Technically Joseph too was an immigrant; as a child

he came through Ellis Island with his parents. The cemetery officials have agreed to bury him on the second day of Passover, if
they can find the space.

A delivery van pulls up to the house with food, flowers, and a
case of wine from Joseph's friends at HBO in New York. The driver
brings everything into the kitchen and hands it off to Susan's girlfriends from Los Angeles. "Sue, where should we put this?"

I don't give a damn where you put it. "Oh, anywhere is fine. How
about taking the wine out to the garage? And I think there's a
vase on the top shelf."

*Everyone is sending over boxes of matzoh. We should be making a Seder, not
planning a shiva. Joseph, do you remember when we were buying this house?
After the Realtor told us about the heating and the roof and the plumbing,
you asked, "Suze, how many can we squeeze in here for the Seder?"*

*Oh, sweetie, your Haggadah. I forgot all about it. The one you just
finished making. Everyone with their assigned parts, all thirty-four of
them. Even the man who delivered the rug two weeks ago! Did anyone
tell our guests not to come tomorrow for the second night? I guess they all
know you are dead. Well, maybe not the rug man. I'm glad I didn't take
your Haggadah to Kinko's to make copies. I was going to, but then Dr.
Lewis called and we went to the hospital. Oh no, the food!*

"Ben, I need you to run over to the Butcherie. I think they're still
open." They'd better be. A large turkey and an eight-pound bris-
ket are waiting to be picked up to supplement the gefilte fish, the
homemade chicken soup with matzoh balls, the chicken breasts,
two whole salmons, vegetable tsimmes, artichokes, the spinach
soufflé, and various other dishes for a group of almost three dozen
participants at a wonderful Seder that won't take place this year.

Susan must attend to a less pleasant task. Accompanied by
her two adult stepchildren, she walks over to Stanetsky, the Jewish funeral home on Beacon Street. The air is heavy with mois-

ture, but at least the rain has stopped and the day is mild. It feels good to be outside and away from all those well-meaning but inane questions about what should go where.

She is carrying one of Joseph's suits in a garment bag. They bought it last year for Jacob's bar mitzvah and had the tailor put in special panels to hold the five-pound batteries that kept the VAD machine going during the service. The alterations were surprisingly expensive. "*Gonif*," Joseph said when the tailor told him what the panels were going to cost. "I'll give you half, or forget it."

"Sweetie, don't start," Susan said as Joseph sat there with an oxygen cannula up his nose. "I don't have time to be sewing pockets in suits." In the end the tailor knocked off a few dollars. *Gonif*, Susan thought as she drove them home.

This is so weird. Four years ago I walked my mother's clothes down the street in the other direction, to the Irish funeral home. It's closer than Stanetsky. Come on, Susan, what does that matter?

"What day is it?"
    "I think Wednesday," says Emily.
    "What's the date?"
    "It's the eighth," says Jon.

They declared him dead yesterday, April 7. He really died Sunday, during surgery, but April 7 is on the death certificate. Mom died on the seventh. Joseph had just gotten out of the hospital and was home in bed. I flew Mom to South Carolina to be buried next to Dad. Hardly anyone came to her funeral. Everyone is coming to this one. I don't think I can do this.

They enter the funeral home. Susan hates this place, which brings back painful memories of other funerals she has been to, like the one a while back for Sol Levine, Pam's father. Raised as a devout Catholic, she worries whether she'll know what to say to the funeral director. The logistics of this death have not been easy, and she sensed some discomfort on the phone when she spoke to him, especially when she explained that Joseph's body won't be released from the hospital until tomorrow afternoon. And Passover starts at sundown tonight.

"I hate this," she says.

"We'll do it fast," Jonathan assures her.

The funeral director ushers them into his office. Emily slumps into a chair, jet-lagged and still in shock. A riot of soft honey-red curls surrounds her face, which is pinched from so much crying. She swoops it up and stuffs it into a scrunchie.

The director tries to appear relaxed, but he must be eager to get home to prepare for his Seder. They review the wording for the obituary. Then Susan leaves with him to pick out a casket. Emily and Jon can't bring themselves to go with her.

It has to be cherry, like my mother's coffin. He loves cherries—no, I guess that should be loved cherries—more than anything, except maybe watermelon. God, he loved watermelon. This is, what, my fifth casket in ten years? I should have bought them in bulk! Joseph, his sister, his mother, and my parents—they're all gone so soon.

"How much is the cherry?"

The funeral director describes the silk lining and the Star of David. "It's a bit steep," he cautions.

"Whatever. The lining . . . I don't care, it's fine. I'll take it."

She stands in a daze, looking around at a room full of caskets.

God, please don't make me come back here ever again.

As they head back to the director's office Susan is barely able to hold herself together. "I picked out one just like Grandma Rachel's," she tells her stepchildren, "except this one's cherry, like my mother's. Grandma's and Auntie Pauline's were oak." They don't remember. Young people don't dwell on those kinds of details.

The funeral director clears his throat. "I have good news. It took a while, but we found a double plot at Workmen's Circle in West Roxbury. They're the only ones who have a double available, and they're willing to do the funeral on Friday." He pulls out a map indicating where the plot is located. He is relieved that it all worked out. He quotes a price and folds his hands on the desk. He looks at Susan, waiting for her response.

"Excuse me, did you say a *double?*"

"Yes, for you and your husband to be buried together."

"But I'm not Jewish." She winces as she says it. She knows that only Jews can be buried in a Jewish cemetery.

"You're not?" He is incredulous. Because she has shown so much familiarity with Jewish death rituals, he naturally assumed . . .

"No, I was born Catholic."

"I see." But he doesn't. He looks down at his papers, suddenly at a loss. All this careful planning, and she's not Jewish? He glances at the clock on the wall. It's almost three. Sundown isn't that far off. No one at the cemetery office will still be there.

"You know," says Jonathan, "my dad was kind of a big guy. I mean, he lived big. He always flew first class. He liked a lot of space."

The director doesn't follow.

"We'll take the double. Put him in the middle."

He nods. "Do you have the clothing for the deceased?" Susan hands him the bag.

"Smooth," Emily tells her brother as they walk out the door. "Really smooth." They all crack up laughing.

Joseph, are you listening? Jon is so fast and so funny, just like you. And so strong. I didn't know he was that strong. I could never have given Jacob the news without Jon by my side. And Emily misses you so. She looks so much like her grandmother and can size people up just as fast. She's so sharp. Whatever are we going to do now?

April 8, 2009, mid-afternoon. Department of Plastic Surgery.
Dr. Pomahac is meeting with his team one last time. The surgery is now scheduled for 6 p.m., right after the Helfgots say their final goodbyes. Pomahac spent some time this afternoon visiting with Jim Maki, who is eager to get started. Then he had a quick talk with Peter Brown from Public Affairs.

"They want to put a wire on you," Brown said, as though it were the most natural thing in the world. "The cameras will be filming throughout." For several months, a television production team has been filming *Boston Med*. Now, a few weeks before they are scheduled to leave, they have been invited to capture this historic surgery. Nobody expected they would find a donor before the crew left.

"You want me to wear it the whole time?"

"Ten years from now we will all be very happy we caught this on tape. And you can turn off the mike whenever you want."

Pomahac remembers a TV interview that Brown arranged for him a few years ago about facial transplant surgery. The resulting publicity became the grease for the slow wheel of approval for this project. Peter has never steered him wrong.

"Okay. You're the boss."

At the final team meeting Pomahac runs down his checklist one more time. Dr. Elof Eriksson is content to sit and listen, to do whatever it takes for this young man to be satisfied that nothing has been left to chance. At this point, he knows, the only thing left to do is to get the donor and the recipient into their respective operating rooms so they can start. Eriksson has been head of plastic surgery since 1986, when he took over from Dr.

Murray. Although he's been part of several breakthrough moments, he has never seen a group of medical people as invigorated by their task as this team is today.

Most of the credit belongs to his protégé, Bo Pomahac. There have been many administrative hurdles. Before they finally gave their approval, the review board came back time after time, requesting certain changes. The finance department had to waive the cost of the operation, and the medical team had to donate their services. He is proud of what Bo has been able to achieve. It was not easy. He knows that Joe Murray is proud as well.

Behind Pomahac's quiet demeanor is a determined soul, as Eriksson discovered a decade earlier, when the twenty-five-year-old sent him a letter requesting an interview. Pomahac, who was about to graduate from medical school in the Czech Republic, wanted to continue his education in the United States. He didn't know anything about Eriksson other than his name. Because it sounded European, Pomahac figured that Elof Eriksson might be willing to meet with a young visitor from central Europe. It was a good guess.

Thirteen years earlier. Brigham and Women's Hospital.
"We have nothing open at the moment," Dr. Eriksson reminds the earnest young man who has come to see him. He had told him the same thing on the telephone. Pomahac has just arrived in Boston, two days after earning his diploma from the Palacky University School of Medicine in the Czech Republic.

"I was hoping you might find something for me to do. I'll do anything you need."

Dr. Eriksson is impressed by the young man's sincerity and pluck. "We've got a lot going on here," he says as they step into Eriksson's Tissue Repair and Gene Therapy Research Lab. Soon they hope to transplant healthy cells directly into patients' wounds to encourage healing. Before long this lab will be a nursery for breakthrough gene therapies. Researchers will soon manipulate viruses to carry genetic material that can invade and disrupt cancerous cells while leaving normal cells unchanged.

Pomahac is dazzled, although he isn't sure exactly what they're really doing here with all this measuring and pipetting and centrifuging. Small electronic monitors are plugged into almost every outlet, and people in lab coats scurry around in a controlled but chaotic-looking dance. He knows only that this is the Boston he dreamed about. In the school where he was trained there was no money for even the most basic lab equipment. They read dated journal articles because nobody could afford a current subscription to a major medical publication. His lab training, he sees now, was a joke. Most people with Pomahac's level of preparation for a place like this would turn around and walk out the door. But his resolve runs deep.

"Stay and watch for as long as you want," Eriksson tells him. "Stop in before you leave."

At the end of the day Pomahac thanks his host and surprises him by asking, "Do you think I could stay?"

"Stay?"

"I mean, could I come back again tomorrow?"

A few months later Eriksson has managed to scrape together some money for a small stipend. Bo has quickly exhausted his modest savings from working on a hops farm back home. He was a laborer in the bagging station, picking tiny pinecones from female plants to make beer, filling fifty-kilo sacks and hauling them to a grinding station before returning to fill yet another sack. It started as a mandatory government summer job for students while he was in medical school.

He met his girlfriend, Hanka, on the farm. She was a medical student at the same school. They have been dating for several years now, e-mailing over CompuServe every day and talking on the phone once a week. They met in 1989, just before the Velvet

Revolution transformed their country and the Berlin Wall fell. By the following summer Czech farms could no longer count on free labor. Pomahac helped a friend organize groups of students to harvest the crop, and he saved enough money to visit the United States. This is his second trip. In 1992, as a foreign exchange student, he flipped hamburgers on the New Jersey shore.

His parents used their savings for the plane ticket that brought him to Dr. Eriksson's lab. But Boston is an expensive city, and the three thousand dollars that he has managed to save won't last very long. Dr. Eriksson has come up with a few dollars for his work in the lab; although it's a tiny amount, it makes all the difference—as long as Pomahac sleeps on a friend's couch.

Eriksson is impressed with his progress. He is a quick study, and he is determined to make the most of this opportunity. Pomahac has much to prove. His father's promising career as a chemical engineer was cut short when he signed on with the anti-Communist uprising in 1968, the short-lived Prague Spring. When Russian tanks rolled in and squelched the revolution, his father was blacklisted and forced into a series of low-level jobs. His mother, a schoolteacher, joined the Communist Party so their two sons would be allowed to attend college. That helped, but not completely. After finishing his grueling course work to become a pilot, his brother was informed at the last minute that he would not be permitted to join the air force. Bo Pomahac took notice. If he was going to be successful it probably wasn't going to happen in the coal-mining town of Ostrava. Now that he has a foothold in Dr. Eriksson's lab he is not about to let go.

April 8, 2009. Department of Plastic Surgery.
The human body is a collection of microscopic spheres, each one holding a blueprint copy of the mystery of life. Each tiny cell is bound together by a membrane of fatty lipids that crowd to-

gether like bubbles in a bathtub, pierced by proteins and sugars. Take just the right mix of a hundred trillion or so of these cells, and you have a human biome.

Tonight the doctors will dismantle a fragment of this microscopic universe from one human being and reassemble it onto another, one tiny piece at a time.

We have come so far, Dr. Eriksson muses. Look at the disciplines represented here: anesthesiology, infectious diseases, pathology, psychiatry, all the surgical specialties, and more. The more complex medicine becomes, the harder it will be to make moments like this happen. Half a century ago, in Joe Murray's day, the big obstacle to experimentation, and to innovation, was the fear of looking foolish to your peers. So much has changed, including committees and protocols and government regulations, all of which slow everything down but add valuable circumspection. When Murray operated on the Herrick twins in 1954 almost nobody owned a television. This surgery is about to be filmed for a TV series. Add runaway health costs to the mix, and it's a wonder anything experimental happens these days.

And yet, in spite of everything, they have made it to this moment. Maybe the miracle isn't the face transplant itself, but the fact that they have been able to cut through the red tape to perform one at all. And the main reason is the quiet, steadfast determination that Eriksson spotted over a decade ago in his young Czech visitor.

He gives his protégé a reassuring nod. It's time they got started.

chapter twelve

Wednesday, April 8, 2009, late afternoon. Helfgot residence.

eturning from the funeral home Susan finds large branches of cherry blossoms spilling from vases in every room. The house has been transformed into a park in spring.

"It's so beautiful," she tells the friend who has been helping her prepare for the crush of visitors they expect after the funeral. The woman is a Boston restaurateur who often travels to Manhattan to spy out the next trend, sometimes taking her daughter and Susan's son, Ben, who have known each other for years. When Joseph was healthy he would tag along too. His business meetings often coincided with these trips, and even when they didn't, he loved going to restaurants, especially in New York.

Susan used to go on similar outings in the early 1980s, when she was involved with a restaurant on Beacon Hill. Joseph was a patron, which was more or less how they met. Now their son, Ben, is interested in the restaurant business—Ben, who has done so well on the science and math AP tests, who has the talent to be a doctor or a scientist. Susan herself never listened to her parents' advice, and Ben seems to be taking after his mother.

Twenty-eight years earlier. Back Bay, Boston.
Susan steps into moist air as the city moves toward a lazy mugginess that threatens afternoon thunderstorms. Her third power breakfast at the Ritz in six months with the same lawyer seems to have finally paid off. The names of two prospects are written on his business card, which jostles beside the lipstick in her small clutch purse. When she gets back to the office she will call them and try to schedule appointments.

As she steps off the elevator the receptionist says, "Some man called you twice. He seems anxious to speak with you." She hands Susan several pink message slips. Two of them are from Joseph Helfgot.

The night before, Susan sat on a bar stool at Jason's, a popular downtown restaurant that would later become a Hard Rock Café. She and a partner from the restaurant were at the bar, pitching a woman from a convention bureau that sponsored fall foliage trips to Boston, trying to persuade her to make their Beacon Hill restaurant a lunch stop on the bus tour. She seemed to be receptive.

Three young men sidled up to the bar, all of them drinking beer in bottles. They wore khakis with loafers and slightly different hues of blue oxford shirts. They had loosened their ties and looked like new recruits for a security firm or insurance company. One of them tried to buy her a drink. "Tell me," she asked him, "do you guys always go out dressed alike?"

A man sitting at the bar laughed. Susan recognized him as someone who had been in their restaurant a few times. She was twenty-seven; he looked to be a little older. *He's definitely Jewish,* she thought, taking in that full head of curly brown hair and the beard. He was thin, with piercing blue eyes.

When the young men moved on, the laughing man slid over to the seat next to hers. Susan's friend, who knew him, introduced them, and he asked Susan about the restaurant business. She explained that she wasn't really in the business. She was their financial adviser. Her real job was selling insurance and investments.

"Social work for the rich?" It was a good line, and she would remember it.

"You'll never guess what I do," he said.

"You look to me like a Sumo wrestler."

"I have a sex talk show on the radio."

Susan suggested, without mincing words, what he could do with himself. "It's getting late," she told her girlfriend. "I've got to go."

He put out an arm and pulled her back. "I'm sorry. I guess that wasn't the best thing for me to say. But I really do have a sex talk show. Mostly, though, I teach sociology at Boston University."

"Get rid of him. Tell him I'm in a meeting."

"Right," says Norma, the receptionist. She buzzes Susan an hour later. "There's a telegram for you."

Susan goes out to the lobby. Who still sends telegrams? She opens the yellow onionskin paper and runs her eyes over the type: "Looking to merge, forming a new corporation. If interested call this number. Sincerely, Joseph H. Helfgot, Ph.D."

She ignores Norma's inquisitive face. Better to say nothing. Susan has just ended a disastrous relationship with someone in the office, and Norma already knows more than enough about Susan's personal life.

Her phone buzzes again. "There's a package for you. It's from *him*."

Good lord. It's an enormous piñata tied up in clear cellophane with a huge velvet bow. The card says, "Open with care."

Inside the piñata are chocolate-covered fortune cookies with little messages inside. They all say the same thing: *I love you.*

An hour or two later it's a dozen long-stem roses, which she leaves on Norma's desk.

Then: "Susan?"

"Oh, God, what now?"

"Piñata man is on the phone. What should I tell him?"

"Damn it, put him through."

"Hi," he says.

"Stop it."

"No."

"I mean it," she tries to say in a stern voice.

"I thought maybe you would like to have lunch."

"No."

"I'm going to be right around the corner, at the Meridien." He explains that he's having lunch with the president of a major company who is working on her Ph.D. Joseph is her doctoral adviser. Susan knows the company. They have a couple of hundred employees. "Maybe you can join us?"

"No."

"She runs a big shop. Maybe you can sell her some insurance."

"That's low."

April 8, 2009, late afternoon. Intensive Care Unit.
As young Rabbi Franken makes his way into the ICU he spots Jacob Helfgot. The boy's face shows his pain. He has grown taller since his bar mitzvah in September. Driving to the hospital just now the rabbi recalled that festive morning.

Joseph and Susan were with Jacob on the bimah, although both the cantor and the rabbi worried whether Joseph could make it through the full two hours. A nurse stood by just in case, but Joseph got through it. When it came time to address his son, he spoke in a weak but proud voice. Joseph even joked that he and Susan, who had been married by a justice of the peace, had at long last been joined together on a bimah in the presence of a rabbi. Everybody laughed.

In the ICU Susan introduces the rabbi to the family members. He knows about the face transplant; Susan told him yesterday, when he came to visit her. Maybe that's the sensation he's picking up. Or maybe it's just this family, so filled with life as they stand together in death. Even with Joseph's body hovering between worlds, the rabbi can feel his energy rippling through the room, through these people who love him, who are part of him.

They gather in a circle around the bed for the last time, holding on to one another as he recites the prayers for the dying. His voice cracks as he looks down at Joseph's face and envisions a different kind of blessing that this face will soon bring to another man.

Joseph, why are you so warm to my touch? Do you have a fever? Joseph, honey, it's time to get up. We have to go home now. Please, Joseph, please wake up. Emily is crying again. I want her not to cry anymore.

"Ribono shel olam," the rabbi says on behalf of Joseph. "Y'hi ratzon milfanecha she'yih'ye shalom m'nuchati, amen. Master of the universe, may it be Your will that my passing be in peace."

As Susan glances up through the glass, the staff steps back. She sees her family superimposed as reflections on the window.

We are ghosts in the glass, just like you, my love. I cannot bear to leave here without you.

Go away, you people standing there outside the glass. You can't have him!

"Em," she whispers, "we need to go. They're waiting." But still they do not leave.

On the other side of the glass the team tries to remain calm. It has been a long day, and night is about to fall. James Maki has been waiting downstairs since mid-morning. A medical team is en route from another city, and their plane is about to land. Helfgot's heart, already transplanted once in the past week, should soon work its magic again. The moment the family leaves it can all begin. But they will not rush the family, even if they linger all night.

They finally emerge, all but Susan. She stays back, alone with her husband. When she opens the door, her eyes are dry. She stops for a moment to embrace a nurse, then the desk attendant. Dr. Couper is on the phone, and he stands up, trying to get off the call so he can say something. But Susan shakes her head. Not now. He nods in understanding.

As the family turns the corner, medical personnel swarm into Helfgot's room. Every minute counts. Recipients are waiting.

I must go back. I can't leave. I have been coming here every day, for so many days. How can it be that I can't come back anymore? Joseph, how can I leave you here, all alone?

She steps into the elevator.

chapter thirteen

Wednesday, April 8, 2009, 11 p.m. Emergency Room.

finally a lull. Bo Pomahac has been on fast-forward since last night, when Esther Charves called with the news that Helfgot's wife had agreed to a face transplant. He was euphoric, barely able to sleep, but now his elation gives way to a small case of nerves. Each of the thirty-eight members of the medical team knows his or her role, but only Pomahac is looking at the whole picture. If anything goes wrong, and there is so much that can, it will be his problem.

Exhaustion has set in, the kind that hits right after a period of intense anxiety, and the past couple of hours have been *really* intense. Earlier tonight everything came to a halt when a routine presurgical chest X-ray picked up a spot on Maki's lung. It wasn't new. A few months ago, when the same spot showed up on an X-ray, the doctors thought it was just a tiny scar, probably from an infection. But tonight, with so much at stake, the radiology team wants to take a closer look.

Now? Are you kidding me? Pomahac can't believe it.

So Maki was sedated for a bronchoscopy, which meant snaking a tiny lavage instrument down into his lung to brush up some epithelial cells that would be examined immediately for signs of possible malignancy, or any other irregularity. During the past few months he has already been through a workup rivaling that of an astronaut training for space flight: a head-to-toe assessment, including X-rays, CT scans, blood work, the whole nine yards, all of it designed to ensure that he is a suitable candidate for a face transplant.

Then came all the prep work leading up to tonight: the earlier dissections of the cadavers in Brussels, the careful 3-D imaging of Maki's head, and specialized training for the many medical people who will be involved in this enormously complicated surgery. The doctors have established a personalized immunosuppressant regimen to give Maki the best possible chance to fight rejection, and it's a good thing they did. They discovered that their patient was deathly allergic to a common medication given to transplant recipients.

A lot of other people are waiting for this operation to begin, including the recovery teams who are hoping to remove solid organs from Joseph Helfgot's body. A medical team from the Midwest has just arrived to take the newly transplanted heart to a patient who, without it, doesn't have long to live. Professionals from the New England Organ Bank will be arriving later to recover tissue. The film crew is here too. Although their presence is unusual, to say the least, both the doctors and the donor family know that having them in the operating rooms will help make the viewing public more aware of organ donation.

In certain transplant operations, an understudy is waiting in the wings in case the intended recipient is sick or unable to get to the hospital in time. But tonight there is no Plan B, no backup recipient for Helfgot's precious face. This surgery is unique to Maki; the team's practice drills have been tailored to his particular facial defects. Someday, perhaps, Dr. Pomahac may be able to perform a face transplant on a backup candidate, but that day, if it ever comes, is years away. If this surgery doesn't happen to-night, Helfgot's face will be lost. And who knows when Pomahac might be offered another one?

He couldn't believe he might have to go to his patient and tell him they'll have to cancel. Maki heard the good news this morning, and, like his doctor, he has been excited all day. What was Pomahac supposed to tell him? "I'm sorry, Jim, but there's a tiny spot on your lung, and the radiology guys don't like it. It's off for today, but we'll probably get another chance." No way was he going to do that.

They'd had a false start just a few days ago. It looked as if a family who agreed to donate a heart might also be willing to give a face, but they balked when asked about donating back tis-sue. "We're not comfortable with that," they said. If they weren't comfortable with back tissue, they would never donate a face. But on Tuesday night Susan Helfgot had said yes.

Pomahac paced around as he waited for radiology and thoracic to decide what to make of the spot. Dr. Phillip Camp, the director of the hospital's lung transplant program, was still in the building, and when Pomahac had him paged he rushed down to assess the situation. After a few minutes he and the other doctors agreed that the blemish on Jim's lung was nothing but old scar tissue after all.

Pomahac takes a deep breath and lets it out. He needs to lie down. He still has an hour before surgery, which has been pushed back to midnight. But there is no on-call room with a bed on the surgical floor, so he makes his way upstairs to the Burn and Trauma Unit.

"Is there somewhere I can rest for an hour?" he asks a nurse.

"How about Mr. Maki's bed?" They are holding a room where Jim will be brought to recuperate after surgery and where he will likely remain for a few months. The room is immaculate, of course, and the nurse anticipates the surgeon's hesitation. "Go on, take it. They'll change the sheets after you leave." They'll have to clean the entire room, because human skin is full of bacteria and viral agents, not to mention mites, fleas, and microscopic pests that travel through the air and latch onto a person's hair and cuticles. Everyone is a carrier, patients and doctors alike.

Pomahac gets into his patient's bed and closes his eyes, hoping nothing else will go wrong. He can't wait to get started.

Back downstairs Dr. Donald Annino, an oncology and reconstruction specialist who is heading up the surgical prep team, is having a hard time getting into the pre-op area. The transplant is still a secret, and the public relations staff wants to keep it that way until Maki is safely awake in recovery. What if a reporter sneaks in? So Peter Brown has asked to have a security guard at the door. As he looks at the names on his sheet of paper, the guard says, "I'm sorry, Doctor, but you don't seem to be on the list."

"What? Let me see that." Annino grabs the clipboard. "Great," he says, chuckling, as he realizes what has happened. He speed-dials Pomahac's cell phone, jolting him out of his all too brief nap.

"Bo, you won't believe this, but I can't get in. Security doesn't have my name. Not just me, it's all of us. We forgot to put our names on the list."

Pomahac laughs. So much for resting. "I'll be right down," he says.

He hops off the bed and heads out. The nurse, seeing him leave, calls housekeeping to sanitize the room for Mr. Maki. Then she goes in to help strip the bed. She doesn't mind. This is a historic event, and she is pleased to play even a tiny role.

Midnight. New England Organ Bank headquarters.
Ever since Tuesday afternoon, when the doctors signed the death certificate, Joseph Helfgot has no longer been a patient of Brigham and Women's Hospital. Hospital personnel are still deeply involved with his care, but direct medical management of Helfgot's

body now belongs to organ bank employees, who appear on the scene and work with the hospital staff whenever there is a brain death with the possibility of organ or tissue donation. Hospitals are legally required to report all deaths to the nearest organ bank office. Many people register as organ donors when applying for or renewing their driver's license, but sometimes potential donors don't drive, or are too young for a license, or come from another country. In these cases the family must make the decision.

Meredith, the organ bank nurse who took Joseph's oral medical history from Susan yesterday, is watching over Helfgot with the help of the Brigham nurses, monitoring his kidney and liver functions, managing the IV fluids and their composition, and maintaining blood pressure and urine output. In a single moment on Tuesday afternoon everything changed from preserving a life, which was no longer possible, to preserving organs, especially Helfgot's new heart.

From the moment Susan said yes to the face donation, the organ bank has been scrambling to get everything in order. Kristina Andrzejewski, the tissue services manager, sits in semidarkness in her office, trying to grab a little sleep while she waits to hear from the hospital. Soon she and her colleagues should get a call telling them the surgeons are ready for them to come and assist in the recovery of Helfgot's organs, and to make sure they are correctly handled for shipment. There has already been a delay, which is not unusual in these situations.

Andrzejewski has another role as well. Solid organs—lungs, kidneys, livers, and hearts—are removed by doctors, but tissue recovery is done by local organ banks. Arriving with their own scalpels and other specialized tools, trained organ bank employees recover skin, bone, and even heart valves that can be used for reconstructive surgery at some other time, such as skin from the back that is used for grafts on burn victims. When she gets to the hospital Andrzejewski will assess Helfgot's body and determine whether any tissue can be recovered for future use. Unlike organs, which must be transplanted almost immediately, properly preserved tissue can remain viable for up to five years, frozen in a protected vault. If only hearts and kidneys could be preserved this way.

For Andrzejewski it all began with a call late Tuesday night from Chris Curran, the organ bank's lead operations manager. Curran is directing this big and complex symphony, managing every nonmedical detail, including the chartered jet waiting at Signature Aviation in a private area of Boston's Logan Airport. One of his jobs is to make sure that as soon as the traveling heart team returns to the plane with the recovered heart, they will get immediate clearance to take off. This is an urgent matter, as hearts can survive no more than five hours without blood flow.

He is concerned about the weather. It has been a wet and ugly week, and on Monday, Opening Day at Fenway Park was rained out. April is often unpleasant in New England, where climate change can be an hourly event. A plane may be grounded at the last moment, which is another reason they try to line up an alternate recipient in the area, just in case. A healthy, available heart is simply too precious to waste.

Curran is also thinking about the complicated legal issues at play here. The gift of a face still lies outside the framework of the United Network of Organ Sharing, or UNOS, the hub that connects the country's organ banks. Today the whole business has been made even more complicated by the documentary film crew, which adds to the general anxiety that the anonymity of the donor could be compromised.

A few weeks ago the television producer sent a crew to Helfgot's house for an interview about what it's like for a patient and his family to wait for a heart. The crew is aware that Helfgot died the other day, but they have no idea that he is the person donating the face. And the few people who do know want to keep it that way.

With his beard gone and four surgeons and their assisting

nurses working on his heavily draped body, not even his wife would be able to recognize him. But the hospital and the organ bank want to be extremely cautious, so the organ bank's lawyer has prepared another set of forms that Susan Helfgot signed on Tuesday, agreeing not to hold them, the television network, or any caregiver or institution responsible if her family's anonymity is compromised.

Curran managed to make it to tonight's Red Sox game in the freezing rain, returning home just as Pomahac was starting his abbreviated nap in Maki's bed. He is glad to get out of the cold and back to his computer and his landline. He looks at his cell phone, half-expecting it will ring, but hoping it won't. Sometimes a donor becomes hemodynamically unstable. If Helfgot starts to crash, Curran will be involved in some quick decisions about which organs can be salvaged, and in what order.

Midnight. Hospital basement.
Dr. Julian Pribaz makes his way along the cold corridor. Midnight may seem like a strange time to start a cutting-edge face transplant, but he doesn't mind. He enjoys nights in the hospital, when he can think his own thoughts without the constant din to distract him. The elevators are empty, the hallways are silent. There is no hint that tonight will be unusual.

Pomahac called him last night. The two surgeons are as ready as they'll ever be, and their donor is right here at the Brigham.

Having Helfgot in the house is a big advantage; there will be no need to orchestrate events between hospitals that could be hundreds of miles apart. This operation will be complicated enough, with other organs being recovered at the same time. The out-of-town heart recovery team is already here for a surgery that was supposed to start at six o'clock. Now it's almost midnight. They can't be pleased about the delay.

Although he and Pomahac practiced in the Brussels lab, with

Pomahac on the right side of the face while Pribaz tackled the left, they have never handled a real face donor whose blood still moves throughout his body. Pribaz relishes this final precious moment of contemplation. As he enters the locker room, where he will change into fresh surgical scrubs, he is overcome with anticipation. He slips off his lab coat, hooks it inside the narrow locker, and slams the thin metal door.

Upstairs Dr. Pomahac enters the operating room just before Helfgot, warm and pink, is wheeled in. The only sound comes from the respirator, which is pumping air in and out of his lungs. Other than Pomahac, there are only two people here: the ventilation specialist and the gurney operator, who is responsible for the critical job of moving a body that is being kept alive by artificial means. This is no simple task, as it involves protecting all the equipment that is being moved along with Helfgot: the respirator, the vital-sign machine, drainage bags for waste, and IV lines. Even the gurney is not really a gurney, but an adjustable bed from the ICU.

Pomahac begins to wonder. Where the hell is everybody? Has the surgery been called off while he was napping? Is it that damn spot on Maki's lung? In a mild panic he calls Julian Pribaz.

"Where are you?"

"In the locker room."

"Oh, good."

"Don't worry. I'll be right there."

Thursday, April 9, 2009, 2:00 a.m. Helfgot residence.
I can't get myself to lie down on this bed. It's a dead man's bed. So I sit on the edge and pick up the picture we keep on the night table. We're together in Laguna Beach, near the water, with another couple. We

look pretty good, the four of us, darkly tanned with healthy smiles in the brilliant California sun. When was that, ten years ago? It feels like a lifetime.

I hear Ben rushing up the stairs. His bedroom door slams closed. He's talking loudly on his cell phone. It feels like daytime, but the clock tells me otherwise. We all know what's going on at the hospital. Who can sleep at a time like this?

Wonderful Ben. He'll leave in a few months for New York. I still can't believe he got into NYU. Joseph was so happy. Thank you, God, for letting him be alive on that wonderful day. Long before we met, Joseph used to teach there. Once we sat for hours in Washington Square Park, across from where he taught sociology, and he told me about those early days. It was freezing, but we didn't care. We had just met. You don't feel the cold when you're falling in love.

So he won't be here to see Ben go off to college. I can't believe Jacob is starting high school, but at least he'll still be home with me. Thank you, God, for not taking all of them away from me at once. I just couldn't bear it.

Please make this night be over. My heart is racing as I keep thinking about Joseph lying on an operating table in a cold room, being pulled this way and that. I don't even know these surgeons, and Couper's not there. What am I thinking? Couper doesn't need to be there. Joseph's dead. They can't hurt him.

But his face—I see it happening. It's too much. I get dizzy and sink down into Joseph's pillow. It smells just like him. Maybe I should have thought about this some more. Pam told me to slow down. But I was so damn sure it was the right thing to do. How can you not give somebody a chance to live a normal life? Dr. Lewis was so excited when he called us about the heart. In the blink of an eye, everything was going to be all right again. And then the other eye blinked.

They've turned on the TV in the kitchen. What is that theme song? Maybe I should go down there and be with them. But I just can't move. Will they call me from the hospital when it's over? Esther promised they would. It's two in the morning. They should be almost done by now. God, please make this go quickly.

It helps to think about this poor man, the recipient. How can you lose a whole face? Esther said it was a horrible accident, that he had been a housepainter. Maybe he fell off scaffolding? She said he can't eat or breathe or talk. Will this surgery save him? I look again: it's 2:10. I'm not going to sleep. There's no way. I close my eyes and see a scalpel tracing Joseph's face. I'm so cold I'm shaking. I pull the covers up over my clothes.

chapter fourteen

Thursday, April 9, 2009, 1:15 a.m.
Helfgot Operating Room.

the first trace is made with a fine scalpel that outlines a portion of Joseph Helfgot's face. In Brussels it took them just over an hour to recover a face when they practiced on their fourth and final cadaver. But tonight will take much longer because they're working with a heart-beating donor with blood flowing through his body. A body undergoes less blood loss after its heart has stopped beating, but that's not true for donors with lifesaving organs. Blood loss must be minimized to prevent Helfgot's blood pressure from crashing, which would compromise his solid organs. If they lose the heart, it would be a disaster.

It doesn't help that the human face is heavily vascularized, with large carotid arteries feeding smaller and smaller vessels, the arterioles and capillaries that take blood to specific locations, each accompanied by counterpart veins that carry it back to the heart. These passageways serve different areas of the face, each with its own sets of muscles that go to work when your brain tells you to chew or swallow, to smile at a baby or wink at a friend, to stick out your tongue or grimace in pain. The phone rings with a friend on the line, and you talk and laugh or cry, holding the phone while sniffing and then tasting the spaghetti sauce on the stove with no thought about the millions of brain signals and controlled muscle twitches that make it all happen. These small, simple, ordinary motions of a face require large amounts of energy gathered from lots of richly oxygenated blood. Ask anyone who has ever had a nosebleed.

The luscious beard that Helfgot had sported since he was a teenager was shaved a few hours ago. It's already trying to grow back, because hair follicles don't need a brain to tell them to go to work. His transplanted heart can also function without his brain: it has a tiny cluster of nerves centered high in the right atrial chamber that will beat, at least for a while, in a last-ditch but hopeless attempt to prolong life. As long as the ventilator supplies the lungs with oxygen that can attach to hemoglobin and travel the bloodstream, keeping the cells healthy, the heart will continue to beat. The human heart is a miraculous instrument, tough enough to fly off to another city and take up residence in the chest of a third person, and keep him going.

Pomahac's team will do everything they can to protect this heart, keeping it strong and healthy during the procedure. Jim Maki probably won't die if he has to wait six months or a year before another donor family offers the gift of a face, or even if a face never arrives. But the man waiting hundreds of miles away may have only a few weeks to live. If a new face means almost everything, a new heart means *everything*. If Helfgot becomes unstable, it's an easy call: the heart trumps the face.

Pomahac knows what's at stake as he works with Julian Pribaz, Elof Eriksson, and other members of the recovery team as they make the first incision. The key is to finish with minimal blood loss. If they fail, and the heart is damaged—well, they all know what that means.

Maybe they should do this in stages. Maybe, Pomahac is thinking, they should stop halfway, after they raise the flap, a section of skin that includes the veins and arteries and carries its own blood supply. They could pause and let the other teams take Helfgot's solid organs before they go deeper into his head to remove the bone that will restore structure to Maki's face. If there is excessive bleeding during surgery it's likely to happen then, when the bone is being cut, which is why it's not unusual for a patient to receive a blood transfusion right before major orthopedic surgery. The transfusion acts as a reserve in the event of major blood loss during surgery and protects the blood pressure from a precipitous drop.

Either way, they will need to remove a large portion of Helfgot's facial bone, cutting into the zygomas on each side near the maxilla intersections. They also need his hard palate and upper teeth, which will require cutting into the pterygomaxillary junctions, then all the way up to the top, through the orbital floor of bone that encases and protects the eye. Maki lost a lot of bone, which was burned off when he was electrocuted, leaving a deep, dark void in the center of his face.

In spite of the massive amount of bone work required, they finally decide to do it all without stopping, unless, of course, the heart becomes threatened. Using a syringe they squirt epinephrine into Helfgot's cheeks to constrict blood flow while they move their scalpels deeper into the face, isolating the nerve bundles they'll need along the way. Then they'll move on to the muscles, watching out for preferred arteries and veins before cutting the mandible bone to expose and recover the facial arteries.

They work through the night, but Pomahac is not tired, not at all. He and his team are operating on pure adrenaline.

• • •

7 a.m. Helfgot Operating Room.

Kristina Andrzejewski from the organ bank enters the OR with two coworkers. Everyone assumed that somebody must have called her during the night with news of another delay, but as dawn broke, nobody had. And so the three of them drove to the hospital to start reviewing Helfgot's chart, a heavy binder with reports of everything that happens to a patient during a hospital stay.

Helfgot's heart doctors are now busy with their other patients, and organ bank staffers assume nothing. Andrzejewski will pore over the chart and make an independent assessment as to whether or not Helfgot's tissue is healthy enough to be preserved.

Because nobody involved in removing Helfgot's face knows anything about his medical history, his chart is essential. Andrzejewski reads through it carefully, looking for red flags that will prevent a potential tissue donation, such as a history of hepatitis C or a latent infection. A recipient can be given antibiotics to fight an active infection, especially when a donated organ spells the difference between life and death. But valves, bone, and tissue that will be preserved for later use are in a different category; they cannot be given antibiotics in a preserved state. Most organic materials are considered medical devices by the FDA and are held to an extremely high standard. If there is any sign of latent infection the material will not be taken.

It takes Andrzejewski more than an hour and a half to go through the three four-inch binders that make up Helfgot's chart. She has read a lot of charts over the years, but she has never seen one this large. She can't believe what this poor man had to endure over the past eighteen months. In the end a previous infection will rule him out as a tissue donor. But she and her team are required by law to stay and observe the operation, ensuring that the heart, kidneys, liver, and pancreas—if they're healthy enough—will all be properly recovered and packaged for delivery. The face too, even though it's only traveling across the hall.

When Andrzejewski enters the operating room she makes eye contact with Dr. Pomahac. They first met two years ago, when he visited the organ bank to ask for their help in expanding the range of organs to include the face.

A monitor on the wall shows the surgery in progress on the table. It's hard to make out what's really happening, but she sees that Pomahac is flipping Helfgot's facial flap back and forth like a pancake, looking at it from the top and then viewing it from the bottom. The flap is still attached to Helfgot's former face by a small pedicle of vessel-rich flesh. Pomahac checks everything one last time until he is satisfied that it's safe to detach.

Andrzejewski watches her staff. Going in, it is impossible to know how they will react to a face recovery. She can handle it, and she's pretty sure they can too, but it's new territory for all of them. She and her associates are ready to document the first New England Organ Bank protocol for packaging a face. There will likely be more faces coming along in the future, and they will almost certainly need to travel more than a few feet.

Andrzejewski has one last task, and this one is personal. She has promised Esther Charves that she won't leave Helfgot's bed-side until he is safely on his way to the funeral home. She is now his protector.

Regina Swanton, who works with Andrzejewski, is tense. The documentary crew have their faces in their cameras, just inches from the operating table and workbenches. She worries that they might accidentally brush against a sterile field, although she can't blame them for trying to get the best possible view. She wonders if anyone else is concerned, but if so, there's no hint of it. These videographers have been filming in operating rooms for months, and they know the drill.

At 7:02 in the morning they are finally done. It has taken almost six hours to detach the parts of Joseph Helfgot's face that will give James Maki a new life. What was a face when they started is now officially known as an allograft. Temporarily in limbo, it awaits its new owner, who will mold it into his own unique identity.

7:30 a.m. Maki Operating Room.
Pomahac needs to check a few things, especially Maki's arteries and veins, which must match up exactly. Maki is completely anesthetized, and for several hours another surgical team has been preparing his face for the transplant, which is just minutes away. Although they have been on their feet and under the operating lights for hours, in a sense their work is only beginning. The next stage will be far more strenuous, because unlike Helfgot, Maki is alive and must remain stable on the operating table for another eight to ten hours, maybe longer.

Pomahac looks down at his patient. He knows every square inch of this face, which bears the terrible scars of the ten surgeries he has performed on it over the past four years. Until tonight a large flap covered the section where Maki's nose and mouth used to be; Pomahac attached it during a previous surgery to cover the gaping hole created by the accident. Now, finally, it has been removed and discarded. But there is a lot of scar tissue, and cutting through heavy scars is never easy. Scars are nature's way of holding damaged skin together, and they will make for slow going.

The allograft is in a tray on a table in Helfgot's operating room, its nerves tagged and blood vessels purged of possible clots before being readied for transport. The doctors must get Maki's blood flowing through it as soon as possible, before the cells begin to die. The aorta brings blood up to the head in swift, strong pulses, through two sets of large carotid arteries located on both sides of the neck. Each set has two branches, the protected internal carotid feeding the back of the head and the brain, and the external one, which feeds the face.

They have decided to use Maki's left external carotid as the main source of blood for the allograft, matching it up with Helfgot's left carotid that was taken along with the allograft tissue. The two sections will be sutured together in a microsurgical technique known as anastomosis. If all goes well, blood should begin to flow from Maki's carotid into the allograft, perfusing it back to life.

If the left carotid fails, they will need an alternative blood supply. Maki's external right carotid is not an option, as it is busy feeding blood to the rest of his own face during surgery. Once they start dissecting his left carotid artery and preparing it for anastomosis, it will be almost impossible to turn back.

Now that his colleagues have exposed it, Pomahac examines Maki's artery to judge its lumen, or circumference. It isn't wide enough to suit him, so he instructs the team to go a little farther along on the carotid to enlarge the vessel cavity. That should help the blood flow more easily.

If we can get this first artery to hold, he is thinking, *we can start to relax.* The biggest danger is a blood clot. There is always the risk of a clot, and clots kill. A few of them snaked their way into Helfgot's aorta while he was receiving his new heart. They traveled through his carotids and into his remarkable brain, shutting down its blood supply and ending his life. In medicine, anything is possible.

7:40 a.m. Helfgot Operating Room.
Before Pomahac returns to Helfgot's operating room, he and Julian Pribaz confer. The brief break allows them to catch their breath before pressing on. Pomahac looks up at the clock. It's close to eight, and there is still a whole surgery left to perform.

He returns to help Andrzejewski package up the face for transport as carefully as if it were going to another city, rather than merely across the hall. The allograft has been bathed in Wisconsin Conservation Fluid, better known as UW solution. This special liquid, developed by two Wisconsin doctors in 1987, matches human pH almost exactly, with a rich mix of chemicals that will keep the cells stable until Maki's blood flows in and takes over.

Pomahac wraps the allograft in sterile plastic and then wraps it again in more plastic. With a gentle motion he eases it into a tray filled with a slurry of ice. Then the package is wrapped in plastic yet again, twice, to ensure that no virus or bacteria can attach itself to the human substrate during its brief journey across the hall. Andrzejewski watches carefully to guard against any mishap or confusion. There have been one or two errors in the sixty years of transplant surgery in other parts of the world, with consequences similar to those of a hospital discharging a baby to the wrong parents. But nobody is really worried that New England's first facial allograft will wind up on the wrong person.

A blue plastic tote container that could just as easily hold a few beers serves as a temporary sarcophagus for the face. Pomahac snaps it shut and carries it out of the room, trailed by the camera crew and a procession of surgeons and residents. He enjoys the feeling of knowing how well protected the allograft is. It will be safe even if he slips.

While Pomahac has busied himself with arteries, veins, and UW solution, another team of surgeons has been busy under the direction of Dr. Christian Sampson. They are removing skin from Helfgot's forearm to be used as Maki's sentinel flap, which is a small piece of skin from the donor that is grafted onto the recipient in a discreet location. (In the French transplant Isabelle Dinoire's sentinel flap was placed under one of her breasts.) At regular interviews after the surgery, doctors will snip off small pieces of the flap and examine them.

These tiny biopsies are studied for signs of possible rejection in the recipient in the form of immune cells that show up and function as early storm clouds, appearing well before the rest of the body reacts to the rejection alarm. If necessary the immunology team can launch a preemptive attack with large quantities of drugs, such as prednisone, that will keep the immune system from rejecting the new organ. Creating a sentinel flap protects the transplanted area from frequent biopsies. Maki's face has already been through far too many assaults, but this one is avoidable. A sentinel flap also provides a second indicator: if it appears that rejection is forming at the site of the transplant, but the flap remains unaffected, it may be just an infection.

For Maki's sentinel flap, Dr. Pribaz has come up with an inventive solution. Maki had severely damaged his right hand when he fell onto the tracks, and his thumb and index finger were badly burned and fused together. Pribaz wants to use a piece of Helfgot's forearm skin to open up the area between Maki's thumb and index finger. This webbing can also serve as the sentinel flap. Sampson's team has been working on the flap, and it too is now being prepared for transport across the hall. On any other day the new hand flap would constitute a significant medical procedure. Today it seems like an afterthought.

9:15 a.m. Maki Operating Room.
Dr. Pomahac brings in the face. The others in the room are excited, but their expressions lie hidden behind their masks. A pity, thinks the videographer, who is trained to pick up the slightest nuance—a furrowed brow, a wince, a bead of sweat. Even the surgeons' eyes are hidden behind large optical lens machines that hang from the ceiling, which can magnify the tiniest blood vessel or nerve. So much is lost when you can't see someone's face.

The surgery starts in earnest, with Pomahac performing the

delicate anastomosis that connects Maki's left carotid artery to its allograft counterpart. More than two dozen people are trying to crowd around the table, but there isn't room for them all. Finally the microscopic sutures are in place. Two years of work have gone into this moment: cadaver trials, protocol issues, recipient approval, and the long wait for the right donor.

Pribaz looks up at Pomahac before he slowly releases the clamp holding Maki's blood supply at bay. This is it, his eyes are saying. The room falls silent. Everything has come to a complete stop as they all watch. And they wait.

Finally, almost imperceptibly, starting on the left side of the cheek where Pomahac has attached the vessel, the pale white allograft begins to turn pink. Gradually color creeps across his patient's new skin, slowing down a little as it struggles against gravity at the bump of the nose, then moving faster down the other side, toward the right cheek. A pink tinge blossoms on a waxen field.

It's as if a spaceship has just landed on the moon. But nobody cheers. Nobody says a word. They are watching a dead man's face being resurrected, which they recognize as a kind of miracle, a Lazarus moment. But there are no shouts of joy, no smacking of backs. The medical team is humbled, and some of them blink back tears. But only for a moment, because there is plenty left to do.

There are more blood vessels to attach—arteries to pump blood and veins to take it away. The surgeons join Helfgot's right jugular to Maki's major facial vein and continue until their work is done, until the allograft is finally vascularized and blood flows in and out of what is now Jim Maki's new face.

The hand surgeons have started attaching the sentinel flap. While they work, Pomahac and Pribaz begin to connect nerves on either side of Maki's face, using a technique known as neurorrhaphy. It is painfully slow going as the doctors knit together

Helfgot's and Maki's tiny nerves, one suture at a time. Nerves near the surface will provide sensation, allowing Maki to feel steam rising from a cup of hot coffee or a light breeze on a warm day. Other nerves, sutured deeper in the face, will one day allow him to chew and swallow. They are piecing together a kind of 3-D jigsaw puzzle, one tiny segment at a time.

Minutes ago, when the allograft first perfused, forcing blood through Maki's facial vessels, the adrenaline pump in Pomahac's body finally began to relax. Now that he can afford it, deep exhaustion starts creeping in. His last normal sleep was on Monday, three nights ago. He leaves the room and scrubs out for a while, allowing the hand team and everyone else to continue without him. He and Pribaz have picked the whole team, and they trust them completely. What he really needs now is a cold Diet Coke, or whatever is still left in the vending machine.

Pomahac's team starts working on the upper palate and teeth, and the rest of the bone. It is attached with titanium plates less than two millimeters wide that will give structure and protection to Maki's underlying facial area. It will allow him to have a mouth again, so he can eat. And smile. And speak.

11:00 a.m. Helfgot Operating Room.

The traveling heart team has been awfully patient. True, they've had no choice, but they've been waiting here since 11:15 last night. They can't sleep, because if Helfgot becomes unstable they'll need to move quickly. For the past hour they have watched the hand surgeons removing a piece of Helfgot's arm for the sentinel flap. Now, at long last, the room belongs to them.

They are feeling some anxiety, which is normal at such moments. Before they flew to Boston Dr. Couper told them that the donor's heart appeared to be sound, but it's impossible to know for sure without actually seeing it. They hope Couper is right.

Back at their own hospital the intended recipient is hoping the same thing.

Helfgot's body is still split open from his heart transplant three days ago. A long, four-inch-wide gash runs down the center of his chest, covered only by a fine-mesh elastomer membrane, a plastic that acts and feels like rubber. Heart recipients are given so much fluid during surgery that it is often impossible to close up the chest right away because their bodies are so swollen. Leaving the body open for a while also makes it easier to go back in if there's a problem.

Sometimes patients are left open because of all the hardware. When Helfgot died he was full of metal: a VAD, a defibrillator-pacemaker unit, and an experimental device for monitoring heart fluid. Only the VAD was removed during his transplant. Couper had intended to remove the rest later.

Working with the heart team, two Brigham surgeons isolate Helfgot's organs by dissection, separating them from their blood supply and flushing them with fluid. The first to be recovered are the liver and pancreas, but a quick biopsy of the liver shows it to be fibrotic. Someone calls the organ bank to tell them the news. The potential recipient will have to hang on a little longer, hoping that another liver will come in time. But Helfgot's liver is not wasted. A staffer at the organ bank calls around and finds a research team that is happy to take it.

The pancreas has been damaged from years of strong medications that led to diabetes, but it too finds a home in a research lab. As Esther suspected, Helfgot's kidneys are compromised from years of diuretics to purge excess fluid buildup from his sorry heart. Helfgot was sixty, but his kidneys look like those of a man of ninety. His corneas are not taken because of early-stage cataracts. And because he was exposed to a recent drug-resistant bacterial infection there can be no tissue donation.

But that still leaves the most important organ of all. At 11:25

a.m. Helfgot's major aorta is cross-clamped. His heart finally stops beating. One of the heart recovery surgeons scribbles the time and day on the pant leg of his scrubs. The five-hour countdown to transplant has begun.

After a quick visual check, the heart is carefully pulled and immediately placed into solution to protect the cells. The ambulance on alert outside the hospital starts up. After a call from the organ bank, the pilot of the small Cessna secures his final flight plan, preparing to depart within half an hour. They have five short hours to get from the Brigham through the Boston Harbor Tunnel to Logan Airport, board their jet, fly to their home city, and race by ambulance back to their hospital. A quick goodbye and they're gone.

It's all over. Everything that can be removed is gone. With a flip of a switch, Helfgot's respirator is turned off. It is now up to other specialists to restore some small semblance of dignity to this man who, until two days ago, was one of the hospital's favorite patients. The body is closed.

chapter fifteen

Thursday, April 9, 2009, mid-afternoon. Helfgot residence.

Y ou fell in love in a bathtub?"

"Joseph was living in Cambridge, right in Harvard Square. He had this tiny, postage-stamp condo with an enormous cast-iron bathtub, where the water stayed hot for hours. You know, the kind with claw feet?"

Susan is on the back porch with two of her girlfriends, who have dragged her away from the insanity inside the house. A man is stacking cases of wine against the wall. There's a beeping sound in the driveway from a truck backing up. The women look up as two men come around the back of the house. One is carrying a round tabletop and the other has several folding chairs hooked under his arms.

Who ordered furniture? We're not having a wedding. And why do we need all that wine? There's still some in the basement from Jacob's bar mitzvah.

"In a bathtub?" asks one of the women. "It must have been a hell of a tub."

"No, it wasn't like that. I went to his apartment to meet him for dinner. We were going to walk around Harvard Square and get something to eat, but it started raining like crazy. So he picked up takeout at the Chinese place next door."

Someone has uncorked a bottle of white wine and poured it into plastic cups. Susan takes a sip. "Ugh, this is awful, it's warm," she says.

"Drink it. So you're in the tub . . ."

"So we're watching a movie and eating takeout. I wish I could remember what movie it was. And then we started making out."

"*Finally*, the good part."

Susan shakes her head. "Not really. My lips started blowing up. And then I couldn't swallow. He had ordered soft-shell crab, and I'm allergic to seafood. Of course I didn't have any, but with all that kissing . . ."

"You had a reaction just from kissing?"

"There was a *lot* of kissing." Everyone laughs.

"Did you have to go to the hospital?"

"I came right from work, and I had my briefcase with me, with my EpiPen. Joseph ran and got it and I gave myself a shot. In a few minutes I was fine."

"Jesus."

"It's a good thing I had my briefcase."

"And *then* the bathtub?"

"I had tiny hives all over, like poison ivy, but worse. I couldn't stand the itching. Joseph was so freaked out. He was flitting around like a bee, completely frantic. It was so sweet." She smiles at the memory.

"He ran to the pharmacy to get me some cream and an oatmeal bath. While he was gone I filled the tub and got in.

He came back with four bags full of stuff. He must have bought one of everything. You know Joseph." They all nod. "He just stood there with those bags, looking at me. His eyes were all red. He had been crying. I didn't know yet that he cried so easily."

"Oh, Sue."

"He said, 'Would you like me to read to you while you sit in the tub?' I just remember looking at him and our eyes locked. That was it. We just knew."

The women sit quietly, the sound of occasional laughter mingling with clinking china coming from the house. The air is heavy with spring. The rain has finally moved on and the sun is fighting its way out of the haze.

Someone pops her head through the screen door. "Sue, telephone. I think it's the funeral home."

"Mrs. Helfgot? We're trying to get an update on your husband. Do you know when he's supposed to get here?"

Susan glances at the clock on the oven door. It's almost four. "He's not there yet? I don't know. Maybe you should call the organ bank."

"Don't worry, Mrs. Helfgot. We'll figure it out. Sorry to have bothered you."

Susan hangs up the phone. The surgery started last night.

They're not done yet? Where the hell is my husband?

Thursday, April 9, 2009, 1:30 p.m. Helfgot Operating Room.
Kristina Andrzejewski is still standing watch. She has been asked to notify Dr. Suzuki when they were ready, and he has arrived with the mask. He examines the space where Helfgot's face used to be, looking with concern and disbelief because the cavity is much bigger than he expected.

Andrzejewski watches him remove the silicone mask from its

container and begin to fit it on the ravaged hole that is now Helfgot's face.

"Much too small," he says. There is a huge gap on the right lower chin of almost five inches.

"I have to fix this. This is not good."

To cut the tension, Andrzejewski asks, "So what's your specialty?"

"I'm a dentist."

A *dentist?* Her surprised look forces an explanation.

"I'm Mr. Maki's prosthodontist. I'm a professor at Tufts Dental School."

"Oh." Because she has never met a prosthodontist in her line of work, it takes her a moment to remember that they specialize in implants, jaw problems, and restorative work. Who better to make a mask, she decides, for a person without a face?

She leaves Dr. Suzuki to his work, but she is fascinated by what he's doing. The mask is creamy white, very *Phantom of the Opera.* It's amazing how real it looks as Suzuki matches it to Helfgot's cheeks.

"It's not right," he says. "Bo said he had to take more than he expected, but I had no idea."

He takes out some liquid silicone and begins working on the side of the right cheek, trying to fill in the space. But it's just too large. He pulls gauze from his bag and packs the area underneath, stuffing it into the cavity so the silicone won't flow back into Helfgot's head when he pours it into the gap.

Silicone takes six hours to dry completely, but he doesn't have that kind of time. Helfgot's body has to arrive at the funeral home before sundown, when the second night of Passover begins. The funeral is tomorrow morning. He watches as clumps of wet silicone fall slowly back into the gauze. Silicone oozes from the sides of the mask and sags down the neck. Frustrated, Suzuki wipes some away and starts to apply fresh silicone.

For the next few hours he repeats the process again and again, becoming slightly panicked as he realizes he is running out of time. It's almost four o'clock.

Andrzejewski thinks the mask looks fine, but Suzuki is a perfectionist. A nurse is working to get the room ready for the next patient. This operating room has been tied up for sixteen hours, and so has Maki's. The faster she can restock it and sterilize it, the sooner they can start to ease the backlog.

With two operating rooms and eight surgeons dedicated to the face transplant for more than a day, several other surgeries have been canceled. A few out-of-town patients arrived this morning only to learn that their procedures will have to be rescheduled. Later there will be phone calls and letters of apology, but no explanation. Although these patients might feel better if they knew all the facts, federal privacy laws forbid any such disclosure.

Suzuki continues to add more silicone. "This is not good," he mutters. He is becoming more and more upset.

The nurse tries to encourage him. "I never thought I could donate my husband's face, but what you're doing makes me reconsider," she says. It's hard to know if he hears her. He shakes his head, his brow now beaded with sweat. "It's no good," he says again. He is angry now. "I promised them I would make it perfect."

"It's really there to dress the wound and show respect," says Andrzejewski. "You've done the best you can. We have to get Mr. Helfgot there before sundown."

It is now 4:45 p.m. Suzuki has been hard at work for three hours, trying to make the mask fit.

"That's it," he finally says. "There's nothing more I can do."

He takes out his paints and begins to work on the color of the mask. Pomahac had told him to make it light, but he is unprepared for how pale the actual skin surrounding the mask has become. He needs to lighten it up, but his paints aren't up to the task. In the end the mask is a bit pinker than the pallor of Helfgot's forehead and neck.

The silicone has still not set, so he takes more gauze and wraps it around the head and face. Helfgot looks a bit like a mummy. It's an unsettling image, but it will have to do.

The hearse has arrived to deliver his patient to the funeral home. *His* patient, because over the past three hours Suzuki has become possessive. He worries about what will happen on the ride to the funeral home. The new part of the mask is like jelly; it still hasn't set. If it is jostled, all his careful work will be destroyed.

But it's time to go. He gently cradles the head and helps lift Joseph Helfgot off the table and onto the gurney for his final ride out of the hospital.

Suzuki leaves exhausted and defeated, although he has been told that the family does not plan to view the body. As he drives home he makes two vows. The next time, if there ever is a next time, he will make a full facial mask and trim it down. And to help set the silicone he will bring a blow dryer.

Kristina Andrzejewski notes the time as she calls her office to say they are finished. It is five o'clock.

Across the hall. Maki Operating Room.
The allograft skin is perfectly tailored. The surgeons have left extra room for the inevitable swelling caused by the tremendous assault to Maki's face during the surgery and from the antirejection meds he will have to take. Pomahac will tighten up the face in a future surgery. But for now, the fusion of two faces is complete.

Around the time the last suture is placed, a hearse pulls into the funeral home with the body of Joseph Helfgot.

chapter sixteen

The team has been guardedly optimistic since finishing up the surgery. It took a long time and seemed to go almost perfectly. Jim Maki passed the first hurdle when he was roused this morning. He was groggy and slightly disoriented, but no worse for wear after seventeen hours under general anesthesia.

The blood flow to the allograft has continued, and so far it is perfusing well. No clots or blown vessels. Maki's nurse, Lorrie MacDonald, is watching him like a hawk for any possible signs of trouble. His vital signs are good. He is remarkably fit for a man of almost sixty with a history of heroin use.

If there are problems after the surgery they are likely to appear during the first twenty-four hours. But one big worry must remain unaddressed for at least a few days. Will Maki's immune system reject his new face? It will certainly try, but to what extent will it succeed? How much havoc will all those cells coded with Joseph Helfgot's DNA wreak on James Maki?

The doctors have matched donor and recipient by age, sex,

and blood type. Their tissue matches well too, sharing several antigens that should help reduce the risk of acute rejection. On paper the two men share a lot. But Maki is half-Japanese, mixed with what appears to be Native American and Caucasian heritage, while Helfgot was a Polish Jew. They come from completely different gene pools, and Maki's immune system will soon be aware of it.

Immunity is a complicated business. Millions of complex chemical signals cascade in perfect sequence as cells combine in a variety of patterns in a frenzied dance to keep invaders at bay. In addition to killing off harmful bacteria and viral agents, the human immune system will also attack foreign tissue and organs. It doesn't know that a new heart or a new face has entered its country legally, that its passport is in order, because until very recently in the course of human history nothing like this has ever happened. Without asking questions, the immune system acts swiftly and without mercy, mounting a full-scale assault to destroy the presumed invader.

There are several branches within the immunity army. T-cells, some of which are known as natural killers, punch holes in virally infected human cells. B-cells morph into antibody-producing machines that tag invading bacteria for destruction by other cells, known as macrophages. Dendritic cells with octopus-like tentacles announce an invasion much like the family cat presenting a dead mouse on the doorstep: they curl up foreign protein into one of their long arms and bring it to a T-cell, marking it for destruction. Along the way these various cells combine with potent chemical cocktails. Some of them seem to remember enemies from previous battles, while other recruits serve as scouts in the hunt for new intruders, like the H1N1 virus or, in Maki's case, a transplanted nose.

The human immune army is on active duty every second of every minute of every day. It knows to ignore visitors with special diplomatic status, like the bacteria in the human gut that aid in digestion. It will detect and attempt to destroy previously healthy cells that have turned renegade and become precancerous. Rejection can occur in an instant. Nobody has yet figured out how to either persuade or trick the body into believing that transplanted organs and tissues are on the guest list.

Someday scientists hope to replace faulty organs with substitutes that are grown from a person's own stem cells. They have managed to grow ears on the backs of mice, but these ears cannot hear. They can grow a human bladder from bovine cells and are learning how to integrate it into the rest of the urinary tract. For now, donated human organs are the only viable option, and rejection is minimized with powerful drugs that slow down the immune system. But there is a steep price to pay. These same drugs suppress the body's response to infection, which compromises patients for the rest of their lives.

Many transplanted patients who would otherwise die live on for decades, but they face higher rates of diabetes, kidney disease, and cancer. Even with these risks James Maki was eager to receive a new face. Dr. Pomahac has several other patients whose injuries are as severe as Maki's, and they too are hoping for new faces. Assuming that Maki continues to do well, they may get the chance to undergo this life-giving surgery. All they will need is a brave donor family, a fair-minded insurance company, and the continued blessing of the hospital's Institutional Review Board.

Because Maki's surgery is new and mostly untried it is considered a research trial and is subject to IRB scrutiny. This is standard operating procedure; every facility in the United States that conducts medical research on human beings must follow strict guidelines established by its own IRB. An independent panel determines what type of research will be permitted within a given institution and defines its parameters. No two IRBs are identical. Dr. Maria Siemionow, who performed the face transplant at

Cleveland Clinic four months ago, has her own set of guidelines, as does Bo Pomahac at the Brigham. Often there are different panels with particular expertise, and the Brigham shares several IRBs with Massachusetts General Hospital, across town. Each of these panels is made up of twenty to thirty members, including not only medical specialists but lawyers, ethicists, statisticians, academicians, business leaders, and other concerned community members.

The goal of these panels is to maximize potential benefits while minimizing potential risks, but achieving that balance is never easy. Originally Pomahac approached the Brigham's IRB panel with the idea of limiting face transplants to patients already on immunosuppressant medication. In other words, only people who had previously undergone a transplant would be eligible. Although this wasn't a practical solution because the population involved was so tiny, it was a good way to begin the conversation.

By 2008 it was clear that Isabelle Dinoire, the French recipient, was thriving almost three years after her surgery. Her outcome was spectacular, affording her a quality of life that was impossible to achieve by other means. She appeared to have suffered no adverse physical or psychological effects. In view of her progress, Pomahac revised his application to offer a partial face transplant to anyone with a medical need, without regard to ongoing immunosuppressant drug therapy. The IRB met again, and Pomahac prepared for the questions he thought they might ask:

Do you have a particular candidate in mind?

What will you do if, unexpectedly, the recipient ends up looking like the donor?

How can you know if someone is psychologically ready for a face transplant?

There would surely be medical questions about rejection therapy, questions about the surgery itself, and perhaps even one or two about protecting the identity of the donor family. No IRB question is inappropriate, but there could be a response by the principal investigator—Pomahac, in this case—that would send him back to the drawing board.

In the end, after much discussion, Pomahac's research trial, opening up eligibility to people like James Maki, was approved by a majority vote.

Now that the surgery is over the IRB will watch and wait. They will monitor Maki's progress and will act as his guardian to make sure he is protected going forward. So far they have given Pomahac permission to perform only this one face transplant.

Pomahac comes in to check on Maki, who is still struggling to clear his head from anesthesia. The surgeon can see something the IRB panel doesn't yet know: the results look promising. He is certain that the man who has endured the horrified stares of children, and who has gone without a real meal for four years, will be pleased.

Pomahac looks at his watch. The press conference is just a few minutes away, and he'll have to stand up in front of all those people. Peter Brown is looking for him; he is relieved that Peter has his back. As he leaves Maki's room he tries to compose himself into a subdued look, to repress the ear-to-ear grin that is trying to take over his face.

Office of Public Affairs, Brigham and Women's Hospital.
Late yesterday afternoon, as the team of doctors, nurses, and specialists wearily finished up what had stretched into a two-day event, Peter Brown sat quietly in his office. He had plenty to think

about. Assuming the patient did well overnight, the hospital's first face transplant patient would wake up to a new day and a new life. So would the hospital, stepping out once again onto a high-stakes medical stage. It is Brown's job to handle that exposure.

His long night has been followed by an even longer day. A press kit, including the formal announcement pictures, is going up on the hospital's website. Two rooms have been set up: a large auditorium and the hospital's TV studio for one-on-one interviews. Security will be tightly controlled to ensure that only credentialed members of the press will be admitted, because you never know who might attempt to crash or even disrupt such an unusual announcement. As a former television news director Brown is eager to get started. It's going to be a great day for the Brigham.

But not, he knows, for the donor family. They are in mourning, ready to bury their loved one, and that knowledge pulls on him as he works.

10 a.m. Boston's North Shore.
Kay Lazar, a reporter for the *Boston Globe,* is scribbling notes at a meeting on the future of health care. She looks at her watch and wonders how much longer it will continue. Her colleague is on vacation and Kay is covering her beat, juggling multiple tasks for this proud but troubled newspaper. To save money the *Globe* recently moved its Health and Science section into Lifestyles and launched "White Coat Notes," a popular online news blog covering the city's sprawling medical community of schools, hospitals, research labs, and medical companies.

Her cell phone vibrates. It's her editor, and she leaves the room to take the call. "Kay, you've got to get down to the Brigham right away. They've just done a face transplant, and they're holding a press conference."

Kay knows the basics, but her colleague, Liz Kowalczyk, has been tracking the story, having interviewed Dr. Pomahac a couple of years ago. "Okay," she says. "I'm on my way."

Kay is not a typical reporter. Never eager to chase ambulances, she specializes in health care reform and has been busy lately with "White Coat Notes." Driving over the Tobin Bridge and into the city she tries to remember everything she can about the subject, which isn't very much. Facial transplants are not her thing. This is Liz's story, and she will be sorry to miss it. After almost an hour in traffic she stops briefly at her office to read Liz's face transplant articles before racing over to the hospital. News trucks and police cars spill onto the street as she pulls into the parking lot reserved for the press. She finds her way into the crowded conference room.

Her editor calls again, this time to tell her that right after the press conference, Peter Brown is giving the *Globe*—is giving *her*, in other words—a one-on-one interview with one of the surgeons. Great. She'll have to think of some smart questions, and fast. She wishes she'd had more time to prepare. She nods at Brown across the room, who is sandwiched between a couple of doctors. Dr. Eriksson steps up to the mike to begin the event. Focus, Kay. Focus.

chapter seventeen

Friday, April 10, 2009, morning.
Stanetsky Memorial Chapels, Brookline, Massachusetts.

I keep thinking the worst has passed, but then the next day arrives. The haunted look in Emily's eyes when she rushed into the hospital room and saw her father dead—how could anything be more awful? But today is even worse.

We are in a room. It is hard to stand up, and someone gently pushes me down onto a couch. Ghoulish pale faces with red bleary eyes are lined up next to me, like the couch where Michael Keaton sat in Beetlejuice. Jacob slumps at the other end. He is wearing his bar mitzvah suit. It fits him better now, I think, maybe.

Somebody wants to know where to sit if they're going to speak. We have a lot of people who want to speak. On the right, says the funeral director. My sister puts her head close to mine and whispers that So-and-so wants to talk with me before we start, but the funeral director shakes his head and says we must begin. He extends his hand to help me up from the couch. No, please, don't touch me.

I try to stand, and Emily grabs my waist and points me toward the door. I love having her by my side.

Susan keeps her head down, walking carefully behind Cousin Ziva, matching each of her footsteps so as not to stumble, afraid to look up and see the pain in the faces of all these people. On an easel next to the coffin is a blown-up black-and-white photograph of Joseph as a boy. He is six, dressed in a suit with a bow tie. The boy has a beaming smile and is looking up with shining eyes, as if to say, "Come and get me, you big world out there!" Susan chose this picture because it captures her husband's extravagant optimism.

A man wearing a bow tie steps up to the podium. He gestures to the picture and says that he and Joseph shared a similar taste in ties. People laugh.

Joseph's nephew, Bobby, breaks down as he describes the man who was more father than uncle to him. Others share stories about Joseph's humor, his compassion, his love of life. A friend fights for composure as he recalls their final conversation. As usual it was about family, and how grateful Joseph was to be alive when Ben opened his acceptance letter from NYU. Pam Levine, Joseph's surrogate daughter, steps up to the podium as she did at her father's funeral, ten years ago in this same room. Then Emily and Jonathan, standing together.

I can't take this. It's too much to bear. The rabbi looks at me from his chair behind the podium and nods in my direction while Emily tries to be brave. He knows the pain this is causing.

The chapel has become very quiet. But then Jonathan says, "My dad would be really pissed that he couldn't be here to listen to everyone saying how great he was." They all breathe again, and the laughter returns.

The cortege snakes its way to the cemetery, led by policemen on motorcycles in tall shiny black boots who expertly shut off intercepting traffic. As they turn a corner Susan sees a procession of cars with glaring headlights right behind.

They pull into the cemetery gates and come to a stop. Ben opens the door to get out but is told by the driver to remain in the car until everyone else arrives. Nobody feels like talking. Exhausted, Susan puts her head back and closes her eyes. She is thinking about the double plot.

She bought one in Los Angeles too, when Joseph's mother died. They brought Rachel to L.A. when her Alzheimer's got bad. A wonderful Philippine woman took care of her, and Susan and Joseph would visit every few days. But by then Rachel no longer knew her son, and she would stare at him with a blank expression. She still remembered Susan and allowed her to comb her hair and rub cream on her dry legs, until finally, one day, she no longer recognized her daughter-in-law. Susan bought the double plot thinking that when the time came she would bury Joseph next to his mother. But then they returned to Boston, and here she is with another double.

She finds it slightly ridiculous that her Polish-born mother-in-law, who grew up in a shtetl and survived Auschwitz, is buried half a world away from her birthplace, just off the 405 freeway near LAX. From a window seat on the right side of a plane she can spot Rachel's grave from the air during the final descent. In L.A. people sometimes vie for a spot near the grave of a star.

God, I hope she isn't mad at me for leaving her all alone near the 120-foot blue-tile Al Jolson Memorial Waterfall.

Twenty-eight years earlier. April 1981. Brooklyn, New York.
Joseph and Susan sail along in his little silver Datsun, the sunroof wide open on this warm spring day. New York City pops into view, and the World Trade Towers in lower Manhattan seem to lean into one another as they reach into the sky. "Joseph, I can see the Statue of Liberty!"

He's seen it enough times. The first time, although he doesn't remember it, was when he was two, coming off a ship with his parents and sister in 1951, a few years before Ellis Island was permanently closed.

They are driving to his mother's apartment in Brooklyn. Passover begins tomorrow, on Saturday night. The only thing Susan knows about the holiday is that ample amounts of bad sweet wine are consumed and no one can eat bread for a week. And just this morning, standing in line at the liquor store with a case of Kedem Concord Grape wine for the Helfgot family Seder, she learned that beer is also off limits. Susan, who grew up in a home of beer-drinking, strict German Catholics, was not aware of this. Going without beer will be a sacrifice, but it doesn't really matter. She just wants to be with Joseph in New York. Life is good, and they are in love.

They spill onto Ocean Parkway and into Brooklyn. After Naftali died it became harder for Rachel to fend off the neighborhood hoodlums. They knew she was alone, and there was little she could do when they came into the store and helped themselves to whatever they liked. In the early 1970s the family left the Lower East Side and moved into a large, post-Depression apartment on Avenue M in Brooklyn. Rachel has been there ever since.

After the store closed, Rachel's knack for making a sale landed her a job at the retail counter of Streit's Matzo Factory on Rivington Street, in lower Manhattan. She would take the train to work every morning and return home several pounds heavier at night. It was an open secret on Avenue M that you could buy Streit's products at Rachel's apartment for half of what they charged on Rivington Street.

As they pull up to the curb, Susan sees a sign on the supermarket across the street: Glatt Kosher.

"What does *glatt* kosher mean?"

"Very kosher." Joseph opens the door to the building. The halls smell of Pine-Sol, garlic, and mothballs, and the air is still stale from the long winter. Everything is painted light green. Even the linoleum floor is a dull green flecked with bits of black and silver.

An old lady at the end of the hallway is pushing bags down the garbage chute. She spots Joseph and calls out, "You call yourself a doctor? I had to buy your poor mother tuna fish today. She has vhat to eat, I ask you? Buy her some food, I *beg* you." Her voice is cracking as though she's on the verge of tears. She shuffles halfway down the hall toward them and disappears into an apartment, her slippers slapping the linoleum.

"Ignore her," Joseph tells Susan. "Bye, Mrs. Pensky," he calls out as she slams her door.

They enter Rachel's apartment. A small woman, barely five feet tall, in a black print polyester dress with white patent-leather high heels stands in the kitchen, fitting Shabbos candles into candlesticks. Her stockings are black and heavily laddered with runs. Over the dress is a yellow and orange paisley sweater, with sleeves scrunched up high past the elbows. As she lights the candles and whirls her hands over the top of the flames, Susan can see the number tattooed on her forearm. Then Rachel squeezes her eyes tightly shut and holds her hands together. Gray hair is stuffed up under a blond hairpiece on top of her head. The heels and hair add several inches to her tiny frame. She finishes the blessing and looks up at her son with a beaming smile.

"Hi, Ma," Joseph says, hugging her and handing her the bag of groceries they picked up on the way. "This is Susan."

But Rachel is busy unloading the grocery bag. "For vhat I need chocolate?" she says, holding a fancy wrapped solid chocolate Seder plate.

"Ma, stop telling Mrs. Pensky you have no money." He goes

to a cabinet and begins to rummage around, shoving aside a few large cans of tomato purée. "Susan, look." Dozens of cans of tuna fish are stacked high up in the back, along with many cans of chicken. The chicken isn't kosher, but Rachel has her own version of the rules, and chicken packed in salt water is good enough.

"*Zi nischt far du,*" she mutters under her breath.

The words sound German to Susan, and she hears the voice of her grandmother, a tall woman from Cincinnati. Susan can't speak Yiddish, but the words seem familiar. What she thinks Joseph's mother has just said is "She's not for you."

Joseph responds with a long string of words that Susan can't begin to comprehend, followed by "Did Pauline call?"

He picks up the phone and dials his sister's house. "Hi, we're here. No fish, right? Susan is allergic to fish." As he talks on the phone to his sister, Susan notices reams of heavy drapery fabric hanging from ceiling to floor on every wall of the apartment.

"For vhat you not eat fish?" asks Rachel.

"Susan is allergic to fish, Ma."

"Alloigic? I never hoid such a thing!" She takes the chocolate Seder plate over to one of the walls and pulls the fabric back, the rings on the curtain rod clinking until several feet of wall are exposed, revealing shelves filled with cans of beans, peanut butter, and tomato sauce. One section, whose shelves are lined with newspaper and doilies, holds boxes of matzoh and macaroons in tins. On a high shelf near the ceiling are shoeboxes with the Thom McAn logo. Rachel shoves the Seder plate onto the Passover shelf with the macaroons and pulls the drape closed.

"Kumn! We're going." She opens the door of the apartment.

"Ma, I don't want the candles burning while we're out."

"Kumn!" He leaves the candles alone.

Joseph stuffs his mother into the front seat of the tiny car. Susan squeezes in back. They drive to Pauline's house in Canarsie, a neighborhood filled with attached homes and fifteen-year-old cars still in active use. At a store along the way Joseph jumps out to buy milk for his sister.

Rachel has been silent since they left the apartment. Now she turns around and faces Susan. "*Schlecht blut!*"

Susan, who has no idea what the phrase means, responds with a smile. Rachel turns back and stares out the windshield.

Later that night, when they are climbing into bed in Rachel's apartment, Susan asks, "What does *schlecht blut* mean?"

"Bad blood. Where did you hear that?"

"Joseph," she says with mock innocence, "I'm starting to get the feeling that your mother doesn't like me."

"She'll get used to you."

April 10, 2009. Independent Workmen's Circle Cemetery.
"Sue, he wants us to come." The funeral director is standing outside, signaling them to get out of the car.

It feels old in this place, sacred and scented with earth. Ben goes first, scooping a bit of dirt into a shovel. He dribbles it over the top of the cherry casket. He hands me the shovel and I go next. Now everyone is taking turns. It seems to be taking a long time.

Now Mick and Dave. They loved Joseph. Their wives and kids come for barbeque in our yard, and Joseph turned them on to merguez sausage. Claude, Sharon's Tunisian husband, started the whole merguez thing back in L.A. We had Shabbat dinner with them every Friday night. We would grill while the kids swam in the pool. God, I miss her. Where is Sharon? There's Claude. Every time Joseph goes to L.A. he brings back some merguez from Claude's kosher butcher. Mick and Dave love merguez now. Joseph grills and we drink beer with shots of tequila while the kids run around. Now we won't have merguez anymore, and damn it, watching Mick and Dave is making me cry.

Mick is holding a huge shovelful of dirt. He once made Joseph laugh when he told him that if it weren't for the circumcision, he'd become a Jew just to be like Joseph. He throws the dirt on top of the coffin.

Now Dave is throwing in a heaping load of dirt. The two of them are scooping up more dirt from the pile and flinging it into Joseph's grave. They are like madmen, and I watch until they are spent, until all the dirt is gone.

Mick pats down the soft mound they have made on top of the grave with the back of his shovel, packing the dirt tight. His face is red and sweaty and streaked with tears. He throws down the shovel and walks away. I am shaking.

It is a warm, sunny day, cloudless at the noon hour. A strong spring wind blows in from the west.

chapter eighteen

Friday, April 10, 2009. Hospital press conference.

This is the kind of event a veteran newsman like Peter Brown lives for. He and his counterpart at the organ bank have been at it for two days, carefully orchestrating the announcement that was issued earlier this morning. In the auditorium, where the energy is palpable as the press conference is about to begin, Brown scans the faces and recognizes almost everyone. He was hoping for a big turnout from the regional media, and they're here. But because this is the country's second face transplant, it's been difficult to gauge the level of national interest.

Exactly four months ago Dr. Maria Siemionow performed the first face transplant at Cleveland Clinic, where she and her team replaced 80 percent of the face of a woman who was the victim of a close-range shotgun blast. Although it was never a race between the two hospitals, nobody at the Brigham will deny that it would have been nice to be first. But the second face transplant is still newsworthy, and Brown recognizes people from CNN, Fox, and ABC.

The hardest part is over; the doctors are pleased, and James Maki seems to be recovering well. The press is convivial, and members of the surgical team are in high spirits. Drs. Eriksson, Pribaz, and Pomahac mill around behind the podium with a few colleagues. A vice president of the organ bank is here too, with the somber task of reminding everyone that there would be no story without a donor, and no donor without a death.

Brown watches a member of the film crew catching a pan shot of the room. He spots Liz Kowalczyk's colleague Kay Lazar. Too bad Liz is away. He gave her the story back in the summer of 2007, when Pomahac received preliminary approval to conduct a face transplant trial. Kay's a pro, though, and Brown hopes she won't mind the documentary crew tagging along to film her interview with Pomahac after the press conference. He nods to Dr. Eriksson: it's time to start.

As Bo Pomahac waits his turn, he is apprehensive. Something could still go wrong. Maki has been out of surgery for less than twenty-four hours, and he isn't even fully awake. Pomahac won't relax for a few more days. He prays that none of Maki's new blood vessels will collapse and that there are no lurking clots.

Mid-afternoon. Helfgot residence.

The funeral is over and the house is filled with mourners. The cherry branches are beginning to bloom on the dining room table. Susan looks around for a place to put a plate of fancy macaroons, but the table marked "kosher for Passover" is completely full. The macaroons are larger than baseballs. *Who on earth would eat one of these things?*

She shoves aside a tray of lasagna with meat sauce on the nonkosher table and nestles the macaroons beside it. The basket of forks is empty again. She grabs the basket and starts toward the

Joseph Helfgot as a first grade student in 1954. A large print of this portrait was displayed next to his casket at the funeral. *Courtesy of the Helfgot family*

Joseph's mother, Rachel Wasserman Helfgot, behind the counter of the family store on the Lower East Side of New York during the 1950s. *Courtesy of the Helfgot family*

A lively Joseph Helfgot in his late forties at a dinner party in a California restaurant. It is the mid-1990s. *Courtesy of Susan Whitman Helfgot*

Susan and Joseph Helfgot on a weekend getaway at Laguna Beach in 1999. This picture ran on the front page of the *Boston Globe* when it was revealed that Joseph had been the facial donor for Jim Maki. *Courtesy of Ellen Aub Doeren*

Joseph Helfgot with his wife and four children at Chef Chang's, a favorite Chinese restaurant near their home in Brookline, Massachusetts, during Thanksgiving weekend in 2008. He died five months later. *Left to right:* Susan, Benjamin, Joseph, Emily, Jonathan, and Jacob Helfgot. *Courtesy of Joan Ganon*

Joseph Helfgot with his son Benjamin in September 2008. A tracheostomy tube aids his breathing and an implanted ventricular assist device helps his heart to beat. The VAD wires embedded deep in his abdomen connect to external batteries hidden in pockets sewn into the lining of a vest. *Courtesy of John D. Brink*

Jim Maki in his backyard at his childhood home in Seattle, Washington.
Courtesy of Jim Maki

Jim Maki in Massachusetts shortly after returning from Vietnam in 1970.
Courtesy of Cynthia Maki

Jim's father, John (Jack) Maki, at age twenty-one, shortly before graduating from the University of Washington.
Courtesy of the Maki family

Newlyweds Cynthia and Jim Maki at Hilton Head, South Carolina, in the late 1970s. *Courtesy of Cynthia Maki*

Jack and Mary Maki playing with baby Jessica. *Courtesy of Cynthia Maki*

Jack Maki with granddaughter Jessica in the mid-1990s, taken at his home in western Massachusetts. *Courtesy of Cynthia Maki*

Jim Maki with Jessica at her high school graduation in 2004. *Courtesy of Cynthia Maki*

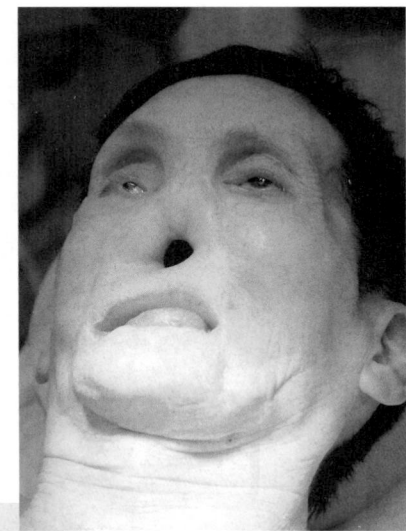

Jim Maki awaiting surgery on the evening of April 8, 2009.
Courtesy of Dr. Bohdan Pomahac

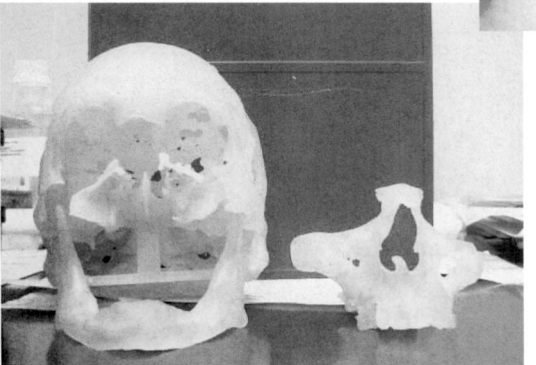

Exact acrylic models of Jim Maki's disfigured head and the portion of Joseph Helfgot's face necessary to restore it sit side by side on a table in Bohdan Pomahac's office.
Courtesy of Dr. Bohdan Pomahac

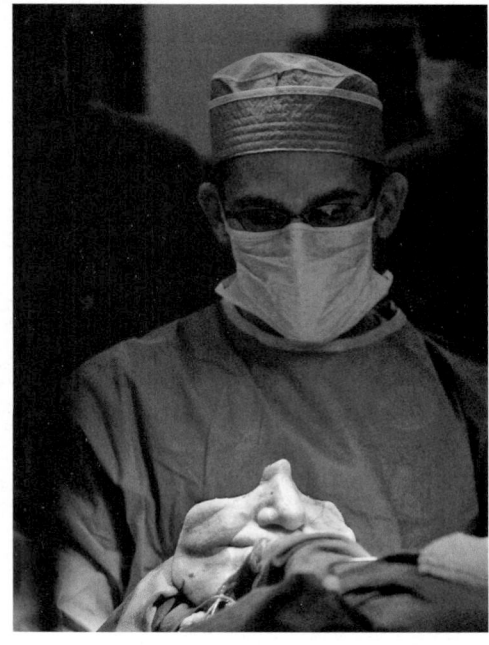

The vascularized composite allograft—skin, blood vessels, nerves, bones, and muscles—moments after being surgically removed from the donor, Joseph Helfgot. Tagged arteries and veins are visible on the lower left.
© April 9, 2009 by J. Kiely, Jr., Lightchaser Photography

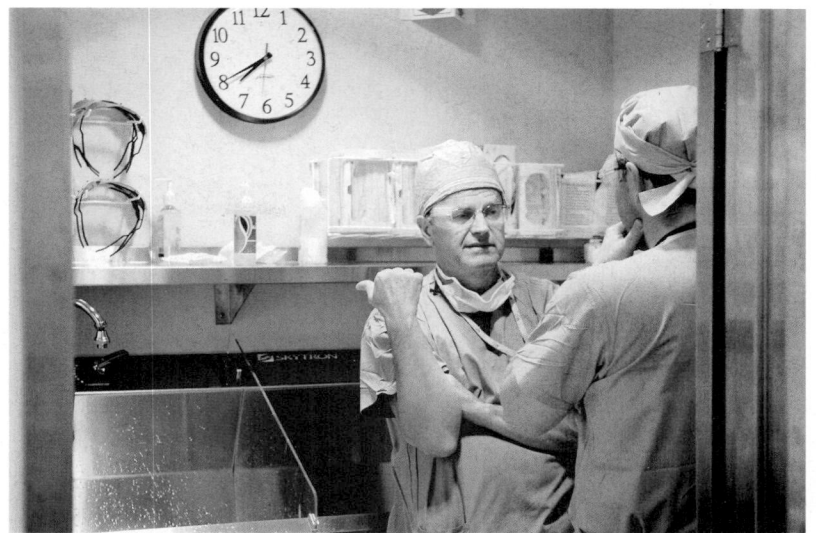

Julian Pribaz discusses the second half of the surgery with fellow surgeon Bohdan Pomahac, his former student at Harvard Medical School. Pribaz motions toward the Maki operating room, where a team is enlarging one of Maki's arteries to better match up with its donor counterpart. The delay provides a brief rest during the 17-hour surgery. © April 9, 2009 by J. Kiely, Jr., Lightchaser Photography

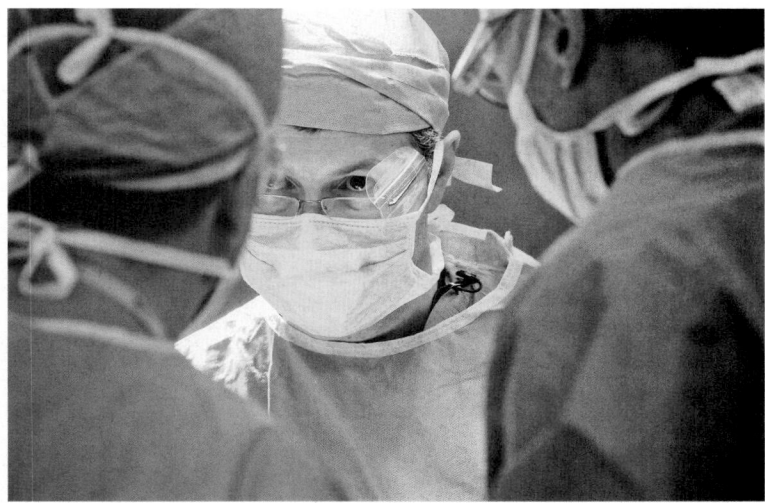

The intensity of the moment is captured in Dr. Pomahac's eyes as he prepares to sever the final blood vessel that will transform a portion of Joseph Helfgot's face into an organ transplant for Jim Maki. Elof Eriksson, Pomahac's mentor and chair of Brigham and Women's Plastic Surgery Department, assists on the right. © April 9, 2009 by J. Kiely, Jr., Lightchaser Photography

Jim Maki and Susan Whitman Helfgot appear together at a hospital press conference six weeks after transplant surgery, on May 21, 2009. *Courtesy of the Department of Public Affairs at Brigham and Women's Hospital*

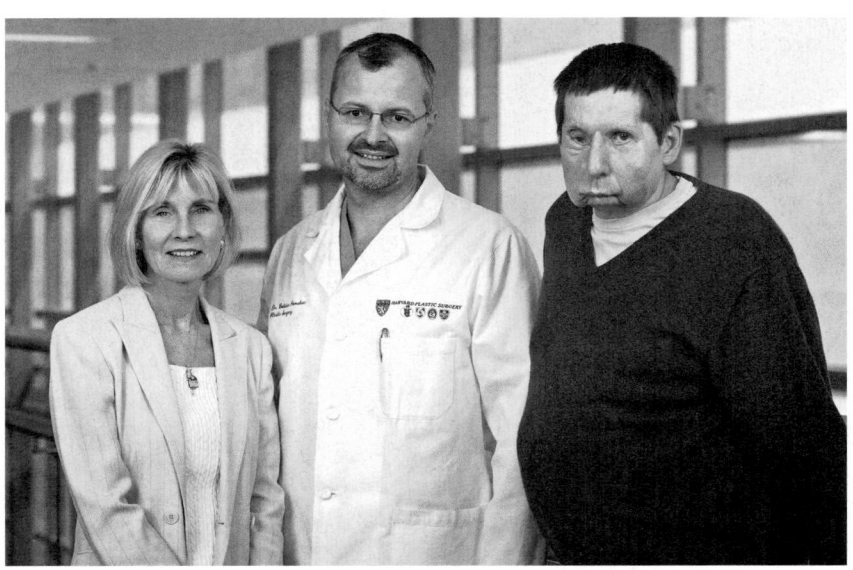

Susan Whitman Helfgot, Dr. Bohdan Pomahac, and Jim Maki thirteen months after Maki's transplant surgery. © *May 13, 2010 by J. Kiely, Jr., Lightchaser Photography*

kitchen. She notices Esther from the organ bank, who is talking to a doctor from the Brigham. She is glad that Esther has come back to the house.

Dave, who shoveled dirt on Joseph's grave, has been handing out pins that look like campaign buttons. A few years ago he and his wife attended a karaoke party with Joseph and Susan to raise money for their children's school. Someone snapped a photo of Joseph singing with the mike in his hand, and Dave has hand-pressed hundreds of buttons with that picture on them, using an old-fashioned machine in his basement. Ben squeezes past his mother in a tan ski cap that he hasn't removed since his father died. There are Joseph buttons all over it.

The kitchen is filled with visitors. Several of them are holding drinks and watching CNN on a TV set on the wall.

"Hey, look," says one of the mourners, who has already consumed several shots of tequila. "They just transplanted some guy's face. Can you believe it? Like that chick whose face got blown off by her husband." Everyone looks up at the TV. Someone in a white coat is speaking. It must be one of the surgeons.

Someone else steps up to the podium, a man from the organ bank. Susan starts to tell her sister to find Esther, but she stops herself just in time. Everyone would know. She listens along with everyone else. "Advances in transplantation only happen when there are individuals and families who can see past their own tragic circumstances and agree to donation . . . It was New England Organ Bank's honor to work with such a remarkable donor family."

They cut away to a video of a surgeon holding a translucent plastic jigsaw model of the facial allograft in one hand. On the table is a head made of the same material, and Susan watches the surgeon insert a stand-in for her husband's face into the plastic skull. They cut back to the doctor who is answering questions from the podium. Susan thinks she will faint. "He's still waking up," the doctor is saying. "He hasn't seen himself yet."

The TV anchor says, "The donor family wishes to remain anonymous, but they have released the following statement: 'To go from being a recipient family to a donor family so suddenly has given us the opportunity to fully understand the power of organ transplants to give and transform lives.'"

It sounds false as I hear it, like a bad actress reciting a lousy script for the first time. And why is she reading only one sentence? Jon and I spent almost an hour composing the statement. Next time I'll write just one line. Wait—next time? What am I thinking?

"Great story, isn't it?" responds the coanchor, smiling widely. Susan's friend from California, who knows what really happened, looks over to her. Susan winks, hoping no one notices how red her face has become.

The hospital.
"Kay?" Brown taps her shoulder. "Dr. Pomahac will be happy to speak with you now. We have some people in town who are filming a documentary. You don't mind if they come along?"

Kay goes down a hall, distracted by the cameras, trying to compose her first question. When she is introduced to Dr. Pomahac, she sees a tired young man, and the questions swirling in her head fade away. "How does it feel?" she asks.

"Surreal."

Helfgot residence.
"Sue, is there someplace we can speak privately?"

Susan leads the general manager of Marketcast into the library and closes the door. Henry Shapiro has had a hand in running Joseph's company almost since the day Joseph and his partners sold it to Reed Elsevier several years ago.

"It's not going to be the same without him," he says.

"I know."

After Joseph and Susan moved the company to L.A. in 1996, Marketcast grew quickly. By 2000 two suitors were vying for it. As a major supplier of business-to-business information, Reed Elsevier wanted to be the top player in providing market research to the lucrative entertainment industry and had already bought *Variety*, an entertainment daily.

The Marketcast acquisition got off to a poor start when the Dutch conglomerate tried to nudge Helfgot aside. Marketcast's president was fun to hang around with, but he was also a loose cannon, saying whatever came into his mind at corporate meetings and arguing with senior management about almost everything. He was also expensive. It wasn't just his salary, but his lavish tastes. Even on short flights he refused to fly coach. He was known to run up five-hundred-dollar business lunches by dragging along favored employees who would normally be munching on sandwiches in the corporate lunchroom.

The executives at Reed were under the impression that movie studios were buying Marketcast's services for its sophisticated statistical analyses of moviegoers' preferences. But they quickly learned that what the studios really wanted was someone who would let them know straight out what the hell was going on with their movie, good or bad. They trusted Helfgot to tell them what all the numbers on those charts really meant, and to be frank with them when they didn't mean anything. Honesty can be scarce in Hollywood, and Joseph Helfgot was brutally honest. It was his greatest asset, although it regularly got him into trouble.

Henry Shapiro was tapped to bring some order to Marketcast, and those who planned Joseph's early demise were soon gone. Shapiro's business acumen blended well with Helfgot's marketing insights, and the company enjoyed explosive growth, its revenues increasing dramatically in just a few years.

"We've been talking it over, Sue, trying to come up with something Joseph might like to be remembered by. We're thinking of endowing a scholarship at UCLA in his name, in sociology. Do you think he would like that?"

Susan is stunned. "I think he would."

Twenty-five years earlier. Cambridge, Massachusetts.
"How'd you do?"

"Not too good." Joseph has just come home after his weekly poker game with the guys, several of whom are independent theater owners. "But I may be able to get into General Cinema. J.D.'s wife told me about a market research study they've just finished. It's completely flawed. I can't believe they paid so much money for it."

For the past few years Helfgot has earned some extra income outside of his teaching job at Boston University by taking on small market research projects around Boston. Fidelity, Lotus Development, Wang, and Polaroid have all hired him to shed light on big decisions, such as the color of a new camera for kids.

"I think Dick Smith likes me," Joseph says after his meeting at General Cinema. "I told him that video rentals were going to be the best thing that ever happened to the movie business." At the time, the exhibition industry was terrified about the impact of VCRs on their business. If people could rent a film and watch it whenever they liked, without worrying about babysitters, ticket prices, or anything else, why would they ever go out to the movies?

Helfgot had a different view. He believed VCRs would actually save the industry. Baby boomers with young children had no time to go to the movies. But if they could rent movies and watch them at home, after the kids were in bed, they would remain interested in what Hollywood was producing and would return to the theaters in a few years. So would their kids, who would be growing up in homes where movies were as familiar as television.

"I told him they needed to do more segmentation on which films they should put in a given market, and that we could poll local audiences. Maybe they could start some sort of frequent moviegoers' club. Then they could segment the avids from the occasionals and the rares." Joseph had already thought about the different kinds of moviegoers. "I'm not sure they like that idea, but who knows?"

Four years later. May 15, 1988. A private club in Boston.
An hour before Susan's wedding, her sister is putting on the bride's makeup. Susan is so tired she can barely stand up. She spent the night stuffing five hundred brochures into envelopes, which she ran over to the post office early this morning. Joseph intended to help, but Rachel was spending the night at their house and she ordered her son to get some sleep. The brochures highlight a study just completed by Helfgot and his two colleagues from the State University of New York at Stony Brook, "Aging Baby Boomers and Declining Leisure Time: Strategic Implications for the Movie Industry." They hope to attract a few nibbles from studio executives and get a foothold in Hollywood.

On their honeymoon in Antigua Susan calls the office. "Anything come in yet?"

"You have two return cards."

"Joseph, we got two cards!" But two trips to L.A. and a thousand dollars in credit card bills later, both leads have fizzled out.

Joseph is undeterred. He continues to consult to General Cinema and then to its rival AMC. At Helfgot's urging AMC launches a program called MovieWatcher, a club for avid filmgoers, which is still going strong.

One day, out of the blue, Orion Pictures calls. They are worried that one of their epic movies is going to flop. Someone has told them about a quirky guy in Boston, and the head of marketing remembers Helfgot from a trade show, where he was brilliant but unfocused. What the hell, they've got little to lose. Kevin Costner in a very long western is going to be a tough sell.

"A western? Three hours long? With *subtitles*? Sure, I'll be there this afternoon."

"Orion wants me to come to New York," he tells Susan.

"Honey, that's great."

"Not great. A three-hour western with subtitles. Are they kidding? No one will see that. I'll give them advice, the movie will bomb, and that will be the end of it. Would *you* see it?"

"I like Kevin Costner."

"Let's hope he likes me. He's directing this thing, and he's supposed to be at the meeting."

"Here's what you need to do," Helfgot tells the Orion executives. "Women love Costner. Forget about the main story line. This is a movie about a man in search of himself, who is struggling to be good. Women love that shit. But he's really brave too, a guy's guy. That's crucial. It's the only way you get the couple's butts *together* in the seats. If she wants to see it, because of Costner, he won't go and they'll see something else. And if you try to sell it as a guy's movie, she won't want to see it. She'll stay home with her girlfriend and he'll go bowling or watch football with a buddy, because guys won't go to a western together. This isn't *The Terminator*. You need to get couples to go *together* or you're dead."

The executives sit around the table as Joseph paces back and forth with a dry-erase marker in his hand. *Dances with Wolves* is just about in the can. It's time to decide how to position it.

One of them sighs deeply. "Okay, let's go with it." At Helfgot's suggestion they market the movie differently in different parts of the country. The East and West Coasts will see trailers with a Native American slant, and the middle and southern areas will get the American soldier slant.

"Suze, what do you think?" Joseph shows her a mockup of the poster. The caption under Costner's face reads, "Lt. John Dunbar is about to discover the frontier . . . within himself."

"I like it."

"They really got it. They got it *exactly*."

"Joseph, stop crying."

Dances with Wolves won seven Oscars in 1991, including Best Picture. At the time, Oscars were not given to movies that nobody saw.

The bedroom phone rings. "Joseph, what do think about a movie starring a guy who eats people?"

"I think you're out of your mind. It sucks." He hangs up the phone. "Suze, I'm going to Orion. They've got something about a cannibal and an FBI lady."

Susan rolls her eyes. "Good luck with that."

Joseph's focus groups tell him that after *Taxi Driver* and *The Accused*, audiences see Jodie Foster as a victim. They don't believe her character can stand up to Hannibal Lecter, a serial killer who targets women. So the studio adds two scenes to the trailer. In the first one, Lecter, locked in his prison cell, terrifies her, and she backs away. In the second, when he tries to intimidate her again, she holds her ground and Lecter backs off. The movie opens on Valentine's Day, although there's nothing romantic about it. It becomes a huge success. A year later *The Silence of the Lambs* wins five Oscars, which doesn't normally happen with horror thrillers.

Twelve years later. Spring 2000. Santa Monica, California.
"Joseph, what's happening?" Susan is in the car, stuck on I-10,

heading home toward the Palisades. Joseph has called her three times today. He's been in meetings with the lawyers to finalize the sale of the company. She has kept her cell phone in a sweaty palm since leaving for work this morning, afraid to set it down for a minute. If this deal goes through, they're set for life. It's hard to believe they had all of fifty-seven dollars in the bank when they got married. They were tapped out, juggling their two businesses and trying to make their payrolls every week. Now, barring a last-minute glitch, they can finally relax a little.

Four years ago they sold Susan's financial services company and their house in Boston and moved to L.A., sinking every dime they had into Marketcast. Joseph opened an office in West L.A. on the Santa Monica border. Two employees came with him from Boston, and he hired several more. But employees are expensive, and so is everything else. Rachel, who died recently, had been in a nursing home. There were college loans for Joseph's older kids, and two younger children to care for. Susan can't believe they had the chutzpah to think they could pull it off.

"Is everything still okay?" she asks. She worries that if the deal falls through it will kill him. Last year Joseph spent several weeks in the hospital after passing out at lunch. They rushed him to UCLA Medical Center, where he was cared for by Gregg Fonarow, a young cardiologist who was trained back east by Dr. Lynne Stevenson at the Brigham. Dr. Fonarow told Joseph he would soon need a new heart, and Susan attended a "What to expect when you're expecting a heart" meeting for spouses of patients. Dr. Fonarow started Joseph on a new regimen of drugs, and the patient seems better now.

"Suze, I'm holding the check."

"Oh, Sweetie."

"I'm at the bank. You know what? The number is too big to fit on the deposit slip. I wish my mother could see it."

"Honey, I'm so proud."

"Meet me for a drink?"

"Twenty minutes."

When Susan arrives at their local hangout Joseph starts crying. "I keep thinking my mother should be here. Your dad too." Susan's father is dying of a brain tumor. For eight months she has been flying back and forth to South Carolina to help her mother. Now they are both crying.

They walk out holding hands into the late afternoon sun and stand by their cars for a moment. Before Joseph opens the car door, he turns around and shouts out, "Unf—ing believable!"

Susan giggles and gets into her car. She follows him home, talking with him on her cell phone the whole way.

April 10, 2009. Helfgot residence.

Someone bursts through the library door, a studio guy from Hollywood. He's holding a pair of jeans in one hand and sneakers in the other. "Susan, I can't stay in this suit another freaking minute. Can I change in here?" He starts unbuckling his belt.

"Jeez, keep your pants on. Let me out of the room first."

"I just keep thinking," he says, pulling off his dress shoes, "who am I going to call now?"

chapter nineteen

The media people pack up their equipment, eager to get back and edit their stories. One of Maki's doctors makes his way over to Dr. Kim.

"Congratulations," she says. "How is Mr. Maki doing?"

"He's awake. Go take a look."

She makes her way into the ICU and looks around for her patient. To ensure privacy, Maki's name is not on the door. The desk nurse nods in the general direction of his room.

From a distance Dr. Kim makes out a facial profile. She is wildly excited and takes a deep breath. Normally there is nothing there when she looks at James Maki from the side—no contours where a mouth or nose would be, just one flat, featureless surface. But now, even from far away, she sees something very different. His face is grotesquely swollen, but there it is: an honest-to-God *face*.

She walks quietly into the room and stares down at the new James Maki. "Mr. Maki?" she calls out softly. No response. Half-sedated, half asleep, he doesn't stir. "You look great," she whispers before she leaves.

Nineteen years earlier.
University of Massachusetts at Amherst.
The seventy-year-old professor quickly makes his way across campus to his office. Spry for his years and looking much younger, Jack Maki is ruminating on his future. He will retire in a few months and will sorely miss academia. It has been his life for half a century, and he finds it hard to imagine any other existence. His department has generously offered to let him keep his office for as long as he wishes, but he has watched other retirees hang around longer than they should have, injecting their opinions and gradually turning into unpopular has-beens.

No, he'll give up the office. So much of his life is on display in this cluttered room—all his awards, pictures, degrees, and, of course, his books.

He has achieved quite a lot for a man who was abandoned as an infant. He sees a picture of his family, and his thoughts immediately turn to Jim, who came to the house yesterday, once again begging for money.

He had such hopes for the boy. He remembers the day Mary received a call that a half-white, half-Japanese baby was available. They had already adopted John and were looking for a girl. But Mary convinced him to go and see the baby. They named him James. He had hoped things would improve when they moved to Amherst. The small college town in the foothills of the Berkshires seemed to offer the hope of a quieter existence. Maybe Jim would settle down and focus on his studies.

• • •

June 1966. Seattle.

"Boys, we have some news."

Jim and John stopped eating and looked up from their plates. Their parents never had any news.

"I have been offered a job to start an Asian studies program at the University of Massachusetts next fall." Jack Maki raised his chin a bit higher and smiled at his family.

It didn't quite sink in at first. "That's good, Dad," John said.

"We'll be moving in July."

"*What?*" Jim gasped.

"In July," his mother said. She looked happy. Mary Maki was always happy when her husband was happy.

"Mom, we can't move. I have to graduate."

"You'll graduate in Amherst. They have a very fine high school," his father explained.

"Well, I'm not going. You can't make me." Jim jumped up from the table and threw down his fork. He ran up the stairs and slammed his bedroom door.

July 1966. Western Massachusetts.

Heavy rain pelts the roof of the white Galaxy 500. Jack Maki leans over the steering wheel, trying to see through the windshield in search of an exit sign.

"This is the one." He eases off I-91 onto a country road.

In the backseat John opens his eyes. Six long days of driving. Will they ever get there? Jim sits next to his brother, watching their father. He drives like an old man, his knuckles white as he grips the wheel.

Dr. Maki has always seemed old to Jim, at least compared to their friends' fathers. Jack feels it too. He even subscribed to *Sports Illustrated* in the hope that it would bring him closer to Jim.

Jim stares out at the New England countryside, which is filled with rolling hills covered in pine trees, green humps on a camel's back. It's so different from Seattle. It's hot in the car, and his father is using tissues from Mary's purse to wipe the condensation off the windshield. The wiper blades barely keep up with the heavy downpour.

"We're almost there," his mother says. "Isn't this exciting!"

Not really. Jim mouths the words silently to the back of his father's head. John stifles a giggle. The boys aren't that close, especially since John started college last fall. The constant tension between Jim and his dad hasn't helped, but this trip has brought the brothers closer. Their world has been turned upside down, more for Jim than for John, who will head back to college in Seattle next month. John feels sorry for his brother, who will have to deal with a new high school during his senior year. That's going to be tough.

The rain abates, changing to a fine mist. Jim cracks his window. Sweet air flows into the car, different from the air in Seattle. Earthier. *This is where I am going to spend my senior year in high school? In the middle of goddamn nowhere?*

John shuts off the engine and they stare at the rental house from the car. It is small and looks abandoned. Newspapers scatter the walk, soaked from all the rain.

The humidity is oppressive as the storm moves off toward the east. They sit in the car for a moment without speaking. Jim has never experienced silence this complete. It's as if the four of them are the only people on the whole planet. A hard lump forms deep in his throat.

"Well, what do you think, boys?"

Jim pulls on his duffel bag and gets out of the car, slamming the door. He heads up the walk to the house.

"Where do you think you're going?" his father calls after him. "Come back here and help us with the bags."

"Maybe he needs a little time," Mary says. "It's a big adjustment."

• • •

1990. University of Massachusetts at Amherst.

That was more than twenty years ago, Jack remembers now, sitting in his office. Where did the time go? They've had a full life, he and Mary. Few real struggles or tragedies. Except for Jim.

Jack hoped things would improve when Jim married Cindy in 1978. Her father was a manager at the General Motors plant in Framingham and helped him get work there. But Jim was fired for using drugs on the job. It was that damned Vietnam War that got Jim on drugs, but Jack didn't realize it until a year or two after he returned.

Jim had done miserably at Amherst High and graduated near the bottom of his class. He had no interest in college, but when Jack insisted that he couldn't live at home unless he stayed in school, Jim enrolled at Greenfield Community College.

On his first day there he signed up for every intramural team they had. He barely had time to make it to class. And then he found a new sport. Jack shakes his head as he thinks about it. Another pointless obsession. Golf.

Spring 1968. Cherry Hill Golf Course, Amherst.

Jim surveys the hole and pulls his driver from his bag. It's a long par 5, with water hazards at the best lies. Tee off too hard and the ball may land in water; too soft, and forget a birdie. The last hazard abuts the green on the approach. Tricky.

He should be in class, but it's a mild spring day, which in New England isn't something to take for granted. Nobody is going to begrudge a guy a game of golf on a day like this. Birdie for sure. He steals one last look down the fairway and checks his wrists, elbows, and feet. In one seamless motion he brings down his left shoulder to meet the tee, sending his arms high into the air. Then, reversing his weight, he brings the club down hard, connecting with the sweet spot in a loud clean crack. The ball lifts very slowly, maximizing the distance.

Two caddies are playing ahead, but now they hang back, watching him. He's going for a birdie. Jim decides on his 3-wood, taking the left approach. Crack, and the ball flies over the next two hazards, coming to rest slightly in the rough. The grass isn't too deep, but it's spongy. He takes his 5 and looks for the flag, but it is obscured behind a bush. One of the caddies runs back and spots it for him. Jim waves thanks and brings the ball cleanly out of the rough. There is no wind today, and it clears the last water hazard, bouncing sharply, rolling downhill past the cup, coming to a stop several feet past the hole.

"Five bucks he sinks it," the caddy says to his buddy. Young Jim Maki has already established himself as a very talented golfer.

Jim assesses the green. It's a nice soft rise up to the cup, a 14-footer, give or take. Not impossible. He hears a car engine start up, but otherwise it is completely still. He loves this part of the game. It's not like basketball, where everybody is screaming at him from the stands, "*Shooooooot!*" After a basketball game he can hardly hear for an hour or two. With golf, you take your time. When a player putts, everybody stays silent. It's just you and the ball.

He checks the hole, repositions himself, and gently sends the club through the ball. It runs up to the cup, slowing gradually until it stops half an inch short. He picks up his gimme.

"Tough luck, Jim," the caddy says. "I thought you had the birdie."

"Next time." He smiles to himself. It's only April.

chapter twenty

October 1968. Amherst, Massachusetts.

ell me you didn't." Dr. Maki grabs Jim's arm.

"Yep, I signed up today."

"Son, do you have any idea what you have just done?" Jim's father is horrified. Mary is pale. She sinks down onto the couch.

"I was going to get drafted anyway. I might as well go on my own terms."

"You might not have had to go at all if you had just stayed in school. You're quitting."

"What am I quitting? Dad, I'm going into the army. There's a *war*."

"Jim, U. Mass has a place waiting for you any time you are ready."

"I'm not going back to school. Anyway, it's too late. I've already signed up."

They took him to the airport a few weeks later. "Call me the minute you get settled in," his mother implored, hugging him. "I'm so proud of you, Jim."

As they drove back to Amherst, Jack comforted his crying wife. "You know, Mary, military discipline may be good for him. Maybe it's what he needs."

May 1969. Fort Lee, Virginia.
A few hundred young recruits stand at attention in the hot sun. "Okay, men. You are now considered one hundred percent AIT— advance infantry trained. I won't lie to you. Most of you will go to Germany, but some of you are going to Vietnam." A sergeant goes through a roll call handing out orders, but Jim's name is not called. There has been a mix-up, but the sergeant assures him he will be going to Germany. He calls his parents to tell them the good news.

That afternoon he heads over to the office to pick up his orders. The desk sergeant checks some papers and then hands him a form. "I got bad news for you, Private. You're going to Vietnam."

Several months later Jim's brother is drafted. Mary is inconsolable. A few months later, when he finishes basic training in South Carolina, John receives orders to leave for Germany. They won't send a family's only two sons to Vietnam at the same time. John will be forever grateful to his younger brother for perhaps saving his life.

A few months later. South Vietnam.
Pvt. James Maki looks out over the country that will be his home for the next year. They should be landing any minute. In the distance he can see the runway strip at Biên Hòa Air Base, where planes full of fresh troops dump their live cargo to sit and wait for orders. Suddenly the plane rises higher and begins to circle. The base is under attack and they can't land. The men make

small talk about the oppressive heat while the plane cuts lazy loops high above the base. Their eyes peer through small windows as the rockets explode below them.

No one will admit it, but they are scared. Jim looks around at the faces of the other men. He signed up; many of the others were drafted. But how they got here makes no difference at a moment like this. The same fear touches them all.

They stay at Long Binh near the air base for a few days. It's a nasty place, dirty, and poor beyond anything they have ever seen. It is one big dirt pile filled with desperate Vietnamese living in squalor, scratching it out while the war drags on—badly, ever since the Tet Offensive.

Several days later he receives his orders and is sent to Phu Bai, a small village at the start of the conflict that has rapidly expanded into a sprawling town built of sandbags. Much of the labor is supplied by children, some as young as eight, who earn a dollar and a half a day standing in the broiling sun, filling up bags.

Heavy construction supplies move through Phu Bai. So do other things, like toothpaste, soap, and cigarettes. A small pack filled with a soldier's toiletries can bring seventy-five dollars on the black market. Pilfered cartons of cigarettes go for close to thirty. A soldier can make a killing in Phu Bai, and some do. What appears on the surface to be an efficiently run facility is more like a western frontier town.

Tempers run high in Phu Bai, and the command hierarchy always seems on the verge of imploding. There is no reason to suck up to those above you, because they too want to get out of here.

Shortly after his arrival Jim stands in a makeshift hospital on an errand for his commanding officer. A North Vietnamese soldier in army fatigues is half propped up on a gurney, moaning. He seems badly hurt. Two Americans are standing over him, shouting in his face. They want to know how many others were traveling with him when he was captured. The man is writhing in pain. One of the soldiers begins to slap him hard across the face without letting up. Jim knows these tactics could save the lives of soldiers who are out combing the nearby roads, but he is sickened by the sight.

When he first arrived he was told to head for the bunkers the minute they came under missile fire. The first time the sirens went off, he and a buddy did as they were told. They were the only ones there. Twenty minutes later, when the all-clear signal was heard, they were still the only ones in the bunker. As they came out into the daylight they saw men sitting on the rooftops of the barracks.

"You idiots, what were you doing down there?" someone shouted. "No one uses the bunkers." A few guys on a nearby roof started laughing at them.

The Viet Cong are efficient. They train their missiles on the hospital and the fuel dump. They don't waste rockets on the barracks, which are too spread out. When the men hear the siren they hop up on the roofs and watch the spectacle, smoking cigarettes until it's all over and they are forced to return to whatever dull job they've been doing. Like a fire drill at school, the rocket attacks break up the monotony. They're a lot easier to take when you know they're not aimed at you.

Except when they are. Jim awakes with a start in the middle of the night. He has somehow fallen onto the floor, across the room from his bunk. But his bunk is gone, and so is the wall. A stray rocket must have landed nearby, and the shock waves sent him flying. Five feet closer and he would have been killed.

Jim doesn't do much of anything until he is finally assigned to be the driver for a senior officer. The officer travels around, supposedly checking up on security, but mostly going wherever he wants. Jim drives him through the small shanty villages that form and then quickly evaporate, depending on where the fighting is.

The villagers make temporary shelters in holes they dig into the ground, covering their "hooches" with leftover cardboard that serves as both walls and roofing.

Driving around is not that bad. It kills time, and the steady breeze in the open jeep is a welcome relief from the constant damp heat. The roads smell a lot better than the villages, which are rank with the odor of rancid fish sizzling on small tin grills.

Spring 1970. South Vietnam.
Private Maki's driving job didn't last long. The captain was reassigned, and they still weren't sending him out to work on anything. Jim hung around camp all day, playing cards and dice. The men played for money, sometimes a lot of money. Why not? Life was short and brutal here, and nobody cared.

One day Jim laid into a buddy who owed him close to four hundred dollars from playing dice. A little while ago, when Jim was on the losing end, this guy had been ruthless about collecting. Now Jim was returning the favor.

"I want it. Now."

"You'll have to wait. I don't have any money."

"I want it right now."

The soldier slipped him a packet of white powder. "This is all I got."

"What the hell is this? I want my money."

"It's all I got. You want it or not?"

Maki wasn't sure what was in the baggie. Heroin was cheap here, not much more than a pack of American cigarettes on the black market. And it was easy to find. Mama-sans, the madams in the villages, always had some lying around. No one was watching you score. No one cared.

Jim had never done drugs, but that night he learned how. He and his dice buddy got extremely high. Jim wanted more. It took a few weeks for the man to repay his debt, but Jim's fondness for heroin didn't end there.

Late one night he returns to barracks. A large sergeant who has been hassling him since his arrival shouts at him. Jim shouts back. It's hot and filthy and they both hate this place. It takes several men to separate them.

"Maki, he's a *sergeant*. What on earth were you thinking? You can't punch a staff sergeant. I don't care what he did to you. He's allowed to beat the crap out of you. You're confined to barracks for a few days while we sort this thing out."

In any other war Private Maki would be court-martialed for having struck an officer and likely sent home to face dishonorable discharge. But this is Vietnam.

"We've decided to transfer your sorry butt out of here. Do you have any idea where you want to go?"

"Maybe on leave?" he says hopefully.

"Don't get smart. You've got ten seconds."

"First Cav?" First Cavalry has a rep for the most kills. It's where you want to be in a war, with men you can count on.

"First Cavalry, Eighth Engineers. You ship out tomorrow morning. Get your gear in order."

"Thank you, sir."

The First Cavalry has been operating just northeast of Biên Hòa, shoring up defenses while it ferrets out enemy infiltration routes. It's an extremely dangerous place. Maki is supposed to be training as a combat engineer, someone who can build things even while short-range missiles and bullets fly overhead. But his new unit leader won't let him go on any missions. He knows why Maki was transferred and isn't sure he can be trusted. Out in the jungle with the enemy, a man has to know his back is covered. He'll watch the new kid for a while and see what he's made of.

● ● ●

Sunday, April 12, 2009, morning. North shore of Boston.
In Cynthia Maki's kitchen her striped tiger cat, eager to be fed, threads its sinewy body between her ankles. Cynthia prepares a bowl of food and sets it on the floor. It's time to get dressed for the trip into Boston to visit Jim.

Jessica is already dressed. She has never been the kind of person who will keep you waiting. Her college graduation is three weeks away. The money for school came from her aunt Michi, Mary Maki's sister. It saddens Cynthia that Michi is in far-off South Carolina, too old and frail to come north for Jessie's graduation. And Mary, who would have been so proud, has been dead for almost twenty years. Her ashes were scattered from a footbridge over the Ohana Pecosh Falls in Mount Rainier National Park by Jim's father and brother. Dr. Maki took a picture of the white powder swirling in the breeze; Cynthia used to look at it when she visited him. Almost five years ago John returned to the same footbridge and scattered Jack's ashes into the flowing waters.

Jessica's middle name is Mari, for Jim's mother. She is getting a bachelor's degree in communication. Jim won't be there either. Just Jessica and Cynthia. After graduation Jessica is going to Europe with her friends. It doesn't seem to faze these kids, getting on a plane and going wherever the winds take them. Jessica has always been confident, just like her father. The audience gave her a standing ovation after she sang "Sit Down, You're Rockin' the Boat" in her elementary school's production of *Guys and Dolls*. Once she got a child's role in a play at the North Shore Music Circus, where Stephanie Mills played the lead. It was a small part, but performing is in her blood.

Cynthia used to sing and dance. Her dance teacher had been a Rockette, and Cynthia dreamed of becoming one too. But she was an inch too short to be considered.

In the summer of 1968 Cynthia and her friends drove to the Newport Folk Festival, where a woman named Janis Joplin sang in a way Cynthia had never heard before. Back at U. Mass she started singing everywhere she went, trying to sound like Joplin. One day a guy asked if she wanted to sing with a group called Clark, Walter, and the Alligators. Later another band wanted her, but Cynthia was in the middle of a semester. She was proud of herself for sticking it out in school. Her mother didn't believe she had really graduated until Cynthia showed her the diploma. Her parents hadn't gone to her graduation.

She sits at the kitchen table for a minute before getting dressed. She likes her kitchen, which is brand-new, with white cabinets and black counters. She has worked hard to get this place. She hasn't had an easy life, and it hasn't gone the way she expected. She and Jimmy have been married for thirty-one years. Unbelievable. They were married in 1978, during March Madness, when Jim was watching one of the games on TV. At halftime they ran out and got married. They had been planning a small wedding, and this wasn't exactly eloping, but still—at halftime during a basketball game, at the house of the local magistrate? It seems crazy when she thinks back on it.

Jimmy loved watching those games. He was a really good basketball player too. And softball, and golf. He could run a pool table several times in a row. In every sport he tried he seemed to be a natural. She can still see him playing shortstop for a slow-pitch softball team in Amherst with a cigarette in his mouth. With his eyes half-closed, he'd scoop up a fast grounder and fire it to first like it was nothing. The memory makes her smile.

On Wednesday Jim called her on his way to the hospital, but she was at work and heard the message only later. On Friday she watched the news about the face transplant. It was strange to hear people talking about her husband on television, but of

course they didn't mention his name. She didn't realize how big
a deal it was until she saw Dr. Pomahac on CNN.

She still can't believe Jim survived the fall. She sat in a room
full of doctors, discussing whether or not he would even survive
the week. The poor guy was hurt beyond recognition, his face
gone, his arm and hand badly burned. At best he would need as
many as a dozen surgeries just to fix his face, Dr. Pomahac told
her. And it wouldn't really be fixed, even then. He might need
a feeding tube for the rest of his life, and a tracheostomy tube
as well. They weren't sure he could breathe on his own if they
turned off the respirator. But they had to try, and he started
breathing through the tube on his own. First the aneurysm, then
the subway fall. Jim was like a cat with nine lives.

"Jessie, you ready to go?" Cynthia wonders what her husband
will look like with his new face. He used to be so handsome.

A single shot rings out. There was no warning of enemy activity
in the area. It came from right outside the perimeter of the camp.
They sit straight up on their cots in silence, waiting for another
round. Jim's heart is pounding.

But it's quiet. They hastily gather together and make a plan,
fanning out in an arc, working their way beyond the fringe of the
base, toward the general direction of the sound. Jim takes slow,
careful steps, his rifle in his hands, until he comes to the top of a
crest running along a deep ditch. He approaches it with caution.
The sound came from very nearby. He inches forward, closer and
closer, craning his neck barely enough to see down into the hole.
He quickly draws back. "Here!" he shouts. "I see something."

They gather around the top of the ditch. At the bottom lies a
fellow soldier with a single shot to his head, his eyes wide open,
his legs and arms splayed at unnatural angles around his torso.
He has killed himself with his M-16. The rifle lies next to him in

the ditch. As Jim bends over the site, someone taps him on his shoulder from behind. He jumps.

April 12, 2009. The hospital.
"Mr. Maki?" Nurse Lorrie is shaking him gently out of his recurring dream. "Wake up. You have visitors."

He looks up. Cindy and Jessica are standing by the bed.
"Jessie? Hi."
"Dad?" Jessie begins to cry.
"Jess, please don't cry. How do I look?"
"Really good, Dad. I never thought you'd look like this again."

chapter twenty-one

a skinny young man with an enormous Jewish Afro stands outside the Manhattan Draft Board office, smoking his third cigarette. He paces nervously back and forth, rehearsing and re-rehearsing in his mind the list of things his Quaker draft counselor has told him. He looks down at his shoes and checks the heels. They are caked with mud. He throws the cigarette onto the sidewalk and grinds it out. Then he bends over and swipes at the ashes with his hand before rubbing his dirty fingers over his shirt and neck.

As he enters the building and passes the MPs, he utters a silent prayer. From out of nowhere the image of Naftali Helfgot appears to him. He rarely thinks about his father, the proud and crippled Auschwitz survivor who was beaten down by the terrible things he had witnessed and endured. Naftali never succeeded in America, and Rachel constantly berated him for his lack of business sense. Marginalized by his wife, he hung around the store, schmoozing with customers and watching for shoplifters while Rachel and Joseph stacked crates of soda and other supplies in the back. With his bad heart Naftali was too weak to lift anything. He did all the banking, in part because it would have been unseemly for his wife to handle the money.

Over time Naftali started drinking more and more. He took out his frustrations on Joseph, sometimes hitting him for his lack of reverence for Judaism. Once, after he found Joseph eating spareribs in a nearby Chinese restaurant, he almost knocked him out. The boy was only ten, but Naftali hit him hard. Rachel intervened, standing between them and protecting her son from further violence.

Joseph hated his father, and he still feels guilty that he was almost unmoved when Naftali dropped dead from a massive heart attack barely a month before Joseph's bar mitzvah. But today, with everything on the line, he misses him for the first time.

December 2, 1961. East Side Hebrew Institute, Lower East Side.
Rachel Helfgot sits with Pauline, her daughter, partially obscured by a screen in the auditorium of the Jewish religious school. Her jaw is firmly set and she stares blankly ahead, knowing the women who sit with her during services are watching to see if she will cry. She can barely endure the injustice of Joseph reading from the Torah without his father standing with him on the bimah. How could her stupid husband drop dead just before the bar mitzvah of their only son?

She holds her prayer book a little higher, pretending to read the Hebrew letters while she mumbles the prayers by rote. Rachel, who grew up poor in a Polish shtetl, never learned to read. When she was still a little girl she watched one evening as her mother prepared to light candles for Rosh Hashanah. She lifted the glass from the oil lamp and lit a small wad of paper from the wick. As she set the glass back into the lamp and turned to recite the blessing, she burned her finger and yelped in pain, knocking over the lamp with her elbow.

Rachel watched as oil spilled down her mother's dress and immediately caught fire. The flames engulfed her mother and Rachel ran from the house screaming for help. But all the men had left for the evening prayers, and there was nothing the girl could do. Rachel's little brother followed, his hair aflame. The house was consumed.

It took three horrible days for her mother to die, screaming in pain the whole time. Her brother survived, but he was badly scarred. The sounds of her mother's screams still haunt Rachel, more than all the other tragedies she has endured. And she has endured a great deal.

After her mother's death Rachel was sent to her father's cousin in Warsaw, a widower with a small infant. As she grew older he began to look at her in a way that made her uneasy. She fled the tiny apartment and lived off the streets, selling candy and flowers during the day and returning at night to the Jewish section, where a kind family gave her refuge.

She married the first young man who came along, and they had two beautiful children, a boy and a girl. They were desperately poor, but Rachel made extra money selling items she pilfered from expensive shops in the heart of Warsaw, returning to the Jewish district at night to sell her bounty. She hired a woman to look after her toddlers during the day. In the summer they would go on picnics at the lake near the palace, with fruit and bread and cheese that Rachel carried in a potato sack.

But the young family's tranquility was short-lived. Germany invaded Poland in 1939 and marched toward Warsaw, quickly surrounding the city. One night Rachel and her husband awoke to the sound of airplanes flying low. Sirens began to wail. An explosion rocked their building as the bombs started falling and continued through the night. The Second World War had erupted on their doorstep.

The Nazis entered the city a few weeks later. Soon the Jews were no longer allowed to leave the northern quadrant of Warsaw except during certain hours, and Rachel was forced to wear a white armband imprinted with a blue Star of David. Food was rationed. She tried to keep her children's stomachs full by giving them carrots. They ate so many that her daughter's skin began to turn orange.

Men were rounded up to work in German labor camps. Her husband was taken, and Rachel never saw him again. A brick wall was built around the Jewish quarter, the ghetto. When it was finished no one could come or go without permission. Rachel's radio was taken, and her son's tricycle. They told her it was for Germany's war effort.

One day a man came and asked for all the knives. He let Rachel keep one dinner knife. She complained that she couldn't cut anything with a dinner knife, but he turned away and began to walk out the door. She continued to rant until he finally turned around. She thought he was going to give her back one of the knives, but he walked up to her and slapped her hard across the face.

While they slowly starved, Rachel heard stories about Jewish children being taken from their homes in the night and sent to refugee camps, supposedly for their safety. She also heard rumors that children were being killed in those camps. Rachel didn't believe these stories, but she was afraid for her little ones, who were seven and five. A nurse came into the ghetto twice a week to check people for typhus. She was known to smuggle small children out in her medical bag, but there were thousands of children and just one woman with one bag. Rachel begged the nurse to take her tiny five-year-old daughter. The nurse said she would try, but there were so many children that Rachel shouldn't expect it.

Late one night soldiers with German shepherds broke down Rachel's door. They swarmed the apartment, hurrying the children into their coats while they shrieked in terror at the barking dogs on short leashes. "Mameh, don't let them take me!" her son cried out. "I don't want to go!"

"We're taking you where it's safe," one of the men said with a gruff German accent.

"No, no!" Rachel screamed. She flew at the soldier, trying to gouge out his eyes. He lowered the butt of his rifle over her head. She awoke the next morning, the sun streaming into the window. A small pool of blood had hardened on the floor next to where she had fallen. When she tried to stand, her right leg gave way. A chunk of flesh was missing from her calf, the dog bite already filling with pus.

She ran limping from the building, screaming that her children had been taken, but the street was full of crying parents. Later that afternoon a man who had slipped out through a dry canal returned with the news that a train had been seen leaving Warsaw earlier that day, filled with Jewish children. Rachel limped back to her apartment and poured her own urine on the dog bite, for there were no medical supplies anymore. She waited until the Nazis finally came for her.

December 2, 1961. East Side Hebrew Institute.
"Ma, stand up." Pauline nudges her mother.

Rachel is thinking about the deposit she gave to Little Hungary Restaurant over on East Houston several months before. They kept part of it, although it wasn't her fault that Naftali dropped dead and she can't make a party for her son. *Gonifs!* In mourning, they will make do with a modest lunch downstairs, after the service, with bagels and cream cheese. The Bremers are taking Rachel and her children out to dinner tonight, because there won't be a party. Mrs. Bremer too is an Auschwitz survivor. Maybe she can get Joseph a summer job at her cousin's place in the Catskills.

Rachel watches her son on the bimah with the other men. As though a thousand knives have not already shredded her heart, she has to endure Joseph's bar mitzvah as a widow. But her eyes are dry. She hasn't cried in years. She squares her shoulders, reciting the prayers she knows by heart, but is unable to conjure up any communion with her Creator. What has she done in her life to deserve this added insult? It's not that she loved her dead husband, who was often terrible to Joseph, especially around matters of religious observance. Sometimes Rachel spots her son sitting alone on a park bench, pulling spareribs out of a small red and white bag. She leaves him alone. It's a small offense, and she is no longer sure God is watching. Or that God even cares.

Spring 1970. Manhattan Draft Board.
Joseph is wearing a tie-dyed T-shirt with the initials SDS emblazoned on the back. Two years ago Students for a Democratic Society shut down a number of college campuses to protest the war. Although the provocative shirt adds to the drama, the SDS can't help him now. Four months ago he sat on his mother's couch in Brooklyn, watching the national draft lottery on television. His birthday came up 123rd out of 366, only a third of the way down. He knew this meant Vietnam. He told himself that he hadn't been born of Holocaust survivors to go halfway around the world and die in a swamp.

The words of the draft counselor rush back to him. "Whatever you do, don't sit. Stand. Pace. Don't get into line. Push your way to the front. Be as rude and as different as possible." For Joseph, who grew up in a tough neighborhood, these instructions aren't hard to follow.

He pushes his way to the front and rips a clipboard from the hands of the man who is giving them out. Leaning against a wall, he checks "yes" to every question on the medical form: bedwetting, night terrors, drug addiction, even attempted suicide.

He throws the clipboard on the table and stands around waiting for the medical exam. The men are told to strip to their underwear. "You won't be wearing underwear," the counselor had said.

As he stands in his altogether holding his clothes, the others give him a wide berth. He is beginning to attract attention, which is good. An MP walks over to him. "You think you're being funny? Didn't your mother teach you to how to dress?"

Joseph spews out a tirade in Yiddish, spitting words at the man until the MP rips the pants out of Joseph's hands and throws them in his face. "Put these back on, now!"

Joseph lets the pants fall to the ground and proceeds to urinate on them. The MP grabs him and pushes him down the hall to the psychiatric service, where he shoves Joseph behind a curtain. A few minutes later a middle-aged doctor with glasses pulls the curtain aside. Joseph, buck naked, lunges at the man, ripping his white coat pocket, scattering his pens, and knocking his glasses into the air. When the doctor tries to retreat and get help, Joseph runs after him, grabbing at his stethoscope.

"Get this f——ing kid out of here!" the psychiatrist shouts. The MPs throw Joseph out of the building.

Several months later he receives his 4F status: unfit for military service.

chapter twenty-two

Sunday, April 12, 2009, afternoon. Helfgot residence.

Visitors have been arriving in a steady stream all weekend.

"Josie," Susan says, "you made something? Really, you didn't have to." She takes a large Tupperware bowl from the woman's hands. Her eyebrows stiffen as she opens the plastic lid and peeks at the contents. *What the hell is this?* "Let's go in the living room. Would you like a glass of wine?"

It's only four o'clock. They still have the evening prayers to-night at seven. Susan isn't sure she can stay upright that long. The phone rings for what seems like the millionth time.

"Sue, someone named Sharon is on the phone." Sharon is Claude's wife, and Susan is reminded of the merguez sausage. She feels tears forming. *What is it about the damn sausage that makes me so weepy?*

She closes the bedroom door. "I am so sorry I wasn't there," Sharon says. "Claude just got home from the airport."

Sharon has been stuck in the last round of financing for a company she recently started. Susan cradles the phone to her ear, looking at the picture of the two smiling couples on the bedside table. Sharon was very pregnant with Daniel. "How is Daniel?" she asks.

"Can you believe his bar mitzvah is next year?"

"No, I can't. I went to his bris only a week ago."

"Listen, Sue, Claude told me about Joseph's face."

"Mm."

"I can't believe you did that. It's amazing."

"You can't tell anybody."

"I won't, but you know it will probably get out. Things like this always do. So many people knew Joseph, and it's a big story, so don't be surprised."

"Okay, I promise if it gets out, I'll let you have the story, okay?" Sharon is an entertainment reporter.

"Sue, how are you doing, the truth?"

"The truth? Really crappy."

They sit quietly for a few moments. "I'm really sorry." Sharon is crying.

Someone calls up the stairs. "Sue, some people are here from the hospital."

"I'll be right down."

"Sue, Joseph suffered a long time. You both went through hell. Don't forget that. He's not suffering anymore. When I saw him last summer with the feeding tube and that purple thing sticking out of his throat . . ." Her voice trails off.

"I know."

Susan heads down the stairs.

Four years earlier. Boston.

Susan stands under a hot shower after spending the day at the Brigham. Things are looking grim for her mother, who has been in the hospital for several weeks. Everything seems to be conspiring against the worn-out body her mother inhabits. But hanging around the ICU is oddly soothing, as the whir and click of medical equipment and the low hum of voices and beepers combine in a soft white noise. No kids, no doorbells, no cell phones.

Being there takes up a lot of her time, and Susan must squeeze the rest of her day into a few hours. She's grateful that Chanukah is already over and that it didn't overlap with Christmas this year. She wonders if her mother will make it to Christmas. She forces her mind to bend toward the positive.

Stepping into the house a few hours ago Susan was greeted by the familiar pile of shoes and backpacks from half the kids in the neighborhood. The shoes always strike her as far too large for teenagers. The dog was on the table, shredding a bag of potato chips. "Freckles! Get down!"

"Mom?" Ben called out. "There's something in my room. *Shit! It's a bat!*"

"What? I can't hear you. And don't say *shit*." She set down a bag of groceries.

"Jesus, Mom. I said *there's a bat in my room.*"

"It may have rabies. Get out and shut the door."

Great. She rummaged around the front hall closet and pulled out her old tennis racket. She had heard that bats can't detect tennis rackets. Would her paltry tennis skills be enough to go up against one measly little bat? If it even was a bat.

Susan carefully cracked the door. It was a bat. She waited until it circled around the room again. On its third pass, she reached out and *thwack!* The bat landed hard against the opposite wall and dropped dead to the floor. It was so tiny. She picked it up with tissue paper and flushed it down the toilet. A mother's work is never done.

"Ben, you can come in now."

"Thanks, Mom." He leaned over and kissed her on the head.

"Now do your homework."

• • •

A shower feels like heaven after a day like today. *What now?* Ben is shouting something through the bathroom door. "What, Ben? I can't hear you."

"I think there's another bat in my room," he yells.

"Get your father," she shouts back, lathering shampoo on her head. Wait, maybe that's not such a great idea. Joseph has a defibrillator in his chest. She turns off the water. "Hold on, I'm coming." She wraps herself in a robe and runs into Ben's room, soap dripping from her hair.

"It's up there." Ben points to a shelf high above his desk. Joseph has one foot on a swivel chair.

"Joseph! Are you crazy? Don't get on that chair. You'll fall and kill yourself." She lugs a large ottoman across the room.

"Susan, put that down, it's too heavy for you." Joseph steps on the ottoman and looks around, pulling books from the shelf. "I don't see anything."

"I'm getting back in the shower. Ben, put the ottoman away." Joseph isn't allowed to carry anything heavier than five pounds.

Susan has finished her shower and grabs a towel. She hears Ben call, "Daaad, I hear it again."

A few seconds later she hears a loud crash.

"*Mom! Dad fell!*"

She slips running back from the bathroom. Joseph is sprawled out on what's left of Ben's desk. His chest has landed on Ben's heavy, boxy computer monitor, and his arms and legs dangle in midair on either side. The swivel chair is on its side, one wheel still spinning. Books and papers litter the floor. Ben is teary eyed.

"I *told* you not to get on that chair!"

Joseph follows her, limping into their bedroom, and gingerly lies down on the bed. "I'm in pain."

"Good!" She tugs on her nightgown and leaves to help Ben clean up the mess.

In the middle of the night, after Joseph has been tossing and turning for hours, Susan hears him reach for the remote. "If you turn on the TV, I will divorce you."

"I thought you were asleep."

"I am *trying* to be asleep."

"What is the dog doing in my bed?" Freckles cautiously eyes his master. The sweet cocker spaniel was picked out by Joseph in a moment of weakness. He has never been able to say no to his children. When Joseph is away the dog sleeps with Susan. When Joseph is home, he tries.

"Please, can't he stay just this once?"

"If I can't watch TV, he can't stay in my bed."

"Freckles, get off the bed," Susan says gently. The dog looks at her and wags his tail.

Joseph lowers his voice. "Off." Freckles jumps off the bed and prances into Jacob's room.

Two hours later Susan is awakened by the sound of the kitchen door closing. "Joseph?" No response. The clock reads a little past 4 a.m. She heads downstairs. The kitchen is cold and smells of winter. Opening the outer door, she sees Joseph limping around the corner of the house, toward a waiting taxi. She runs after him in her bare feet, her nightgown no match for the bitter wind. "What are you doing?"

"I'm going to the hospital."

"I'll take you."

"Go back to bed."

"Don't be ridiculous."

"Go back inside. It's cold."

"Joseph, please don't—" But he is already in the cab, which is speeding away. She runs inside and calls his BlackBerry. "I'm coming to the hospital."

"No, stay with the kids. I think I might have broken something. I'm in a lot of pain."

"Are you sure?"

"About the pain?" There is a pause. "I'm sorry. I shouldn't have stood on the chair. It was stupid." He sounds like a penitent teenager.

"It's all right."

"I love you, Suze."

"I'm coming." She is worried.

"Take your time. Get the kids off to school. Nothing moves fast in a hospital."

She shakes Ben awake.

"Is Dad okay?"

"He's fine. They're just checking him out to be sure. I'm going over there now. Make sure Jacob gets to school."

"Okay."

In the Emergency Room Joseph is connected to a morphine drip. He has two broken vertebrae and three broken ribs.

Back home he spends New Year's Day on the couch, watching the Rose Bowl game and reading movie scripts, popping painkillers as though they are candy. They seem to be having little effect. "Do I remind you of House?" he asks Ben. Gregory House is a fictional TV doctor with a drug habit.

"Dad, give me a break."

Six weeks later. Helfgot residence.

Joseph hasn't slept in days. He is still in intense pain from the fall and is very cranky. Susan is heading over to the Brigham to visit her mother, who has recently been admitted for the third time in six months. She too is cranky. Dan the plumber has just arrived to fix a leaky shower, and she feels better knowing that someone is in the house with Joseph while she runs over to the hospital.

Joseph looks like hell. Too much pain medication and too

little sleep is a bad mix. And her mother is back in the ICU, wearing a hockey mask that pushes air in her face, keeping her alive. She looks like Jason in *Friday the 13th*. Susan looks in the closet for something to wear.

Joseph calls out from the bed. "Suze?"

"What?"

"Why is my computer floating in the middle of the air?"

"Whaaaaat?" She turns around. The laptop is sitting on the bed.

"There." He points to a spot in the middle of the room. "I don't feel well," he says in a weak voice.

"That's it. We're going to the hospital. Dan!" Rubbing his dirty hands on his pants, the plumber walks into the bedroom. "Help me get Joseph into the car."

"Susan?" Joseph calls out into the air. "I can't see you."

Oh my God! She considers calling an ambulance, but she can get him there faster herself. They half-carry him downstairs and push him into the back of the car. Susan races over to the Brigham, two miles away. As she pulls up to the Emergency entrance, a cop comes over. "Ma'am, you can't park here."

"Heart attack!" she shouts. Experience has taught her that these are the magic words. A wheelchair appears out of nowhere and someone takes her keys.

"Please page Dr. Stevenson," she calls out to the receptionist. The woman sends them straight to triage. The Helfgots have been here more than a few times.

"I can't get a blood pressure."

"Where is Dr. Stevenson?"

Soon Carol Flavell arrives. A nurse practitioner with the heart failure group, she has been following Joseph closely for years. Not her most compliant patient, he once apologized for his bad behavior by sending her an enormous fruit basket. The cellophane began to rip as she lugged it down the Pike, and a cascade of grapefruits and tangerines began rolling in various directions. It was quite a sight, watching senior physicians chasing after the runaway fruit.

"What's happening?" she asks Susan.

"They can't get a b.p."

"Take it again," Carol tells the attendant.

"Fifty over ten."

"Let's get him into the crash bed. Susan, did you bring his interrogator?" In addition to his defibrillator-pacemaker, Joseph has a new, experimental device in his chest that records fluid pressures in the heart. The device is read by an "interrogator," an instrument they keep in a drawer next to the bed.

"I just wanted to get him here as fast as I could," Susan says apologetically.

"It's okay. Could you run home and get it? It would really help us to have it."

"I'll be right back."

On her way out the kitchen door, interrogator in hand, Susan hears her cell phone ring.

"It's Carol. We're losing him."

"Oh my God."

"Come fast. Can you get someone to drive you? Sue, are you there?"

"Yeah." She is opening the car door.

"Get someone to bring you."

"Okay." She throws the phone on the floor and reverses the car out of her driveway, narrowly missing a garbage truck driver who has just hopped out of his cab.

A young social worker is standing at the emergency room entrance and escorts Susan back to the crash room. Curtains enshroud the space, and the two women stand anxiously on the other side, listening for clues. People suddenly start running toward the curtain as Susan hears the words "Code Blue, Code

Blue" on the intercom. *Oh my God!* A young doctor runs up. It's her mother's cardiologist.

"What is your mother doing down here?" he asks, confusion on his face.

"It's my hus——" Susan starts to say, but he has already pulled the curtain closed.

"Three thirty-seven." Lynne Stevenson finally pulls back the curtain. Susan feels her knees buckle. The social worker steadies her.

"Three thirty-seven what?" she asks.

"Minutes he was dead—this time. The first time he was out about a minute and a half."

"That's when I called you," Carol says.

Her mother's cardiologist says, "You aren't having a very good week, are you?" She shakes her head and looks at Joseph, who has a tube down his throat.

As they prepare to take him up to the ICU, Dr. Stevenson whispers, "I got his body back, but I don't know about his mind. He was out a very long time."

A few days later. Intensive Care Unit.
A woman comes to the door of Joseph's room where Susan is sitting. "Mrs. Helfgot, I'm Dr. Kim. I'm a psychiatrist working with the Trauma Unit." They shake hands.

Joseph is sitting up in bed, watching TV. "Mr. Helfgot, you had a little trouble last night?"

"Who are you?"

"I'm Dr. Christine Kim."

"Susan, can I talk to you?"

"Sure, honey."

"Without *her*."

Dr. Kim moves out of earshot. Joseph pulls his wife down to the bed and whispers in her ear. "Last night they turned everything into a reality TV show. About doctors. It's a Japanese show. All the doctors are Japanese. They come in at night. That teleprompter," he says, pointing to the monitor on the wall, "it's in Japanese. See?"

Susan looks up at the monitor. Joseph's heartbeat is being graphed in a green line, his EKG showing irregular shallow heartbeats. She nods her head, and he pulls her even closer. "The people who come at night to do the show? I think *she* is one of them." He gently nods in Dr. Kim's direction.

Yesterday Joseph convinced the nurse on duty to call the house. Susan was down in the hospital coffee shop. Jacob picked up the phone. "Dad, is that you?"

"Jacoby, when are you coming? This boat is really cool."

"Dad?"

"I told them the ship can't leave on the cruise until you and Ben get here too. I keep waiting for you. I'm really mad at you for not coming."

Ben grabbed the phone from his younger brother. "Dad, it's Ben. We'll call you back, okay?" Jacob had started to cry.

Dr. Kim steps outside with Susan. "Sometimes it just takes time. Your husband is on a lot of medication."

"But he'll eventually be all right, won't he?"

"It takes time, but usually, yes."

Usually?

A week later Dr. Kim steps into Joseph's room. He is sitting on the edge of his bed, papers scattered everywhere, banging away on his laptop like a madman. As he shifts his weight to look up, papers slide off the bed. "Do you go to the movies?" he asks her.

"Sometimes. How are you feeling today, Mr. Helfgot? I heard you had a bit of trouble again last night."

"Yeah, I guess I did. I thought they were making a movie here, and the ICU was a casino." He gets up and walks around the room. One of his heart leads pops off, triggering an alarm. His nurse comes in and admonishes him. "Stay in bed."

"It was so real. But I remember the whole thing exactly."

"So the next time this happens, when something doesn't seem to make any sense, you can remember that you've had an experience like it before. Can you try to do that, Mr. Helfgot?"

He gets up off the bed and walks to the door of his room. All the leads snap off. He points to the row of monitors next to the nurses' station. "I thought those were teleprompters. They were so *real*. Can you believe it?" He returns to the bed and starts re-attaching the leads. His nurse rushes in. "Joseph, you're killing me. Put your hands down. They have to go a certain way."

"Mr. Helfgot, often when people see things that aren't real, they see something that they can relate to, something they're already familiar with—in your case, the movies."

"But it was *so* real. So do *you* know when I'm going to get out of here? No one seems to know."

"That's up to Dr. Stevenson. How is your pain?"

"Lousy. I need you to write something to help me sleep."

"We'll work on that."

"I thought Kevin, my nurse yesterday, was Fat Gerry, this drug dealer I used to know. Kevin doesn't even look Italian. He's not even fat."

"I'll see you tomorrow." Dr. Kim scribbles something in her notes. "Remember, you're here at the Brigham. We're a hospital. We don't make movies here." She smiles at him and points to a sign Susan has put up on the wall. YOU ARE AT BRIGHAM AND WOMEN'S HOSPITAL. "You're doing much better, Mr. Helfgot. See you tomorrow."

• • •

Monday, April 13, 2009. Trauma and Burn Unit.
Dr. Pomahac has just finished checking James Maki's face. The microvascular connections seem to be holding, and the swelling is abating. So far there has been no rejection.

"When do you think I can look at myself?"

"Let's see what Dr. Kim thinks. Do you feel ready?"

"Definitely. Let's do it."

"Do you think he's ready?" Dr. Pomahac asks Dr. Kim later that day.

"Mr. Maki says he's ready. I think he is."

"Then let's set up a time. Peter told me the crew wants to film it."

"But it's such a private moment," she says. "No one knows how a person may react in that situation."

When Dr. Pomahac began to review potential candidates for facial allograft surgery, he needed to satisfy two important criteria. The first was this: of all the patients in his care, who would benefit the most from such an extreme surgical procedure? Everyone agreed, including the surgeons at the meeting in Brussels, that Pomahac's most seriously disfigured patient was James Maki. Another dozen surgeries would do little to improve his ability to speak, swallow, or breathe, basic functions that the rest of humanity takes for granted.

The second criterion was whether Jim Maki was physically and emotionally strong enough to endure the marathon surgery and its long aftermath. Dr. Kim spent months meeting with him and concluded that he wanted the surgery, understood what it entailed, and would be able to cope with his new identity.

Jim has been desperate to learn what life with a new face might hold. Four years of sobriety in isolation, enduring surgery after surgery from the subway fall that should have killed him, left him a changed man. He knew he was on God's time now and wanted to make good on the years he had left. Without a face, that would be impossible.

His doctors believed he deserved a chance. And so James Perry Maki became the first person listed for a facial transplant at the Brigham.

Dr. Kim is appropriately concerned for her patient. It can't be easy to wake up with someone else's face. Whose nose am I wearing? Whose teeth are in my mouth? Mr. Maki has asked more than a few times who died to give him this face. It is good that he registers the enormity of this gift, that someone had to die in order for him to be whole. It will take time before he merges with his new face.

Dr. Kim understands what he's going through. She too still sees another man's face when she visits with Maki. And like her patient, she wonders about that other man and what kind of life he might have led.

chapter twenty-three

Monday, April 13, 2009, late morning.
The offices of the Boston Globe.

friday's face transplant announcement was big news. After her
interview with Dr. Pomahac, Kay Lazar spoke with four other
sources and sifted through the literature on face transplants, including an article on Dr. Pomahac's three-year quest to perform
the operation at the Brigham. Somehow she managed to put it
all together before her deadline. A quick Friday afternoon post-
ing on White Coat Notes ran ahead of her full story in Saturday's
paper. She has already moved on to her next assignment.

The blogger for White Coat Notes is reading e-mail comments from readers, including quite a few about the transplant.
She spends a lot of her time screening the postings that stream
into the newspaper's website, filtering them for believability and
a semblance of decorum. Today, as she skims the postings, she
reads one that stops her cold.

"Kay, look at this." A man has written that during a Torah
study session at his Brookline synagogue, the rabbi cited the face

transplant as an example of the ultimate mitzvah, an act of kindness that can never be directly repaid. And apparently he mentioned that the donor had been a member of the congregation.

Really?

Kay stares at the words for a few seconds. So the unknown
donor may not be so anonymous after all. She quickly types a
response to the person who submitted the comment: "We
would very much like to share with our readers, in a sensitive
and thoughtful way, this exceptional person's story of generosity.
Could you contact me?"

"I don't know the man," the correspondent responds. "The
rabbi didn't mention him by name. Let me check with him and get
back to you. I can tell you he is from Brookline and in the entertainment business." The day drags on, but Kay hears nothing further.

She starts scanning the local death notices. Nothing rings a
bell, so she checks the websites of *The Hollywood Reporter* and
Variety. In *Variety* she finds an obituary that begins, "Joseph
Helfgot, the sociologist who founded the media market research
firm Marketcast, died from complications of a heart transplant
Wednesday in Boston. He was sixty."

In Boston—died from a heart transplant!

Kay's mind races. She replays the press conference in her
mind, trying to recall exactly what was said about the donor. She
pulls up the statement released by his family. "To go from being
a recipient family to a donor family so suddenly has given us the
opportunity to fully understand the power of organ transplants
to give and transform lives." *The donor was a transplant recipient.*
Kay pulls up her newspaper's obituary for Joseph Helfgot.

"Brookline resident Joseph Helfgot is survived by his wife,
Susan, and three sons and a daughter. He was sixty years of age."
Brookline. The Torah study session was in Brookline. And there
are condolence notices from Los Angeles posted on the funeral
home's website. The donor has to be Joseph Helfgot!

Kay and her associates mull over what to do next. "Just drive over to the house," someone suggests. It takes them only a moment to find the family's address.

"I don't know," Kay says. "Wouldn't that be a complete shock to his widow?" Kay isn't a hard-nosed reporter, and she hates interviewing grief-stricken families. She has chiseled her career around an aversion to situations like this. Now she has to make a tough choice. "Is the family still sitting shiva?" she wonders aloud. They look at the obit again. Visiting hours were Friday through Sunday.

A few hours go by. The man from the temple still hasn't responded. "Kay, just get in the car and go." But she decides to wait out the day.

Later that day. Trauma and Burn Unit.

Dr. Kim's concern for her patient is evident on her face. Maki's hospital room is crowded with people. She hopes he won't be overwhelmed by so much attention for what should be such an intimate moment. Dr. Pomahac stands ready to watch his patient's reaction to his handiwork. Maki's nurse checks his face. She attends to some minor detail and runs a comb through his dark hair so he will look his best.

Cameras whir quietly in the background. Although Dr. Kim's concern is well intentioned, it turns out to be unnecessary. Jim Maki is so excited that later he will be surprised to learn that a TV crew was in the room. What he will always remember is what awaited him when he finally saw his new face.

Dr. Kim shares a knowing look with the nurse practitioner in charge of Maki's care. Maki is only four days out from surgery, and normally they would never allow so many people in a patient's room. They are crowding around the bed, and Maki's nurse will be glad when the hubbub dies down and she can get back to taking care of him.

"Are you ready?" asks Dr. Kim.

"Yep."

"You're not nervous?" Dr. Pomahac asks.

"No, I'm excited. I want to see my face."

The room falls silent. After a moment someone asks, "Who has the mirror?"

People look at Dr. Kim, and then at the nurse, and finally at Dr. Pomahac, as though a surgeon might have a mirror at the ready. A few chuckles finally explode into full-blown laughter. No one has thought to bring a mirror. A nurse runs from the room, frantically making her way around the floor. But there are no mirrors in the Trauma and Burn Unit. Some patients need time to get used to their deformities. Others never do.

She bursts back into the room. "Here," she says, and hands Dr. Kim a tiny makeup mirror.

"Oh, we can't give Mr. Maki that," says Dr. Kim. "It's too small. Can we find something a little larger?" The nurse leaves again and the room falls silent once more. The climax they have been waiting for still eludes them, and for the most mundane of reasons. But they all see the humor in this moment: in a building filled with some of the most sophisticated and expensive equipment ever devised, they are waiting for someone to bring in one of the oldest, simplest tools in recorded history.

Finally the nurse returns with a large square mirror she has pilfered from another floor. She hands it to Dr. Kim. "How's this?" she asks.

"Perfect. Are you all set?" asks Dr. Kim.

"Yes."

She holds the mirror for him as he takes his first long look at himself, moving his head carefully left and then right to inspect his cheeks, and then raising his chin to see under his neck. Looking at Pomahac, he says, "The guy who orchestrated this did a good job."

Maki's facial nerves, which were microsurgically joined to Helfgot's, have not yet healed, so he feels no sensation of any kind, not even pain. He touches the new skin, his fingertips lightly resting on something soft and slightly rubbery. Although the mirror suggests that this new material somehow belongs to him, this isn't *his* face, he realizes as he studies the image. The nose belonged to another man. These teeth and upper lip used to live in someone else's mouth, drinking his water and speaking his words.

The monstrosity that was his previous face is gone. He has returned to life as a full human being. He is humbled. He is shocked. And he is grateful.

There are long, red suture lines running from the corners of his mouth down along the jowls and back up the sides of his face over the cheeks. The lines join perfectly at a common point in the center of the bridge of his new nose. His whole face is still swollen from the surgery, the puffiness exacerbated by large doses of antirejection drugs. But to a man who has lived with a hole in the center of his face for four long years, these imperfections are almost invisible.

He looks up at Dr. Pomahac and says, "My new face looks just like my old one."

It's quite a stretch, but nobody is going to argue.

Bo Pomahac returns Maki's gaze as one man to another, rather than as doctor to patient. Maki's features are swollen and scar ridden, but otherwise normal. The faith that this damaged and fragile man had placed in him, and the risks they had all taken—the naysayers were wrong. This *was* the right thing to do.

And there are plenty of others out there, Pomahac knows, including soldiers with terrible injuries and people with severe burns or tumor disfigurements, who will take hope from the news of this groundbreaking operation. Each one of them, like James Maki, is desperately hoping to feel human again.

chapter twenty-four

Tuesday, April 14, 2009. Helfgot residence.

S ue, there's a call for you from the *Boston Globe.*"
Susan is pulling clothes out of the dryer and putting
them into Jacob's outstretched arms. A week ago today she was at
the hospital, where Dr. Rawn was having trouble looking her in
the eye. *Has it been a week already?* Earlier this morning Jacob tact-
fully reminded his mother that he was out of clean underwear
and that maybe she should think about doing some laundry.

"I'm not ordering the *Globe.* Tell them someone died or
they'll keep calling back."

"She says she's a reporter."

*A reporter? Why on earth is she calling? A local paper ran a story on
Joseph when we first moved back here from California. Or maybe it's for
an obituary.* "Can you tell her it's not a good time?"

"She says it's important."

"I'm coming." Susan picks up the phone.

"Mrs. Helfgot, this is Kay Lazar with the *Boston Globe.*"

"Yes?"

"I want to tell you that I am very sorry for your loss."

"Thank you."

"I would like to speak with you. It's about your husband."

Something isn't right. Her voice–I can hear it in her voice. "What about my husband?"

But Kay is too experienced to say any more on the phone. And she hates this kind of call. "May I come and see you?"

This woman sounds mortified. Something is wrong.

"I guess this afternoon would be fine," Susan says.

As I stand with the phone to my ear, I watch the people in the kitchen moving all around me. They are in one world, the one I used to live in. This reporter is taking me into another place. I think she knows.

Emily's and Jonathan's luggage is on the kitchen floor. I can't bear that they're actually going home. While we are all together I feel safe. This woman is coming over and something is about to happen. And they won't be here with me.

She picks up the phone. "Sharon?"

"Sue, how are you doing?"

"Listen, I think they know. What am I going to do?"

"Who knows?"

"The *Globe*."

"What do you mean you *think* they know?"

"A reporter called. She's coming over to see me about Joseph. Her voice, Sharon, she knows something. What should I do?"

"If she already knows, there's nothing you can do. You could ask her not to break it, but I don't think she would sit on a story like this. It's too big."

"I know."

"Sue, it's your story. You should do whatever you think is right."

"If she knows, I'll tell her to hold it so you can run it first, okay?" Sharon runs a website called TheWrap.

"Thanks. Try to make her do all the talking. She may not know as much as you think."

Kay Lazar is annoyed with herself as she rings the doorbell. How could she show up late for this interview? She called Susan from the road, but this could be a difficult conversation and she doesn't like starting off on the wrong foot.

A petite woman whose age Kay can't quite estimate opens the door with a smile.

"Hi." Susan extends her hand in greeting. She studies the reporter's face and is met with empathetic eyes. Kay's sympathy for the new widow is genuine, but Susan detects something else. Discomfort, certainly. Anxiety, perhaps. *I think she knows Joseph was the donor.*

Susan has been hoping that Kay was coming for some other reason. Unlike Joseph, who enjoyed public attention and loved mixing with celebrities, she has long had an aversion to fame. In the ninth grade she watched as her classmate Chris Evert became an overnight tennis sensation. The paparazzi showed up at school and the poor girl couldn't go anywhere. It was, Susan felt later, a little like having leprosy in reverse. Even now, as she guesses that the jig is up, she still hopes to remain anonymous.

Kay follows her into the living room. In a Jewish house of mourning it is customary to cover the mirrors, as vanity is temporarily suspended by grief. In this house the mirrors are crowded with photographs, thickly layered, one on top of another and strung together with cellophane tape. Frozen memories of a man who died too soon.

"Mrs. Helfgot," Kay starts.

"Please, call me Susan." They sit a moment too long. Finally, "It's all right, Kay. I think I know why you're here." Lazar nods a yes.

• • •

Shortly after Kay leaves the phone rings. "Mrs. Helfgot? I'm Peter Brown from the Public Affairs Office at Brigham and Women's Hospital. I want to express my sympathy for your loss."

"Thank you."

"Kay Lazar from the *Boston Globe* called me just now. She told me that you two discussed your husband's gift. She was wondering if you would mind her speaking with a few people here. Dr. Stevenson is willing to share her thoughts if that's all right with you."

"I . . . sure, that's fine." After the call Susan walks outside into the yard. She sits in the garden that is still struggling to fight off the long New England winter. A friend comes outside to check on her.

"Sue, are you okay? Who was that on the phone?"

"The hospital wanted my permission for the *Globe* to talk to Joseph's doctors. It hit me during the conversation that the man who got his face is going to know who we are."

"Is that bad?"

"It could be. I suddenly remembered that he has the right to decide whether he wants to know about the donor. What if he doesn't? There's a whole system in place, and I forgot all about it. Donor families and recipients can write letters to each other, but without giving their names. The organ bank passes the notes along, but if either side doesn't respond, everyone stays anonymous. When I spoke to the reporter I wasn't even thinking about that. Maybe I should have called Esther from the organ bank before I spoke to the *Globe*. I've made a big mistake. What am I going to do?"

"Sue, calm down. You have been through so much. Don't beat yourself up. It's not as if you do this every day." She pauses, and then adds, "Thank God."

"I feel terrible."

"Don't. You did a wonderful thing, giving that man Joseph's face. I bet he wants to know whose face he's wearing. I'd want to know. Wouldn't you?"

"I guess."

"Stop worrying about it. You have enough to deal with."

"I have a bad feeling this whole thing is going to blow up in my face. Why are you looking at me like that?"

"Did you hear what you said? Blow up in your face?"

"Oh, very funny."

"Sue, did *you* call the *Globe*? No, they called you. Take it one step at a time. It'll all work out."

Later that day. Office of Public Affairs,
Brigham and Women's Hospital.
Peter Brown is with Dr. Lynne Stevenson. They have Kay Lazar on speakerphone.

"Kay, I have Joseph's cardiologist here with me. Mrs. Helfgot gave us permission to speak with you."

Kay asks Stevenson, "Do you remember when Joseph Helfgot's son had his bar mitzvah last year? His wife told me that's all he lived for."

"He always talked about staying alive for that day. He was so extremely ill. We really weren't sure if he would make it with a trachea tube and the VAD." She pauses for a moment. "Joseph lent me a videotape of his mother describing her experiences during the Holocaust, and how she would sacrifice anything for her children, and her sorrow that Joseph's father had not lived to be at their son's bar mitzvah. That's what Joseph was living for, this passage for *his* son. He said the additional time was worth all the suffering he had endured.

"The last time I saw him, he told me that the few months

when he was able to leave the hospital for the bar mitzvah and his sixtieth birthday party were the richest of his life." She clears her throat. "He told me that even if there was nothing after that, it was all worth it." She stops speaking, unable to continue.

Wednesday, April 15, 2009, early morning. Boston.
Christine Kim relaxes in her favorite chair, checking her laptop to see whether anything new has been written about her patient. Although he remains anonymous at the moment, he is far from unknown. And whatever she reads, he will likely read someday.

She worries: How will he cope if his name is revealed? What if people think he isn't worthy of this gift when they find out about his past? Between heroin and jail, it's not a pretty story.

Mr. Maki has spent four long years recovering from his hellish fall onto the subway tracks. At first, he told Dr. Kim, he didn't even realize that his face had been destroyed. It took a long time for his mind to clear after the accident, which wasn't surprising. After being severely electrocuted he was on life support for several weeks. He was also withdrawing from the effects of the various illicit drugs he had been using while taking strong medication to stave off the pain of his injuries. His mind was foggy. He wasn't fully conscious the first few times Dr. Pomahac took him to the operating room.

After three or four months in the hospital he woke up one day fifty miles south of Boston in the severely disabled unit of the New Bedford Rehabilitation Hospital. During the two years he spent there his mind gradually began to clear. The first time he saw himself in the mirror, he couldn't believe the image staring back at him. Everything from his lower lip up to his eyes was gone. No upper teeth or mouth, not even a nose. All that was left was a hole so deep it was black inside. He was looking at a man without a face. It would be shocking enough on another person, but to see *yourself* like that?

The unit was mostly filled with people whose brain injuries had left them virtually mindless. But Jim Maki slowly climbed his way out of that facility, and then rotated through other institutions before moving to a veterans' home outside of Boston. During that time he had multiple surgeries back at the hospital. A trachea tube and a feeding tube were his constant companions.

He knew the accident should have killed him, and he spent four years as an invalid, most of it flat on his back, reflecting on that fact. Dr. Kim knows that patients in such dire situations typically experience one of two reactions: either complete despair or a Herculean resolve to find some meaning in the madness.

Mr. Maki was in the second group, and his optimistic outlook has amazed her. She met him well before anyone had even mentioned the possibility of a face transplant. But even then he was calm, taking each day as it came and never complaining about his situation. In one respect he was actually lucky. He was rushed to the Brigham the night he fell, straight into the hands of Dr. Bohdan Pomahac.

She clicks on Favorites and waits for "Boston Face Transplant" entries to pop up. She scans the first one. The *Boston Globe*: "Gift of a face a testament to donor's enduring values." What? The donor's family has spoken to a newspaper? Hadn't they asked to remain anonymous? She knows the donor was a Brigham patient who died in transplant surgery, but that is the extent of the information given to the press—and to her. Mr. Maki is eager to know who gave him this precious gift, this second chance, but she hasn't been able to help him. The ramifications begin to swirl in her brain as she clicks on the article.

The front page of the *Globe* leaps out at her, with a picture of a handsome, middle-aged couple. They are dressed casually, and the woman's arms are entwined in her husband's. Their heads touch as she leans into his shoulder. They are smiling, and they seem intimately connected to each other and completely at peace. They are obviously in love.

Dr. Kim's mouth turns dry. She cannot take her eyes away from the man's face as he stares back at her from the screen. She knows this face. It's Mr. Helfgot!

Frozen in her chair, she gazes at the picture. As she reads Kay Lazar's article, she cries.

7 a.m. Helfgot residence.
Susan's alarm goes off and she wakens with a start. She is sitting up in bed with the light on, a book half-open in her lap, her glasses on her face. She was up at four, thinking about the recipient and missing her husband. She finally dozed off while reading the same page for the second time, or maybe the third. She can't get the recipient out of her mind. Will he be upset when he finds out she has shared her story? She's been half-expecting the organ bank to call. Kay Lazar was in touch with the hospital. She must have spoken with the organ bank. Maybe they're angry with her for speaking to a reporter.

She reaches for her cell phone and shuts off the sound as it begins another high-pitched scream. Jacob must have been fooling with her ringtones again, because the alarm mimics a test of the Emergency Broadcast System. *God, get me through this day.*

While she waits for the coffee to brew, she pulls up the *Globe* online. The picture her friend Ellie took several years ago looks so different on a computer screen. It sits above the headline, and she and Joseph seem to be other people. But she likes the story Kay has written. It is intended to make a person weep, and it does.

Although it's barely seven, the kitchen doorbell rings. It must be Jacob's friend picking up his backpack. The kid leaves his stuff all over town; it's a wonder he finds his way home at night. Still in her pajamas, backpack in hand, she opens the kitchen door.

But instead of the boy with the missing backpack, she finds a man on her stoop with a microphone in his hand. Right behind him is a another man, holding a camera. The man with the mike looks at her expectantly and opens his mouth to speak, but she slams the door.

Freckles runs into the family room and jumps on top of the couch, barking wildly in front of the window. Susan follows him and cracks the shade just a bit. Several dark SUVs are parked along the curb with people inside. A few people stand on the sidewalk, including the man with the mike and his friend. He takes a long drag from a cigarette while he talks on a cell phone. *Holy crap. I didn't see this coming.*

Trauma and Burn Unit.

Nurse Lorrie MacDonald takes a practiced look at her patient's nose and gingerly places a tissue against his nostrils, using a few quick dabs to sop up a bit of nasal drip. In this case the dripping is a good sign, indicating that Maki's new nose is working. A week after the surgery Lorrie is still tentative in her ministrations. The last thing they need is for any of the tiny, newly attached vessels or nerves to tear. It's all a bit nerve-wracking, she thinks, and smiles to herself at the unspoken pun.

Jim is trying to say something. It is remarkable that a man who was unable to have a real conversation just a few days ago can now speak in coherent sentences.

He is thirsty. Dr. Pomahac has given her permission to start him on a tiny amount of water, although in this case the water is first mixed with a thickening agent. Jim's throat is recovering from the trauma of the surgery and the trachea tube is still in place, so swallowing is a bit of an effort. They don't want any liquid to slip past Maki's weak epiglottis and into his lungs. That could lead to pneumonia. Thickening the liquid first helps keep everything headed in the right direction, down the esophagus.

Thin liquid can squeeze past even the tiniest crack and into his windpipe.

"I'd like real water, not this stuff," he says to her, sucking on something resembling wallpaper paste.

"You're doing really well, Mr. Maki. Try to be patient."

After his drink Lorrie brushes Jim's teeth for him again. Ensuring that his mouth is pristine and free from excessive germs is critical to keep infection at bay. He is still on very high doses of two antirejection drugs, and his immune system is compromised. But Jim Maki couldn't care less about the germs in his mouth right now. He has something else in there that is far more interesting. Teeth! Not dentures, but a full set of real teeth on the top of his mouth. The donor sure had a lot of fillings. Maybe he didn't brush enough. Jim will still need a bottom set of dentures to match the real teeth on top. Dr. Suzuki will make them once the swelling goes down.

11 a.m. Helfgot residence.
Susan's brother, John, a lean southerner who restores antebellum mansions, has been in Boston ever since Joseph's death last week. He's in no hurry to leave. The homes he works on aren't going anywhere, and he likes being here for his older sister. He sits at the kitchen table, sipping a beer and fielding the flurry of phone calls that started early that morning.

"No, she can't come to the phone right now. Can I take a message?"

"No, she isn't doing any interviews."

"No, I don't know if she will be doing any interviews."

"No, she doesn't."

"No, she won't."

"No, she can't."

Every now and then he gets up and ambles outside in his

black boots and skinny jeans, his long hair tucked under a bandana like Steven Van Zandt from Springsteen's band. He sticks his head into car windows and suggests, not too politely, that they might want to consider moving on.

Sometimes the cars leave. Sometimes they don't.

chapter twenty-five

Tuesday, May 5, 2009. Helfgot residence.

On the TV a small woman is facing a room full of reporters. Her face is unnaturally swollen. With a slightly impaired voice she tells the cameras, "Well, I guess I'm the one you came to see today. My name is Connie."

Susan's hand stops, her paring knife halfway into a potato as she watches the nightly news. *The woman from Cleveland!* She shoots a look over at Jacob, who sits at the kitchen table pushing around his homework. She tries to gauge his reaction to seeing a woman on television who received a face transplant. He doesn't seem interested. "Is this upsetting you, sweetie? I'll turn it off."

He shrugs a no in that trying-to-be-cool way adolescents have. "Mom, can you write a note to my math teacher?"

Susan has written a lot of notes to teachers in the past few years. The kids' concern for their sick father has taken precedence over the names of the major rivers of the world or all the state capitals. But Joseph has been dead for almost a month. Maybe it's time to get back on track.

Back on track? What goddamn track would that be?

"Sweetie, it's time to get back on track." She tries out the line in a voice she hopes will sound right, somewhere between *I know it's really hard because you miss your dad so much you think you will burst and how could a mother possibly expect her kid to do homework at a time like this?* and *Do your damn homework, Jacob.*

His eyes brim with tears. "I miss Daddy."

"I'm sorry, Jakey, did I sound harsh?"

"Mom, I don't know how to do this part. I wasn't there when the teacher went over it."

His brother comes through the door. "Mom, what's for dinner?"

"Ben, please help Jacob with his math." Susan shuts off the television. They have moved to another story. She goes back to the potatoes.

Two days later she watches Diane Sawyer interview Connie Culp on *Good Morning America.* Culp tells Sawyer that her injuries occurred when her husband shot her in the face. Sawyer wipes the woman's tears with a tissue. Connie Culp says she can't feel the tears. *Good Lord.* They cut to Connie in her bathroom putting on mascara. *It must be easier to be a woman than a man if you have had a face transplant. You can do a lot with makeup. Susan, stop. You're being ridiculous.*

The phone rings and she turns down the sound. Esther Charves has stayed in touch, and now she's a little concerned because Connie Culp has been in the news for several days.

"Yeah, I'm watching it now."

A few weeks ago, when Susan shared her story with the *Boston Globe*, some members of the press wanted more and camped out at her house for three or four days. Katie Couric's people called, asking for an interview, and so did Diane Sawyer's. And now, less than three weeks later, Connie Culp has gone public. It doesn't feel like a coincidence.

Peter Brown from the Brigham told Susan last week that the

man who received Joseph's face would soon hold a press conference at the hospital, and perhaps she might like to participate.
Would she be willing to read a statement?

Before he asked Susan, Brown had a long conversation with
the organ bank team. "Mrs. Helfgot has already expressed an interest in meeting the recipient," they told him. "And Mr. Maki
knows her identity because of the newspaper story. All that's left
is to introduce them." There was another legal document for
Susan to sign, granting permission for the television documentary crew to film her first meeting with James Maki.

"Mrs. Helfgot, we would like you to meet Mr. Maki a day or
so before the press conference. That way, you will have a chance
to get to know one another."

She is eager to meet the man who received her husband's
face. The country needs more organ donors, and Susan wouldn't
be in this miserable place if Joseph hadn't had to wait so long for
a heart. The press conference will be a good platform for talking
about organ donation.

Susan wonders about the man she will soon meet. She still
knows nothing about him. What does he look like? In her mind
she keeps seeing a male version of Connie Culp, but with Joseph's prominent nose. When she describes the images to Esther
they both burst out laughing. "Susan, he can't possibly look anything like that woman. His surgery was different, for starters."

Esther is right. Although both recipients were given new
faces, their wounds occurred in very different ways and their surgeries were completely different. Unlike heart or lung or kidney
transplants, every facial transplant is unique.

"Don't worry," Esther tells her. "You'll see what he looks like
soon enough." Peter Brown has arranged for Susan to come in
to look at pictures of Mr. Maki before she meets him. Esther and
some of her colleagues from the organ bank will be there to support her. They want her to get used to Jim's reconstructed face

before they meet. No one, including Susan, knows how she will react.

They talk about Connie Culp. "I still can't believe what she went through," says Susan. "Twenty-seven operations, and the whole time knowing it's because your husband shot you in the face." *My God. The lives some people are forced to endure.* Susan thinks of her own marriage and how much Joseph loved her.

Esther asks how the kids are coping. "And you, Sue?" Esther is one of the very few people who really understand. She knows what families go through when they must delay the funeral until the organs are removed, how they think about their loved one being worked on by surgeons even after death. It is brutal for these families, but it brings purpose to the grief they must endure, which is often much worse in the days after the funeral.

"How am I doing? I don't even know. At the moment I need something to wear to this thing in L.A." Susan and her two boys will soon fly to California for a memorial planned by Joseph's company. Emily and Jonathan will meet them there. They will all be together again, and although it is for a sad occasion, Susan is looking forward to having everyone back together, even if only for a day or two.

"You'd look good in a sack," Esther says.

"You're the best, Esther. Bye."

"Freckles!" she calls out. The dog jumps up and down wildly. "Sit." She clips on his lead and steps outside into the cool spring morning. A man needed a face, and he got one. It had seemed so simple at the time.

May 10, 2009. Trauma and Burn Unit.

"Hi, Jim." John Maki hasn't come east yet to visit his brother, but he saw the press conference with all the doctors and he read the story by Kay Lazar. "How are you doing?"

"I'm all right."

It's the first time in four years that Jim has been able to have a real phone conversation with his brother. John is thrilled. For a long time the only way he knew how Jim was getting on was through sporadic conversations with Cynthia. Sometimes he spoke with Jessica, but she was mostly off at college, mercifully away from the horror that has been her father's life since the accident.

"So, how do you look?"

"A lot better than I did before." John hears him trying to chuckle.

"When are you getting out of the hospital?"

"Pretty soon. I'm going to be on TV. They're having a press conference for me. I'm writing a speech."

He's writing a speech? John never thought he'd hear anything like that coming from Jim. Writing a speech! Well, why not? He's always had a way with words. For years Jim carried around a thick packet of white note cards bound with a rubber band. He told people he was writing a dictionary, one that everyone could use, whatever their education. His plan was to take complicated words and explain them in a simple way. He used to read the *New York Times* editorial pages. Even after Vietnam, when he was in tough shape, he still read the paper every day. He'd underline words he didn't know, and he'd make a new card for each one.

He also used the cards to hustle money for drugs. He'd ask people to think of a difficult word and bet them he could spell it. Everyone knows drug addicts aren't that smart. But Jim could spell almost any word, and he could tell you what it meant too. That brother of his was really something.

"Are you in any pain?"

"I'm starting to feel something here and there, like after you go to the dentist." Every few days they remove a bit of skin under his chin with a tool that looks like a miniature apple corer. They also take some from the new skin they have attached between the thumb and index finger on his right hand. It's beginning to tug a bit in an uncomfortable way. It's good that he is getting back some nerve sensation, but that doesn't make the procedure any more pleasant.

"You'll have to come out to Seattle and visit me when you get stronger."

"Yeah, I'd like that."

In the early 1990s Jim moved in with John for a while. After his wife died, Professor Maki gave John some money to remodel the space above his garage into a bed and bath. His plan was to come out to Seattle every now and then and spend time with John. Jim helped with the remodeling while John went off to work. Before long Jim hired a plumber, an electrician, and a carpenter. He met them all playing poker. John was a bridge player. Both brothers loved cards and were good players. Jim gambled. John entered tournaments.

John had to admit that Jim hired good workmen. And when Jim was around, the house was often full of people. John led a quiet life, but Jim was gregarious, and people tended to follow him everywhere he went. He could walk into a room and the mood would change. "Do you know that a baleen whale weighs . . ." And Jim would rattle off some obscure fact and get everyone going.

John thinks maybe he should send his brother some new clothes, but he'll wait until Jim goes home. He still remembers the clothes that were stolen at the rehab hospital. That was a tough place, filled with people suffering from severe brain injuries. He thought Jim would never make it out of there. John crossed the country five times to make sure they were taking good care of his brother. Now they're talking on the phone as if nothing had happened. Life can really surprise you.

In spite of everything that Jim has put the family through,

John still loves his brother. "I'm going to be looking for you on TV. I'll come east soon to visit."

When he hangs up, John hopes he and Jim can spend time together again. They had fun when Jim came to Seattle to help fix up the house. But drugs put a wedge between them, and one day Jim drifted on. Maybe now they'll have the relationship that has eluded them their entire adult lives. It would have made their parents happy.

John pictures his mother, a smile framing her soft face. It's a pity she isn't here to see the change that has come over Jim.

Trauma and Burn Unit.
Jim slides out the mirror in his tray table and inspects his face. The scar lines running over his cheeks are beginning to lighten a little. He gently runs a finger over the new skin but feels nothing. Dr. Pomahac has told him it will take time for the nerves to start working. But his fingertips feel something. He traces a zigzag across his cheek. It's definitely scratchy. He peers more closely into the mirror and rubs his chin.

"Lorrie!" She runs in.

"What's wrong?"

"I think I'm growing a beard!"

"Of course you are. You have a real face now."

"I could never grow a beard before."

"Since the accident?"

"No, I never could."

He takes a long look at the stubble that has sprouted across his entire lower face. "I really have a beard."

The other day he received a visit from Isabelle Dinoire, the world's first face transplant recipient. She came to see him during her trip to the United States. She was a small, shy woman who spoke in French through an interpreter. But she didn't really need to say anything. Her face said it all.

A month after her transplant Dinoire had gone through a bout of rejection. She was alone in the hospital, feeling grateful for her new face, but not seeing many people and wondering when her life would ever resume. It was a difficult time for her. Now her scar lines are virtually invisible, and here she is, traveling overseas.

Jim knows that his face will improve with time, but the waiting is hard. "But you will do better and better," Dinoire reassured him, a smile on her lips. He believes her. She knows.

chapter twenty-six

Wednesday, May 13, 2009. Los Angeles.

S usan is bored, and the long flight to Los Angeles is wearing on her frayed nerves. From Boston it takes longer to fly to L.A. than to London. She has finished a mystery novel and is looking over their hotel reservation, mostly for the pleasure of seeing that welcoming, soothing phrase one more time: *early check-in.*

She is exhausted and can't wait to get to the pool. She isn't looking forward to Friday's memorial program. When she gets home, she will meet James Maki, followed by the press conference. She needs to pace herself, just as if she were running.

She plans to run tomorrow morning, taking her old route from the hotel north to Chautauqua and over to the next cliff, past their first house on Via de la Paz.

She and her father used to walk down the street to the cliff's edge, talking about the kids and Joseph's business. Her father always said that he was thinking of renting a small apartment next time, so he and her mother could stay longer. He and Susan would look for dolphins as they stood under the swaying eucalyptus trees, whose heady scent mixed with the smell of the sea. But then, one day, there was no next time. He was gone in less than a year from a deadly brain tumor.

The cliff dissolves and the face of James Maki appears under her eyelids. Not his new face, but the one that was horribly disfigured. She can't get that haunting image out of her head. Last week she and Esther sat at a computer screen in Peter Brown's office. *Boston Med* filmed it. Susan was wearing a wire, and the crew asked her a lot of questions. Now she can't remember a word she said. Who remembers anything at a time like that?

One week earlier. Office of Public Affairs.
Susan gazes at the first picture and then quickly turns away from the camera. She doesn't want them to film her reaction. Her husband's nose is now on a man with brown eyes and dark hair. And it really *is* Joseph's nose on Maki. Wait, *Maki?* His name is on the picture. *He's Japanese? A Jewish nose on a Japanese face: Does that even work? I guess it has to.* She forces herself to focus on the details in the photograph. His face is huge, swollen and puffy, just like Connie Culp's. One eye is sewn shut. She can see the black stitches.

"Here's what the gentleman looked like before," says Peter Brown as he clicks on another file. Susan sees a shell of a face with a large crater in the middle. *Holy Jesus.*

"And this is what he looked like when they brought him in." He shows her several black-and-white pencil drawings.

"There aren't any photos from the time of the accident?"

"Dr. Pomahac doesn't want anyone to see them. They're too gruesome."

May 13, 2009. Los Angeles.
As the plane begins its descent toward the Los Angeles basin, the Salton Sea glistens like aluminum foil in the late morning sun.

Susan can't stop thinking about meeting James Maki. She still isn't sure it's the right thing to do. She appreciates the drama of their meeting, but what about the four Helfgot children? Right after the *Globe* story, when the press was hanging around, Jacob and Ben had to leave the house by a side door. And how will Emily and Jonathan feel about the attention that may come their way? There aren't many situations in life with no real precedent, but this surely is one of them.

Now they seem to be over Joshua Tree National Park, a bit farther south than usual. When she and Joseph lived in L.A., she sometimes spent the night there, lugging her telescope and pup tent into the desert with other stargazers, braving the nightly temperature plunge. As dawn approached they would hunker down to sleep and awaken in the bright sun, broiling like lobsters in their sleeping bags. Joseph never understood her passion for twinkling dots in a dark sky, but then, she never understood his fondness for sea urchin. There are certain things that couples just can't share.

As she checks her seat belt, an image of the astronomer Tycho Brahe pops into her head. It's been like this since Joseph died, random thoughts coming and going more often than usual. She studied Brahe in school. His student, Johannes Kepler, used Brahe's precise observations to come up with the laws of planetary motion in the early seventeenth century. It has always irked Susan that Kepler got most of the credit when it was Brahe who performed the painstaking observations. And he made them without a telescope, which wasn't invented until eight years after his death.

That's the way it works in science. It took three dozen people in two operating rooms to perform the face transplant, but Dr. Pomahac is getting all the attention. Although he deserves much of the credit, he is the first to acknowledge that without Dr. Pribaz and Dr. Eriksson, it would never have happened.

But that's not the real reason Susan is thinking about Brahe. She has often seen his portrait in her astronomy books, the handsome young Dane sporting an outlandish neck ruffle with a singular oddity jumping off the page. Tycho Brahe had no nose, the result of a duel when he was a student. Fortunately for Kepler, and for science, Brahe survived the incident, sporting various artificial noses made of silver and gold, and sometimes copper. (Copper was lighter, but it turned his skin green.) To keep his prosthetic nose in place, he kept a small jar of glue in his pocket. Medicine has come a long way.

"We're stopping at In-N-Out, right, Mom?" Ben asks as the plane turns north.

"Benny, can you see Grandma Rachel's grave?" He is in a window seat on the right side of the plane.

"I can't see anything," he says. "It's too murky." A thick layer of ocean fog envelops the plane as they approach the Pacific.

"So we're stopping, right?"

"Let's see what the driver says." Every Helfgot trip to L.A. begins at the In-N-Out Burger next to the airport.

Someone from Marketcast is sending a car for them, but Susan isn't sure if the driver will make an extra stop.

When Joseph was too sick to drive, a man named Cesar took him around to meetings and the three places her husband deemed essential: In-N-Out Burger, Tito's Tacos, and Pink's Hot Dog Stand.

Cesar was kind, decent, and hardworking, carrying Joseph's bags when Joseph was too weak to lift them. He would pick up medicine and seltzer water to quench Joseph's burning thirst from the strong diuretics he had to take. Cesar would meet Susan and the kids when they flew in from Boston and take them anywhere they wanted to go.

Once, when Joseph was still driving, he passed out and flew through a stoplight and up onto a curb, smashing into a parked car. He was lucky he didn't kill anyone, or himself. Dr. Gregg Fonorow, his physician at UCLA Medical Center, patched him up with some additional drugs that worked for a while.

She has no idea how to get in touch with Cesar. The company asked her to send back Joseph's BlackBerry when he died. It was company property, they said, but to Joseph it was his life. She downloaded Cesar's number along with those of some other friends, but Cesar's cell phone number was disconnected when she tried it the other day.

Cesar doesn't know Joseph is dead. He may think we just decided to blow him off for no apparent reason. It's been two years since Joseph was here. And now he will never be here. She remembers the night they dined on a patio rooftop looking out at the Hollywood Hills and made the decision to sell everything and move to L.A. to give Marketcast a fair chance. It was a gamble, and it worked. They got lucky, but only financially. As the plane lands she starts to cry, and she peers out the window so her boys can't see her face. *I wish I could call Cesar.*

"Hi there. I'm Kirk." A not-so-young man with bleached blond hair flashes a wide smile. His overbuffed arms are holding up a sign that says DIVA Limo. Under the logo is written H-E-L-G-O-T. *People always miss the F.* Susan can hear Joseph saying, *No, that's not right. It's H-E-L–F as in Frank–G-O-T.*

"I'll be your driver while you are here."

"Can you take us to In-N-Out?" asks Jacob.

"Sure."

It takes forever to crawl the one block from LAX to In-N-Out Burger. Kirk tells them all about the house he is planning to buy with his dad if they can scrape together the down payment, and how his daughter in high school wants to be an actor, just like him. He's had some parts on daytime soaps and had a good run as an extra on *Baywatch* that turned into a few lines here and there, but the show ended a while ago and things have been slow since then.

Now he's trying his luck in films. From their itinerary, he's assuming the Helfgots are connected to the movie business. On Friday morning he will pick them up from their hotel and take them to the ArcLight Theater. He's trying to figure out if they are on the business side or the talent side. He's guessing the business side.

"Mom, do you want anything?" Ben and Jake have decided the drive-thru line is too long. They are going inside to order.

"I'll have a number three with a chocolate shake."

She watches her boys run into In-N-Out, their first visit without their dad. This time they'll skip the fish taco place before heading over to the hotel. Joseph loved fish tacos, but he was the only one who did.

They always stay at the Miramar in Santa Monica, always in Room 209. The suite has a counter with bar stools, and when the boys were little they would play bartender and make their parents pretend drinks before going off to swim in the hotel pool. Susan can't wait to get there.

Today they are booked in Room 309. There's a problem with the plumbing in 209, but the woman in reservations has assured her that the rooms are identical. "Everything is exactly the same, Mrs. Helfgot, just one floor up." *Exactly the same? If only!*

"I'm sorry, Kirk. What were you saying?"

"I tried out for a small part in *The Mist* . . ."

"Kirk?"

"Yes?"

"Did anyone tell you why we are here?"

"No, ma'am, they don't tell us anything. They're not allowed to."

"My husband passed away a month ago."

"Really? I'm sorry."

"He worked here in L.A., but we live in Boston. We're here for his memorial on Friday."

"Oh, is that the thing at the ArcLight?"

"Yes. So we may not be a very happy bunch. If we seem to be, you know, ignoring you or something, I just wanted you to know."

He turns around to face her. "How would you like me to act?"

"I'm sorry?"

"How do you want me to act?"

"Just be yourself."

Late afternoon. Trauma and Burn Unit.

"Are you sure?" Lorrie MacDonald isn't sure. Not at all, but Dr. Pomahac seems to be.

"C'mon, let's just do it." Jim Maki is eager.

Dr. Pomahac's beeper goes off. "It's okay," he says. "He's ready." And he disappears.

Lorrie stands with her hands on her hips, her head sideways, assessing her patient. She shakes her head. "I don't know."

"Dr. Pomahac says it's okay."

"All right, but if anything goes wrong, it's not my fault."

"Can we just do this?"

Lorrie puts on gloves and gently places a towel over Jim's face. She presses her hands down on either side of his nose, making sure her fingers are covering his scar lines. "Okay, are you ready?"

"Yeah."

"You're sure?"

"I'm sure."

"Okay, then. Go ahead. Slow and gentle. *Real* slow."

Jim takes a deep breath and then slowly releases it through his new nostrils. He gives a final hard push.

"*Slow!* I said *slow!*"

"I think it worked."

"You did it!"

James Maki has just blown his new nose for the first time.

chapter twenty-seven

Tuesday, May 19, 2009, early morning.
Trauma and Burn Unit.

a t the start of their training, medical students learn how
to spot the three classic signs of infection that have been
known since medieval times: *calor* (fever), *dolor* (pain), and *rubor*
(inflammation). Rejected tissue turns an angry red when the immune system detects foreign cells, treating them exactly like an
infection. Large doses of chemicals are released at the site, dilat-
ing blood vessels and causing blood and plasma to leak into surrounding tissue, making it swollen and red.

Dr. Stefan Tullius, chief of transplant surgery at the Brigham,
stands at James Maki's bedside. What he sees isn't pretty: the allograft has become inflamed. Is it being rejected? The flap from
Helfgot is now a bright crimson, right up to the scar line that joins
it to Maki's own face. The inflammation signifies something, but
what, exactly? Dr. Tullius and his colleagues aren't quite sure. Whatever is going on doesn't seem to fit normal rejection parameters.

The latest biopsy from the sentinel flap on Maki's hand is

clean; there are no signs of rejection. The biopsy from his face is inconclusive. But the skin on his face is definitely inflamed, a classic sign of rejection.

Skin can turn red for several reasons: embarrassment, exercise, even love. Cold and heat have a similar effect. It's not clear that Maki's inflammation indicates rejection, but they can't afford to ignore it.

Lorrie MacDonald, his nurse, says the redness is transitory. It sometimes fades when the doctors leave, or after a biopsy is completed and Mr. Maki has had time to rest. It's peculiar, no question about it. For now they will classify it as mild rejection and treat it with a steroid bolus.

Brookline, Massachusetts.

A videographer walks down the street, past large old homes and a few brick apartment buildings, shooting background of Joseph Helfgot's neighborhood. Large trees buckle the sidewalk, making his steps difficult. But he's getting some good footage and is engrossed in his task.

Brookline High School sits at the end of the block, and the street is alive with traffic. Teenagers, punks and nerds alike, are walking to school in the cool morning air. Skateboarders whiz along and cars crammed with kids smoking cigarettes race down the narrow street. He thinks his producer will be pleased.

He is careful not to film any of the kids' faces. That's a huge no-no. He takes a final long shot and heads back into the house to see if his buddy has finished wiring Mrs. Helfgot for her historic meeting with James Maki.

They stand around the kitchen sink making small talk as they prepare to go to the hospital. Susan slips a last spoon into the dishwasher. In the distance police sirens wail. The sounds get louder until they come to a stop right outside the door.

Susan worries there's been an accident. Maybe a skateboarder has been hit. It's happened before. She peers out the window. The police cars are right there, with lights flashing. *What on earth is going on?*

As she opens the kitchen door a policeman pushes it open from the other side, stepping past her into the house.

"Ma'am," he says in a severe voice. "Do you know these men?" He takes a stern look at them.

"Yes," she says, the word coming out like a question. "What is—"

"Are you *sure?*" he interrupts.

"Yes."

"Somebody called the station about a guy taking pictures of boys going to school." He's staring at the large video camera.

"Oh, I see." She shakes her head. "No. That's not right. They're here about my husband's face." *Joseph's face? What kind of stupid thing is that to say?*

The officer looks at her, trying to process what he has just heard. His partner arrives.

"You know, the face transplant a few weeks ago?"

"Your husband got a face transplant?"

"No, he gave the face. They're making a show about it."

"Which of you guys is her husband?"

"Nobody's my husband."

"We're making a documentary," says the videographer.

"So where is your husband?"

"I told you, he donated the face."

The policemen are confused.

Hoping it will help, Susan adds, "He's dead."

They stare at her. She decides not to say any more.

"We okay here?" the second officer asks his partner. He is surveying the scene outside. A group of kids has spilled off the sidewalk onto the street, blocking traffic. They are curious about the police cars and happy for any reason to be late for school.

"You sure you're okay?" the first policeman asks Susan one last time.

"I'm fine, really."

The moment the door closes, they start laughing. "It's too bad we don't have *that* on camera," one of the men says. "Damn, that was good."

Office of Public Affairs. Brigham and Women's Hospital.
"Before you meet Mr. Maki, he asked me to share something with you."

Susan is spent, but her eyes are alert. She and Peter Brown are alone. Her wire is turned off.

"It's about his past."

"I don't know anything about him, really."

"I know. But the fact that you're meeting with him suggests you are supportive. People will take it to mean you are endorsing not just the transplant, but also the man. At least, that's how it will be perceived at the press conference."

"But I *am* supportive of him." *What is he getting at? He seems uncomfortable.* "I'm not really sure what you're getting at, Peter."

"Mr. Maki has a history of drug use."

"Oh."

A sixty-year-old who has used a lot of drugs. He probably got arrested for drugs. "The organ bank said he was sixty, so I'm not that surprised. You don't reach sixty as an addict without some run-ins with the law, right?"

"Well, it's a bit more than that."

"Oh." *More than that? What else is there that I don't know? Did he murder someone? Please tell me he didn't kill anybody.* "What *exactly* did he do, Peter?"

"Mr. Maki has a criminal record. He has spent time in jail."

He sits back, waiting for her reaction. He clicks his pen once or twice. It sounds awfully loud. He waits some more. Will all hell break loose? He certainly hopes not. It is unbearably quiet in the room as Susan digests the news.

Thank God he hasn't killed anyone. This is so intense. My heart feels like it's going to jump out of my body. What was I thinking when I said yes to all this? Breathe, Susan.

"Oh," she says.

Joseph, are you okay with all this?

Suze, you still would have said yes. You know you would. No one should live without a face. He didn't do anything that bad. He was a drug addict, feeding his habit.

Brown waits awhile, expecting something more. Finally he says, "Are you okay?"

"I guess so. Thank you for telling me. I really appreciate it. It doesn't change anything."

An hour later:

Do I really want to know this man? Yes, I have to talk with him. I have to tell him it's okay, and he shouldn't feel guilty about getting Joseph's face. God, I'm scared.

The organ bank sent Susan a pamphlet about transplantation. There was a section on survivor guilt. As if Susan doesn't know all about survivor guilt. She could have laughed out loud. *Let me introduce you to the Helfgot family.*

Guilt consumed her mother-in-law like a slow-moving cancer. As a young child Rachel watched her mother burn to death, unable to do anything to help. Her husband was forced into a work camp and her children snatched from her in the Warsaw ghetto. She tried to make the men stop, but she was helpless against them. Then she endured day after day in Auschwitz. Will today be the day they take me to the gas? She always made it to the next selection, but millions did not. Mountains of guilt pressed down on her heart. It's a wonder that she never went completely mad. Her past made her eccentricities seem tame.

Rachel passed her guilt on to Joseph, and Susan has often wondered how he avoided becoming crazy. She knows that he would have wanted Jim Maki to move on without guilt. *I have to talk to this man.*

Trauma and Burn Unit.
Jim Maki walks around his room, pacing. He sits down and immediately stands up again. What's taking them so long? All he wants to do is thank the donor's wife the moment she walks through the door.

Finally Susan and the camera crew arrive. There are more people and cameras in the room. Jim Maki sits in a chair, wearing a Brigham baseball cap. A "Precautions" sign is posted on the door. Susan opens a small closet on a nearby wall, pulling out a yellow gown and a pair of gloves. *I wore a yellow gown every day when Joseph was here.* She slips it on in one quick motion.

James Maki stands up. There seems nothing more natural than to just go over to him and shake hands, but it feels contrived. *I feel like I know this man. Blood really is thicker than water.* The handshake instantly dissolves into a hug.

"Thank you," Jim says to her.

"It's okay, really it is. It's what Joseph would have wanted." She is thankful that she has seen the pictures. "You look good, Jim."

"You have children?" he asks Susan.

"Yes." They sit down. She tells him about the four Helfgot kids. It is difficult to speak. There is so much she wants to say. But not here.

"You have a daughter."

"Yes, Jessica. She may be going to South Korea to teach English. She just graduated from college." He is proud.

"How are you feeling?" Susan is already worried about his health. It is important that he stay healthy, that Joseph's life will somehow continue through this man. If Maki should die, her sole consolation will die with him, and the anguish she is barely able to keep at bay will break her in two.

"I feel great."

"I'm really glad." It is hard to have a real conversation. Susan shoots a quick look at Peter Brown, remembering their talk. She sends an almost imperceptible smile in his direction.

He has Joseph's nose. It is his nose. I can tell. They talk a while longer. As she stands to leave, he stands too and offers his arms. As they embrace she suddenly plants a kiss on his cheek. *I just kissed Joseph's face. This is unbelievable. I have to get out of here before I start crying.*

He calls out after her, "See you Thursday for the press conference."

She is halfway down the hall with no idea how she got there.

"Mrs. Helfgot, do you mind if we ask you some questions on camera?" a woman asks her.

"What? Oh, sure." The questions come fast and furious. Later Susan remembers almost nothing from the interview. She hopes that when the show finally airs nothing she said will embarrass her children.

chapter twenty-eight

Thursday, May 21, 2009, late morning.
Brigham and Women's Hospital.

J im Maki and Susan Helfgot are hidden from view outside the auditorium, where the press conference has already started. Dr. Elof Eriksson is speaking: "One month ago . . ."

One month ago my husband had been buried for eleven days. I am on funeral time. They're keeping real time.

"I want to thank Susan for the gift she and her husband have given to James, the gift of a new face. And I want to thank James for taking that important first step."

God, it's hot. I wonder if Jim thinks it's hot. His face seems flushed.

"He is a pioneer," Dr. Eriksson continues.

Susan has noticed that Jim instinctively touches his finger to his neck whenever he speaks, a habit he developed when he had a trachea tube. When a tracheostomy is performed, air flows through a tube inserted below the vocal cords, making speech impossible. To allow a person to speak, the tube must be capped, forcing air up across the cords. In the 1980s David Muir, a man with muscular dystrophy, invented a small button valve that can be placed over the trachea tube, making speech possible. Until then most patients just covered up the tube with a finger when they had something to say.

Joseph had a bright purple valve. It's still in the night table drawer on his side of the bed. Sometimes it popped off when he coughed, and would skid across the floor, forcing her to crawl around to find it. She would wash it, making sure she dried it completely before giving it back to him. If a drop of water found its way into Joseph's lungs, it could have killed him.

God, I hated that damn thing. Jim probably had one too and just said the hell with it. A lot of people do. They just cover up the hole with their finger and . . . Susan! Get it together. You are about to be on national television and you're thinking about a stupid purple button.

Jim and Susan wish they could just go in and sit down with everyone else and get on with it. Susan is a wreck, and she worries that she'll start to cry when she reads her speech. Nurse Lorrie is with them, trying to help Jim stay relaxed, ready with tissues to mop up the incessant drool from his mouth. Eventually, when Jim's facial nerves start working in his new upper mouth, he will instinctively begin to swallow the twenty or so droplets of saliva his body produces every minute. For now, much of it escapes if he doesn't keep his mouth closed. It's hard for him to control his new mouth. Until last month he didn't even have a mouth.

Susan watches as a slight line of drool starts to drip down the front of Jim's starched blue shirt. Lorrie catches most of it in a tissue. *I hope the spot on his shirt doesn't show up on television.*

It's their turn to speak. Peter Brown leads them into the auditorium. Jim goes first.

"The first part of my life was nothing but trouble." His voice cracks. He thanks the nurses and doctors for giving him another lease on life. "I will be forever grateful."

Now it's Susan's turn. *Don't cry. Whatever you do when you start talking, don't cry.*

She talks about how Joseph worked on movies like *Iron Man* and *Spider-Man*, but that the people who sign up as organ donors are the real superheroes. She asks everyone to visit the website of Donate Life America and register to become a potential organ donor. "Most of all, my thanks go to Jim, who through tenacity and sheer bravery has come so far in such a short time."

As she sits down, Jim looks into her eyes. She smiles at him, and his good, left eye crinkles into feathery lines. Although his face doesn't move, she knows he is smiling back.

They return to Jim's room and sit for a while with Dr. Kim, who asks, "So, how do you think it all went?"

"I'm glad it's over," Susan says.

Jim is tired, his shoulders stooped. "I think we did all right." He takes off his white baseball cap and rubs his scalp.

The film crew is outside. They have been taping throughout the event, but this is a private meeting. Susan has heard from a few scouts who think their story might make an interesting book, and they are trying to figure out their next steps.

Dr. Kim and Susan have already discussed whether a book is a good idea. Dr. Kim thinks it might be helpful for Jim to reflect on his past in a structured way. "The two of you will have a lot of work to do if there's a book, won't you?"

"What else do I have to spend my time on?" Jim is making a joke, and a crinkle forms in the corner of his left eye.

Susan shrugs. "Who knows?" She doesn't expect there will ever be a book. The movie business is filled with options that are never exercised, and she suspects that the world of books works the same way.

"How are you feeling?" she asks Jim.

"Good, I'm going home."

Home? Where? She isn't sure whether it's appropriate to ask him where he lives. *I wish this weren't so awkward.* "You mean now?"

"Yep, today. They're going to take me in a car. I'm ready."

Who wouldn't be? Hospitals are hell. "Do you live far from here?"

"Not too far."

She searches for something to talk about. "How are your daughter's plans to teach overseas coming along?"

"She's in Europe right now, traveling with her girlfriends. She hasn't heard back yet about South Korea." He looks at Dr. Kim. "You're from Korea."

She laughs lightly. "I came here when I was thirteen. And I wasn't traveling alone."

"That must have been a big change. Was it hard?" Susan asks.

"We lived with different relatives for a time."

"I lived in several different places when I was a child," Susan says. "We moved to Boston in my senior year of high school."

"Me too," Jim tells her.

"Really, you moved in your senior year?"

"To Amherst. My dad was a professor at the University of Massachusetts. I grew up in Seattle."

"My husband was a professor when I first met him."

"I heard that."

"Was it hard, moving from Seattle to Amherst?"

"I didn't like it."

"I didn't like moving either," says Susan. "We came here from Fort Lauderdale. My dad bought a house on the South Shore near the water. He thought it would be like Florida, but Massachusetts is nothing like Florida."

"I know what you mean."

Someone from Security escorts Susan to her car, keeping her away from the press. It seems silly, because they know where she lives. The phone is ringing as she walks into the house. It's probably a reporter. *Maybe I should change my number.*

"Hello?"

It's one of Joseph's doctors. Maybe he saw the press conference. "Sue, I have a question for you. Do you know anything about rosacea?"

"It's a rash, isn't it?"

"Yes, on the face. Do you know if Joseph ever had it?"

"I don't think so."

"You're sure?"

"If he had it, I would probably remember. Wouldn't you have known? You guys knew every square inch of his body."

"Well, rosacea's not always easy to diagnose. It can be very mild and go undetected."

"Why are asking me this?"

"I'm not allowed to tell you."

This must be about Jim. His face was flushed at the press conference.

"Do you think Jim Maki might have rosacea?"

He doesn't respond.

"Is it dangerous?"

"Generally, no."

Susan sits down at her computer and logs on to the National Rosacea Society website. The disorder has no definitive cause. It is a vascular condition in which blood vessels dilate, causing severe and sustained blushing. Over time blood vessels may break, and skin tissue, particularly around the nose, may thicken. *Joseph never had any of that. Did he blush a lot? Not really.* She reads on. Flare-ups of rosacea can be tied to emotional stress. "Emotional stress?" she reads again, this time aloud. *Like maybe getting a face transplant and having a press conference? Wouldn't that qualify as stressful?*

• • •

The car chugs up the steep hill away from the river and passes the church. The woman who runs the veterans' home knows that Jim Maki is only a minute away, and she stands on the porch steps, watching for the car.

He eases himself out of the backseat. She smiles at him and they enter the house. "We missed you," she says.

Everything looks pretty much the same. A few of the guys sitting in the dining room look up and tell Jim it's good to see him again. One of them tells him that his face looks great. After a few minutes he heads upstairs to his room. He's happy to be back here with his television and computer. He really missed his computer.

He sits on the bed. The last time he was in this room it was barely spring, a chilly, rainy day. The Red Sox had lost their season opener the night before. Tonight they play the Blue Jays, and Jon Lester is on the mound. In 2006 the Red Sox pitcher was diagnosed with a rare form of non-Hodgkin's lymphoma, but he has made a remarkable recovery. Jim closes his door and turns on the television.

Early July. Department of Plastic Surgery.
Dr. Pomahac fiddles with his PowerPoint presentation. He has been improving it sporadically since the spring, when he put it together to demonstrate the remarkable results they have been able to achieve with a facial allograft. Each time he runs through it he thinks of another nuance or an observation that might be helpful to others who are interested in this groundbreaking reconstructive surgery.

Everyone wants to hear about the transplant, and by now he could probably give this presentation in his sleep. He's become pretty good at speaking with reporters, too. They always ask the same question: "What was the most exciting moment?" He always answers the same way: "When blood began to spread through the allograft, turning it pink. Surreal."

An acrylic model of Maki's skull sits on a table in the corner of his office. Next to it is a model representing the section of Helfgot's face that was used to repair his patient. He clicks away, one slide at a time. A surgeon operates on the notion that there is always room for slight improvements.

James Maki has been doing remarkably well, notwithstanding the rash they have cautiously begun to call rosacea. If that's what it is, it's not a big problem. And if it's a mild form of rejection, they will know soon enough. But the skin guys think it's rosacea, and they turn out to be right.

With Maki doing so well, Dr. Pomahac has begun to allow himself the luxury of focusing on the future. He thinks about the patient who will most likely be their next candidate. He and Julian Pribaz took several patients' files to Brussels last year. The problem is finding more families like the Helfgots.

By the summer of 2009 just two face transplants have been performed in the United States and only a handful of others in the rest of the world. Along with a curious public and a more critical medical community, another constituency has been watching these events with keen interest: the U.S. Department of Defense.

Every war leaves its own stamp of brutality. In Vietnam it was Agent Orange, which led to horrific birth defects and many deaths. In Kosovo it was land mines. In Iraq and Afghanistan it's improvised explosive devices, known as IEDs. First used by the Irish Republican Army in the 1970s, these roadside bombs have been responsible for 40 percent of American military casualties in Iraq; some estimates put the number at 60 percent. Those who are fortunate enough to escape death from IED explosions often lose limbs. The Army Office of the Surgeon General recently reported that well over a thousand American soldiers have undergone limb amputations as a result of IEDs during the Iraq and Afghanistan conflicts.

IEDs are brutal, and sometimes they shear off portions of faces and skulls as well. In 2006 the television journalist Bob Woodruff was almost killed by an exploding IED in Iraq. Surgeons used an acrylic implant to restore a large portion of his skull. He also underwent extensive plastic surgery.

Some wounded warriors have suffered facial injuries so severe they cannot be helped by traditional methods. Two hundred or more American veterans, many living as virtual shut-ins, could benefit from this life-giving surgery. The Department of Defense is determined to find a way to help them. Several successful hand transplants for military amputees have set the stage for the possibility of face transplants. Quietly, on an obscure government website in the spring of 2009, the Department of Defense sent out a request for proposals to conduct clinical studies.

Bo Pomahac filed the Brigham's grant application a few weeks ago. If funding is approved the Brigham's pathology, radiology, immunology, and psychiatry departments will all play a role. Juggling the work will be complicated, but it's the kind of problem that Pomahac and the Brigham would like to have.

He would like to help the soldiers who have given a face for their country. Other than giving one's life, it's hard to imagine a greater sacrifice. He looks up at the model of Jim Maki's skull and imagines several more lined up beside it.

chapter twenty-nine

September 2, 2009. East Madison, New Hampshire.

S usan sits motionless in a tiny kayak, a silent intruder on the shallow pond that is separated from the main part of Purity Lake by a thin strip of lilies. She studies the beaver lodge, searching for signs of life. It seems to have grown higher since last summer. Twigs and small branches jut out in every direction, weaving upward toward a central peak ten feet high.

She scans the surface of the clear, still water, content to wait. Beavers are nocturnal and the day is waning. The fish have forgotten she is here; they jump up along the small molded boat, sucking in stray water bugs. A dragonfly alights on the tip of her yellow paddle. It perches inches from her knee, slowly waving its long wings up and down, basking in a small shaft of sun that splits the boat in half. There is only the sound of the fish and the occasional call from a bird high in the trees.

She closes her eyes and almost falls asleep, lulled by the gentle motion of the boat. For sixteen years she has been coming to this pond, ever since the day they drove back from Mount Washington and Joseph saw a sign for a small resort and decided to have a look. This place is humble and old-fashioned, with Jell-O pudding desserts and a well-used pool table in the basement of the dining hall. Joseph said it reminded him of the Catskills resort where he worked as a teenager. He was fired for serving scalding hot coffee to an important patron who had complained repeatedly that his coffee was always cold. On a bet Joseph had dropped the man's cup into a pot of boiling pasta water and left it there. He fished it out with tongs and cooled the handle with an ice cube before placing the cup on a saucer. He filled it with hot coffee and raced out to the man's table, where he presented it with a flourish. "This cup of coffee is *hot*, Mr. Weinstein. Please be careful."

The other busboys watched in horror as the man screamed in pain. Joseph felt terrible and apologized to everyone, but he was dismissed anyway. It was one of those stories he liked to tell during games of gin rummy under an umbrella at the lake, while his boys begged for quarters for the pool table.

The Helfgots love this place. Even when they lived in L.A. they still made the annual trek back to southern New Hampshire during the final week of the summer, always reserving the same room for the following year before they left. They have forged deep friendships with other guests who come that week. During the year they occasionally come together for a wedding, bar mitzvah, or funeral.

When they were here two summers ago Susan rushed Joseph to the local hospital. The defibrillator in his chest had fired in the middle of the night, not just once but several times, shocking them both awake. The local doctors did what they could. The ambulance driver refused to take Joseph to Boston, claiming he was too unstable. Maybe so, but Susan knew that if her husband went into cardiac arrest, a small rural hospital might not be able to save him. Two years earlier it took a room full of doctors more than three minutes to bring him back when his heart stopped beating, and that heart was even weaker now.

So she wheeled him out to the car and had him lie down in the backseat. She raced toward Boston, praying he wouldn't die before they got there.

Joseph stayed at the Brigham for a month. He went home with a permanent IV that pushed a continual drip of a powerful heart drug called milrinone directly into his heart muscle.

He never went swimming in Purity Lake again. The IV line had to stay completely dry to prevent an infection, and the battery-operated pump could not get wet. A few months later the IV line became infected anyway. By then Joseph's heart was so weak he had to stay in the hospital, waiting for a heart. Dr. Kenneth Baughman came in one day and plopped down on the edge of the bed. Joseph called him Clint, after Clint Eastwood. He was known to be a straight shooter, just like Clint's character in *Dirty Harry*.

"I'm afraid we're not going to get you a heart in time, my friend," he told Joseph. "Your numbers are so bad that I don't think we should wait any longer. Maybe it's time for a VAD."

Seven months later, and missing all the toes on one foot, Joseph left the hospital with a ventricular assist device in his chest. It was already summer again, and the Helfgots returned to the lake, their car filled with medical supplies, including a respirator, because Joseph couldn't breathe at night without artificial support. When a person lies down, his lungs have to work hard to overcome gravity. And Joseph suffered from sleep apnea, which made breathing even harder.

He managed to last for two days and didn't argue when Susan decided it was time to head back. At least he had made it to the lake. The kids stayed on with friends, and everyone waved goodbye as Joseph took a last, longing look at the water. Susan backed the car away from the farmhouse inn and drove away from the mountains. Seven months later he was dead.

This summer she is here with Jacob, having dropped off Ben

at NYU the week before. It's not easy to be at the lake without Joseph and Ben, but not being here at all would be worse.

She spots a bubble trail forming along the surface of the water and sucks in her breath. The bubbles move toward the kayak, coming closer and closer until she finally spots the beaver through the clear water a few feet below the surface. It curls around the boat, investigating for a moment, and then dives deep, pushing its mighty tail against the water, forming small eddies on the surface that dissolve back into glass almost immediately.

The beaver moves to the other side of the kayak and disappears beneath the lily stems. Then it shoots away, leaving another trail of bubbles. Susan follows them until they fade and turns toward the main lake. A loon cries out. She will return here tomorrow. She might see the blue heron.

Guests are scattered on lounge chairs as she nears the wooden bridge. She is glad she can't see the pity in their eyes as they look up from their books. She can almost hear the whispering: "Remember that face transplant? Yeah, did you know Joseph Helfgot was the donor?"

"No, really?"

"Yeah, that's his wife in the kayak."

As she pulls the tiny craft onto the beach, their heads are back in their books. When she plops down on the grass they pretend to notice her for the first time.

The next day Susan and a friend go into town to buy backpacks for their kids at one of the outlet stores. They stop at a gas station where cell phones can get a signal. The scent of pine trees mixes with gasoline as she sits on the curb checking her messages. Jim Maki has called four times in the past couple of days. Why?

"Hey, Jim. What's up?"

She hears the sound of a TV baseball game. "Peter Brown called me. He says Dr. Oz wants to do a show with me."

"Really? That's great, Jim. What do you think about it?"

"They want you and Dr. Pomahac too. You need to call him."

"Okay." She has never watched Dr. Oz.

"Dr. Kim left yesterday."

"I knew she was leaving soon." Susan and Dr. Kim exchange e-mails from time to time. Dr. Kim has decided to spend a year in Germany, working with troops coming off battlefields in Iraq and Afghanistan. She continues to have a close relationship with Jim, who has been relying on her for encouragement and guidance.

Susan almost says, *You're really going to miss her, aren't you?* But something holds her back. She doesn't want to presume to know him that well. They are still almost strangers.

"Well, I guess that's all I had to say."

"Bye, Jim."

In the kayak the next day she replays their conversation. Has Jim ever been to the White Mountains, she wonders, or seen a beaver lodge, or heard a loon's call shatter the silence? Did the Maki family ever spend time in New Hampshire? She stares back at a turtle the size of a fist who is sunning himself on a dead log a few feet away, his long neck pointing in her direction.

Jim has been in the hospital recently, with a small amount of rejection that requires medication. He seems to be doing well, but HIPAA regulations prevent her from knowing anything other than what he chooses to tell. The lack of information about this man sits uncomfortably with her. Maybe it's the abrupt break from the past few years, when she functioned as her husband's personal nurse. No more cleaning sterile VAD wounds, or looking for purple buttons, or taking blood pressures, or arguing over the phone about the delivery of medical supplies to the house, insisting loudly that she can't wait until Monday for sterile gloves because she needs them *now!* In an instant that whole world came to an end.

Is it withdrawal from her daily fix of medical minutiae? Or perhaps a growing concern for the man who wears her husband's face?

"I am having some rejection," Jim told her a few weeks ago.

"They're going to keep me at the Brigham for a few days." That's all he said. Maybe he didn't know the details. Or maybe he's just not a detail kind of guy.

Joseph certainly wasn't. When he and Susan were learning about life on a VAD machine, he didn't want to know too much. And there was so much to know: so many instructions to keep track of, so many warning signs to attend to. She stepped out of his hospital room to call their electrician to make sure the outlet in their bedroom was properly wired for the VAD machine. Later the nurse practitioner told her that the moment she left, Joseph had asked, "Have you seen my wife? Will we be able to have sex with that machine in me? I love my wife."

"Mr. Helfgot, you're having a major medical procedure in an hour. This is a really big deal. Do you have any *other* questions?"

"Yeah. Will I be around for my son's bar mitzvah?"

Maybe James Maki is a bit like that, not one for medical details. *What the hell! It's none of my business.* She laughs at the absurdity of it all, causing the kayak to shake a little, sending ripples across the pond and scattering the nearby fish.

Occasionally she asks Jim, "Are you sure you're okay?" She is itching to ask *What's really going on? Are you running a fever? How high? What meds are they giving you? What do the doctors think? What does Dr. Pomahac say? Is he worried?* But she doesn't probe. After all, she barely knows James Maki. He is still almost a stranger, and maybe always will be. And yet—

Jim is excited about doing a book. He has been telling Susan his life story in fits and starts. He's had a difficult and lonely life, and Susan always feels drained after they have spoken. How do people endure so much?

Two weeks ago Dr. Kim sat with Jim and Susan and talked into a tape recorder. She kidded with him about his reluctance to consider himself brave.

"You weren't ever scared?" she asked.

"No."

"You *are* brave."

"Not really," he replied. But Susan sensed that he was proud of himself. He had signed up to go to Vietnam. James Maki is no coward.

As Dr. Kim spoke passionately about her new challenge in Germany, Susan could feel Jim's anxiety. He seemed lost in thought. This petite woman with long, shiny black hair has a special calm, an inner peace that seems to flow out of her and into the room. Her softly accented Korean voice emerges from a mouth perpetually caught in a small half-smile. Her warm eyes meet you fully, inviting you to share your secrets, but never pushing. Of course Jim will miss her terribly. Who wouldn't? She has guided him through the early weeks of life with a new face and prepared him for his interview with Diane Sawyer in June. She was there for the press conference. But what about *The Dr. Oz Show*? Who will help him through that?

Susan had called Peter Brown from the gas station in New Hampshire. "Dr. Oz used to be a regular on *Oprah*. He has his own daytime show now, with a studio audience. They tape in New York."

"Does Dr. Pomahac want to do it?" She doesn't feel comfortable about it unless he'll be there too.

"I think he does. It's a medical show, and Dr. Oz is a real doctor, a heart surgeon. And it's not for a while."

A syndicated television show in front of a live audience? That's a big deal. Is Jim even strong enough to travel? Do they know how many germs there are on a plane? That can't be good for someone on immunosuppressant medication. Maybe we should take the train. *We? Susan, are you crazy? Let the hospital worry about Jim!*

She starts to shiver. Clouds scuttle across the sky, forming a loose blanket that hides the sun. The water bugs and fish have disappeared. There is no peace here today, and she smacks the yellow paddle hard into the still water, making loud, angry splashes as she glides back to shore.

chapter thirty

October 2009. Helfgot residence.

Jacob shoves textbooks into his overcrowded backpack.

"You're sure you don't mind the monogram?"

"Mom, you've asked me a hundred times. I don't care! I gotta go. Can I have my allowance?"

At the outlet store in New Hampshire, Susan and her friend found monogrammed backpacks on a half-price table. They were identical to the full-price ones, except for the letters. A black bag that Susan liked had the letters WBC faintly visible on the front flap. Susan guessed that Wendy Boyd Chase had asked for black initials on a pink bag, which is why this one was discounted. "Think the kids will notice?" she asked. Half-price meant saving close to forty dollars for each one.

"We could always cut out the threads."

Susan has started shopping the sale racks, which she never used to do. She stood at a window the other day, looking at an expensive pair of Italian shoes, until she finally turned away. When she can't sleep at three in the morning, which is often, the future lurks dark and hazy. Her hands become clammy when she goes online to pay her bills, the same bills in the same amounts that she has been paying for years. All she can see are eight years of college tuition checks. Although there is money to cover those expenses, she feels panic whenever she thinks about it. Jacob heads out the door with Wilhelmina Beatrice Crazowski's backpack slung over his shoulder.

The other day he watched her stuff ten cans of tomato purée into a kitchen cabinet. "Mom, you're starting to act like Dad."

"They were ten for ten dollars."

"Now you're scaring me."

The two of them are still making their way through a case of unsweetened grape syrup that Joseph bought online two years ago. Each bottle makes twenty gallons of juice, and a case makes 240 gallons. Even with a bar mitzvah and a funeral this year, a family can drink only so much grape juice. The case sits on the floor of the pantry under an oversized box of dog food. Other items collect dust in the pantry, thanks to Joseph's shopping habits, including seven packages of tube socks and a box of a hundred Tea Bags of the World that he brought home from Costco. Susan used to cringe whenever he pulled into the driveway after a trip to Costco.

It became more difficult when he had to stop driving. "Can we take a quick run to the grocery store?"

"We went yesterday. There's no more room in the fridge."

"Suze, there's nothing to eat," he would say, scanning the refrigerator with the door wide open, which drove her crazy.

At the supermarket Joseph would work his way through the different brands of olive oil, reading every label.

"Just pick one, *please*. I need to get home and start dinner."

"This one has chardonnay in it."

"Yes, and it's fourteen dollars!"

"You don't care what you eat."

"I do. That's how we met, remember? The restaurant? Joann introduced us? Which I now am beginning to regret."

"This one has cornichons in it."

"How much?"

"Nine-fifty."

"I'm going out to the car. Let me know when you get to the checkout. And put back one of those crates of clementines. We can't eat two crates. They'll go bad."

"They're on sale."

"Put one back."

"You can make juice."

"I'm going out to the car."

Now I'm buying discarded backpacks and hoarding tomato purée. She tidies up the bathroom. Jacob has left toothpaste goop in the sink and the people from *Dr. Oz* are on their way over.

The segment producer leans against the kitchen counter, sipping a cup of coffee while the camera crew sets up. They are here to film background scenes. B-roll, they call it.

The producer is telling Susan how sorry she is that her husband has died. *You think you're sorry? Imagine how I feel.* Susan thinks this every time people say they are sorry. But she found herself saying the very same thing to a friend whose wife had recently died. What else is there to say?

Susan hesitated when she was asked if they could shoot Jim's interview at her house. This will take their relationship to a new level. He will see the family pictures with Joseph hanging on her walls. He will have lunch at her kitchen table.

She started to say, "I don't know . . ." Then, "Sure, why not?" She can't think of a good reason not to do it, but the question made her uneasy. She sits at her computer in another room while they interview Jim.

"I thought people would act different when I had this surgery, but they still stare at me," he says with regret. He is only six months out from surgery, and Dr. Pomahac and Jim's dentist, Dr. Suzuki, have both assured him that his looks will continue to improve. There will be some dental work and more surgery. Although he looks a lot better than a man without a face, he still has a long way to go.

Mid-November. Back Bay Train Station, Boston.
Susan and Lorrie, Jim's nurse, stand in line for donuts and coffee while waiting for their train to New York, where they will tape *The Dr. Oz Show.* Lorrie is worried about Jim's sugar intake. He asked for six packets for his coffee. "How about just a couple this time, Jim?" she says. But they are in a festive mood and she says it with a light touch. It's hard to be a disciplinarian at a time like this.

"Do you want me to put the sugar in your coffee?" Susan asks. His right arm is useless, and he can't rip open the packets.

"Put in four," he says.

Susan stirs in two packets.

When their train is called, they take the elevator downstairs because he can't negotiate the escalator with his limited sight and lame right arm. As the train approaches, the passengers crowd around. Jim cranes his neck as he hears the train and steps closer to the edge of the platform to get a better look. Lorrie's and Susan's eyes lock in a moment of terror.

"*Jim!*" they scream out in unison.

"What?"

"Don't stand so close to the edge," Lorrie says.

Susan lets out a lungful of air as the train pulls up.

chapter thirty-one

November 18, 2009. New York City.

hey leave the train and walk into a sea of people at Penn Station. A greeter from *The Dr. Oz Show* leads them to a car. Jim climbs into the front and stares out the window. "New York is a big town," he says to the driver.

"You've been here before, right?" Susan asks. Everybody gets to New York sooner or later.

"Yeah, lots of times. Forty-second Street is pretty neat."

"You like Broadway shows?"

"Yeah."

He is thinking about Avenue C, but he doesn't want to mention it. Back in the 1970s, right after Vietnam, he would visit New York often. It was easy to score here. Parts of the Lower East Side, especially near Thompkins Square Park, had dissolved into informal anarchy. Buildings were abandoned as competing gangs vied for control of the tenements that had turned into crack houses. Few remnants of the old days remained. Eastern European immigrants like the Helfgots, whose store was on Avenue

B, had settled across the river in Brooklyn or over in Queens.

It was easy to score in Times Square too, but there was less chance of getting arrested on the streets of Alphabet City, as Joseph's old neighborhood was called. But it was much more dangerous.

As they crawl along Sixth Avenue they pass a branch of Chase Manhattan Bank. Susan smiles as she remembers the day she took Joseph's mother into the bank's main branch near Wall Street. Rachel had wanted to get her passbook stamped. She never believed the interest was hers until they stamped it in her little blue book. She might have been the last person in New York to still have an actual savings passbook.

They had been to Katz's for lunch. The famous deli has served pastrami and corned beef at the same location for more than a hundred years. Billy Crystal sat there with Meg Ryan in *When Harry Met Sally*, which led to the famous line, "I'll have what she's having."

Twenty-six years earlier. Lower East Side.
"Ma, leave it. Let's go."

"I'm taking it." Rachel pulls paper napkins from a dispenser on the table and tries to wrap her half-eaten pastrami sandwich. Katz's makes a big sandwich, and these napkins aren't up to the task. She closes the messy package and shoves it into her purse.

Joseph drives through the financial district, avoiding FDR Drive on his way to the Brooklyn Battery Tunnel.

"Stop! Stop! I vant to get my book stamped. I need to do dis anyway." Rachel is almost illiterate, but she knows the words *Chase Manhattan Bank*. "I go in here."

"Ma, I can't stop here. I'll take you to the bank in Brooklyn tomorrow."

"Please, I beg you. I vant to go out now. She can go with me." Rachel always refers to Susan as "she." Without waiting for an answer, she opens the passenger-side door and gets out.

"Christ," Joseph says to Susan, who jumps out and follows Rachel into the bank. "I'll drive around."

Rachel and Susan queue up in line. The bank is old and cavernous, with marble everywhere. Businesspeople in expensive suits with impeccable manners wait patiently for the next available teller. Rachel thinks it's taking too long. Each time a teller window opens up, she pushes on the back of the man in front of them. "Hurry up, mister."

He gives them a dirty look the second time she shoves him along. "Rachel, stop it." Susan smiles at the man. He scowls back at her.

As they inch up, Rachel starts searching for her passbook. Various items come out and Susan holds each one: a hairnet, crumpled papers, an empty bottle of celery soda from Katz's.

"Dus is it. I find it." As she pulls out the passbook, a small piece of pastrami falls into the cuff of the man's trousers. He senses something is amiss and looks down. "*What the hell?*"

He furiously shakes his leg in a hokey-pokey motion, trying to dislodge the deli meat. It finally falls off and he kicks it away. A large glob of mustard sticks to the summer-weight wool.

"*Christ*, lady!"

"I'm sorry, Mister." A security guard has noticed the commotion and steps over. He leads Rachel and Susan away from the line and sits them down at a desk.

"But I vas next," Rachel says insistently. She holds out her arm and shows the security guard her numbered tattoo. She always gets a lot of sympathy when she shows the tattoo.

"Rachel, shh!" Susan says. She traces circles next to her head with her finger, and the security guard nods his head.

"Could she just get her book stamped, please?" The security guard stays with them while a manager takes the passbook to a teller window.

"Thank you, you are a very fine man, very fine," Rachel says, pumping the manager's hand before they leave.

"How much is dus?" She hands the book to Susan.

Susan reads the number to her. "Vhat? Such a *bissel?* I think I get more in the Brooklyn bank. Vhere is Yosel?"

"He's driving around the block."

"He should be here."

"He can't park here. He'll be right back."

"Something bad is happened, I know it." She starts screaming Joseph's name. "Yosel!"

"Rachel, please!" Susan yells at her. "He's driving around the block. You're making me crazy, now stop it." Joseph pulls up. "See, here he is. Everything is fine." She pats Rachel on the back and opens the front door for her.

"I am sorry." She looks at Susan. "You need to make a baby for mine son. You are a very fine poi-son."

"How did you make out?" Joseph asks as they climb into the car.

"I vas so scared something bad happened to you."

"Did you get the book stamped?"

He looks at Susan through the rearview mirror. She smiles at him and points a finger to the side of her head, pretending to shoot herself. It's not the first time she has made this gesture in this particular threesome.

"I love you, Suze."

"I know."

November 18, 2009. Rockefeller Center, Manhattan.
"Jim wants sushi for dinner," Lorrie says. "Do you know any places nearby?"

"Let me call my son. I think he may know a place." He does, right around the corner, and Susan feels a little less guilty leaving Lorrie and Jim alone as she jumps into a cab to meet Ben for dinner.

The next morning they meet for coffee before the show. As Jim sips it, some of the coffee dribbles down his chin. After six months his facial nerves have begun to waken, but they aren't yet fully functional. Hundreds of electrical impulses rush from his brain: *Now, purse your lips. Place the rim of the cup on the lower part of your lip; then put the top lip down and take a tiny sip. That's right. You've got it.* The nerve signals run across his face, stimulating the many small muscles that must work together to take one simple sip of coffee.

Susan notices a man at a nearby table staring at Jim. She gets up to grab more napkins and sees that he is still staring when she returns. She shakes off an urge to walk over and say, "What the hell are you looking at?" Instead she stares him down until he becomes uncomfortable and looks away.

A few moments later she takes a quick look to see if the man is still staring. *Are you kidding me? Maybe I should punch him.*

Lorrie has been watching. She says quietly, "A man came up to our table last night while we were eating sushi. He stood there, right in front us, staring."

Susan can't believe it. Maybe he was a doctor and was genuinely curious. "Did he say anything?"

"No."

The Dr. Oz Show tapes next to *Late Night with Jimmy Fallon*, and as Susan sits in makeup she hears the Fallon audience laughing through the thin wall. They are working on her hair. Three other women are also on the show today, and they sit in a row beside Susan. They have until New Year's Eve to lose two dress sizes. Dr. Oz is encouraging them.

"Why are you on the show?" one of them asks Susan. The hairdresser is putting something gluey on the woman's head to give her spikes.

Susan watches, fascinated by the spikes. *So that's how they do it.* "I'm here to talk about face transplants."

"Oh, yeah, I saw your husband in the hallway. He looks pretty good."

"He's not my husband." Susan decides to forgo saying any more. When she tried to explain things to the policeman who barged into her kitchen when the film crew was there, it didn't go all that well.

Jim is in another room getting a quick haircut. Then he changes into the new clothes he bought especially for the show. Peter Brown, who has just arrived with Dr. Pomahac, tells the makeup artist to watch out for Jim's sensitive facial areas.

A woman brings Jim's other clothes into the greenroom. "These are your husband's," she says, handing them to Susan. Lorrie swallows a laugh.

"Sure, thanks, but he's not my husband." She wonders why they think that she, rather than Lorrie, is his wife. Then she realizes that she is still wearing her wedding band.

December 21, 2009. Helfgot residence.

Susan sits down with a cup of hot coffee and brings up the *Boston Globe* online. A picture of Jim and Lorrie sitting together in a hospital room appears on the screen. Susan reads the caption: "Brigham gets $3.4m for face transplants." The grant application went in a while ago, and everyone has been keeping their fingers crossed. Now Pomahac can proceed with more face transplants. Jim told Susan the other day about a man who was in the next room when he got his transplant. He had been shot in the face. He was hoping to get a new face, and maybe now he will.

The hospital will receive funding from the U.S. military to pay for several more face transplants for veterans, and perhaps a few civilians too, who have catastrophic injuries and need this life-giving surgery.

Susan smiles. *Merry Christmas, Dr. Pomahac.*

a personal plea

more than 100,000 Americans are currently waiting for a phone call informing them that an organ they urgently need has become available. Unfortunately, the majority of these people will die before that call ever comes.

Most of them, close to 80 percent, are waiting for a kidney. It's not that kidneys fail more often than other organs, but that dialysis buys extra time. Other members of this group are hoping to receive new lungs, a liver, a pancreas, or most urgently, a heart.

Unlike those who wait in vain for an organ donor, most of us have no idea when, or how, we will die. But in certain cases our death can prolong another person's life. If you haven't yet signed an organ donation card, won't you please do so today? Becoming an organ donor could be the final good deed you perform on earth.

As a popular bumper sticker puts it, "Don't take your organs to heaven. Heaven knows we need them here."

For more information on organ donation, please visit:

Donate Life America—www.donatelife.net
United Network for Organ Sharing—www.unos.org
U.S. Government—www.organdonor.gov

—Susan Whitman Helfgot

acknowledgments

I am indebted to all those who helped and encouraged me during the writing of this book, especially Ellen Ball, Ellen Doeren, Julie Leitman, Julienne Martone, Paula Silver, and Sharon Waxman. My family, especially my sister Maryann Brink, and my friends around the country have been a constant source of support.

I will never forget the medical caregivers at Brigham and Women's Hospital who fought valiantly to save my husband's life. Among many others, I am grateful to Carol Flavell, Leslie Griffin, Lisa Kelley, Kevin McWha, Kristin Morrissey, and Drs. Lynne Werner-Stevenson, James Rawn, Greg Couper, and Gerald Weinhouse.

Esther Charves saves lives every day. She and the many others who work for organ banks around the world are like angels among us.

For their comments on sections of this book, I thank Jennifer Roecklein-Canfield, Stefan Tullius, Julian Pribaz, Bohdan Pomahac, Andrew Selwyn, Marcelo Suzuki, Christine Kim, Kay Lazar, and Sean Fitzpatrick at the New England Organ Bank. John Maki Jr. offered important comments and suggestions. I now know why authors often add a line at this point absolving their early readers of responsibility for any errors in the book. I feel the same way.

Peter Brown has been a thoughtful guide throughout. I am grateful to Lisa Quinn for arranging countless meetings and telephone conversations with hospital staff.

Many other people generously shared their memories and perspectives. I especially thank Joseph E. Murray, Elof Eriksson, Francis Delmonico, Anne Fulhbrigge, Hanka Pomahac, Lorraine MacDonald, Pamela Albert, Kristina Andrzejewski, Christopher Curran, Tenaya Wallace, Joan Ganon, William Doeren, Michael Schwartz, Lucy Wollin, Frank Stryjewski, Katherine Mitchell, Terence Wrong, Carl Hansen, Denise Batchelder, Gregory Ferland, David Lapidus, Jay Hardiman, and Cynthia Maki.

I am indebted to my agent, Ike Williams, who encouraged me to keep writing and persuaded William Novak to read an early proposal, and to Hope Denekamp and Katherine Flynn, who helped the book along in countless ways. Johanna Ehrmann provided excellent advice when I was starting out. I am grateful to Priscilla Painton, my editor, and her colleagues at Simon & Schuster: Victoria Meyer, Danielle Lynn, Mara Lurie, and Michael Szczerban. I am indebted to Judith Hoover, a superb copy editor. Special thanks to Cathy Saypol and Johanna Ramos-Boyer.

Two wonderful friends were lost during the writing of this book. I remember Kenneth Baughman and Brenda Selwyn with deep affection.

This book could not have been written without Bill Novak. In the most sorrowful year of my life I experienced many moments of joy and laughter in his company as he guided each page toward home. I am humbled and filled with gratitude for his friendship and generosity.

To my family—Emily, Jon, Ben, Jacob, Bobby, and Pam—thank you for your bravery.

I applaud Jim Maki's fierce tenacity and his courage in sharing his story. I am honored to call him my friend.

I am indebted to the late John Maki Sr., whose privately published memoir, *Voyage Through the Twentieth Century*, was most helpful.

Some events described in this book happened long ago, and with the guidance of those involved I have re-created a few of them.

Other events took place more recently, often under extreme circumstances. Invariably there are tiny spaces in the collective memory of those who shared a moment of history on a rainy spring night in Boston. I have tried to fill those spaces honorably.

—*S. W. H.*

The Pursuit of Happiness,
and Other Sobering Thoughts

George F. Will

The Pursuit of Happiness, and Other Sobering Thoughts

HARPER & ROW, PUBLISHERS

NEW YORK, HAGERSTOWN

SAN FRANCISCO

LONDON

This work was originally published in *Newsweek* and *The Washington Post.*

FIRST EDITION

Designed by Sidney Feinberg

Library of Congress Cataloging in Publication Data

Will, George F.
    The pursuit of happiness, and other sobering
thoughts.
    "These articles previously appeared in Newsweek
magazine and the Washington post."
    Includes index.
    I. Title.
AC8.W613   1978   081   77–25956
ISBN 0–06–014663–X
78 79 80 81 82 10 9 8 7 6 5 4 3 2 1

To Louise Will
prima magistra

Contents

2. *Issues*

3. *Manners*

4. *Campaigning*

5. *Governing*

6. *Foreigners*

7. *Personal*

Acknowledgments

Thanks are due to William F. Buckley Jr., editor of *National Review*, and to Meg Greenfield, deputy editor of the editorial page of *The Washington Post*, who do not invariably think alike but who almost simultaneously suggested that I write a column. To Priscilla Buckley, managing editor of *National Review*, and Philip L. Geyelin, editor of the editorial page of *The Washington Post*, for early encouragement and counsel. To William Dickinson and his associates at The Washington Post Writers Group, and to Edward Kosner, editor of *Newsweek*, who get what I write to those for whom I write. To Erwin Glikes of Harper & Row, who made it not only possible but pleasant to publish this book. And to my wife Madeleine Will, the first and best editor of what I write.

Introduction

My two sons soon will be old and rude enough to demand to know what it is I do all day in my office at home. It is not very informative to say I am a columnist. And fortunately the perverse national genius for turning respectable nouns into disagreeable verbs ("to loan"; worse still, "to critique") has not yet produced the verb "to columnize." To explain what I do, and hence what is in this volume, I can begin, as is my wont, with a historical swoop. My craft has a distinguished pedigree.

The periodical essay was a popular and critical success in English journalism in the eighteenth century, especially after the work of Addison and Steele in *The Tatler* and *The Spectator.* Samuel Johnson, during two of the years when he was compiling his *Dictionary,* wrote a twice-weekly essay, *The Rambler,* which for convenience can be considered the first column of the sort that today is a fixture of American journalism. I note this with some diffidence, aware that what Henry Adams said of the succession of Presidents from Washington to Grant (that it disproved the theory of evolution) may be said of the succession of columnists from Dr. Johnson to Will.

For centuries The Cheshire Cheese, a London pub located a few steps from Dr. Johnson's house, has been favored with the custom of journalists and other writers, and for that reason has been the subject of innumerable essays celebrating its ambiance and clever clientele. But I regard as a hero of his profession the writer who wanted his epitaph to be: "He Had Many Faults, But He Never Wrote an Article About The Cheshire Cheese." I have made it an aim of my life to die without ever having written a column about which presidential advisers are ascending and which are descending. I write about the "inside" of public life in another sense. My subject is not what is secret, but what is latent, the kernel of principle and other significance that exists, recognized or not, inside events, actions, policies and manners.

By manners I mean "conduct in its moral aspect," the way people address one another in conversation and through culture, the way they rear children, and educate, inform and entertain themselves. The agreeableness of a society, and of the people who bear its impress, depends as much on the manners that prevail as on the politicians who prevail.

Without a recurrent concern for principledness, nations, like individuals, are guided by vagrant impulses and imperious appetites. This is especially true of a nation that has moved far and fast from its early conditions, material and moral. Modern Americans travel light, with little philosophic baggage other than a fervent belief in their right to the pursuit of happiness. But there is more to the political enterprise. There is what Professor Michael Oakeshott calls "pursuing the intimations of our tradition," using customs and institutions that are "the footprints of thinkers and statesmen who knew which way to turn their feet without knowing anything about a final destination."

This volume begins with essays about people, thinkers and statesmen and others, and ends with an essay about one of my favorite people. In between are essays about campaigning and governing, the pursuit and exercise of power, here and elsewhere. The recurring theme is that politics should be about the cultivation and conservation of character. The chemistry of character, and what can and should be done to effect it—this is the central problem of political philosophy, and political life, especially under popular government. Canning, a man of many measures, exclaimed impatiently, "Away with the cant of 'measures, not men!'—the idle supposition that it is the harness and not the horses that draw the chariot along." The drama of life in a republic is that there are so many horses.

A column is not an adequate format for the full, orderly deployment and defense of a political philosophy. But it is a fine format for an argument. And a collection of columns can present the contours of a philosophy. According to Professor Oakeshott, the pedigree of every political philosophy reveals it to be "the creature, not of premeditation in advance of political activity, but of meditations upon a manner of politics." These columns are meditations, my attempts to examine issues and events through the lens of principles that, I am confident, constitute a coherent conservative philosophy.

I cheerfully concede what some *soi-disant* conservatives charge: my conservatism is not theirs. Some of what passes for conservatism is a radically anti-political ideology, decayed Jeffersonianism characterized by a frivolous hostility toward the state, and lacking the traditional conservative appreciation of the dignity of the political vocation and the grandeur of its responsibilities.

To John Adams, the most conservative and least comforting Founder, this much was certain: "A society can no more subsist without gentlemen than an army without officers." Needless to say, Abigail's husband did not need reminding that ladies are at least as formidable and essential as gentlemen. But the nation needs to be reminded, by proper conservatives, of the following.

Men and women are biological facts. Ladies and gentlemen—*citizens*—are social artifacts, works of political art. They carry the culture that is sustained by wise laws, and traditions of civility. At the end of the day we are right to judge a society by the character of the people it produces. That is why statecraft is, inevitably, soulcraft.

The following essays deal with that and other sobering thoughts. But a pleasant meal should include sherbet as well as meat, and this volume contains both. It would be a shame, and probably a sin, to write as much as I do and not provide some light fun. Solemnity is the occupational disease of people who commit punditry, and as Chesterton warned, "Satan fell by force of gravity."

PART 1

People

Cool-Hand Jerry Brown

The winds of political doctrine have shifted slightly. Realism, sometimes called conservatism, is enjoying a modest vogue. More than a few
politicians will admit to seeing social imperfections that are as immune
to governmental remedy as they are undeserved. These politicians
preach governmental Niebuhrism—acceptance of human finitude, the
tragic sense of life, the politics of contracting expectations.

The most interesting of them is California's young (thirty-seven)
Democratic Governor Edmund G. (Jerry) Brown Jr. The cool son of a
warm father ("Pat" Brown was elected governor in 1958 and 1962),
Jerry Brown, unlike most politicians, pleases his constituents more
while governing them than while courting them. In ten months in office
he has ravished their hearts, something he failed to do during his 1974
campaign, when he barely beat a bland Republican. A recent poll gave
him an 86 percent "good" or "fair" rating, only 8 percent "poor."

Elected after, and because of, "Impeachment Summer," he is alert
to the fact that the public thinks politicians are a bit hazy about the
difference between right and wrong. So Brown has made a show of
looking at the world with a blue-pencilly squint, and his life is ostentatiously ascetic. He shuns limousines and private aircraft. He sleeps on
a mattress on the floor of a spartan apartment, rather than in the new
$1.3 million governor's mansion. He has been rigid about refusing the
small gifts that pour in upon public officials. His office even bragged
about spurning such offerings as a copy of *Peter Rabbit* in Latin. Three
years at a Jesuit seminary should have equipped Brown to recognize the
sin of pride in feeling so virtuous about forgoing so little.

These are gestures, but politics always is in large measure theater,
and the gestures are practical. The gestures give the media what they
want most from government—something to talk about. And the gestures give the voters the pleasing assurance that, for the mighty, power
must be its own reward.

Brown has worked that pedal on the organ about as much as it can
be worked, so, increasingly, attention will be paid to his ideas, like this
one: "I want to be governor of the 54 percent of the people who didn't
vote at all last year." It is a rare politician who, given an opportunity
to wring his hands and deplore nonvoting as a sin, instead says that most

people who decide not to vote do so "not out of . . . laziness but out of a clear choice, that whatever we are doing isn't worth commenting on one way or the other."

Brown's decision to be governor of nonvoters puts him at the head of the silentest of all majorities. By adopting it as his special constituency, he adopts a theory of representation that is not just novel but convenient, too. The task of fashioning programs for people who express no mandate leaves the politician free to improvise. He even seems to dislike the idea of "programs," which implies a menu of treats for every appetite.

Not surprisingly, Brown's most determined improvising on behalf of nonvoters has been in the form of rigorous budget austerity. He gave lynx-eyed attention to the details of this year's 194-page spending bill. As a result, his budget involved an increase of less than 6 percent over the previous budget, less than the rate of inflation and less than the average annual increase of 12.2 percent in Governor Reagan's budget.

But it is wrong to think that Brown is, or sees himself as, just an embankment against the continuous overflow of government into citizens' lives. Brown seems to question the premise, not just the excesses of the modern state. The premise is that the state should regard citizens as consumers and should be "responsive" to whatever appetites they acquire. He gave a glimpse of his distinctive self on a recent broadcast of NBC's "Meet the Press." It is not every day that a politician tells a national television audience that we need to return "to a traditional view of human nature . . . to a sense that human nature is constant, it is weak, it needs a type of government that recognizes that mankind is really brought down by its own instincts. . . ." Not only does he have a view of human nature that is less than Franciscan in charity, he also has—not coincidentally—a view of government's function that is less than Rooseveltian.

I believe that the crux of his philosophy, as I piece it together from somewhat cryptic fragments, is this.

What ruins individuals and nations is overdeveloped appetites, which are stimulated by the illusion that mankind has escaped the constraints of scarcity. The government has nourished this illusion. It has tried to be all things to all people, or at least as many things to as many people as possible in order to spur consumption generally, and to satisfy the most voracious interest groups.

Liberals fault government as unresponsive. Actually, government today has a hair-trigger responsiveness to intense, organized interests.

Conservatives claim that government is too strong and overbearing. Actually, government is fat but pathetically weak. It does not have the strength to say "No!" to determined petitioners.

Modern government—spending more than it taxes, subsidizing and regulating and conferring countless other blessings—is a mighty engine for the stimulation of consumption. Every government benefit creates a constituency for the expansion of the benefit, so the servile state inflames more appetites than it slakes. It has fostered a perverse entrepreneurship, the manipulation of government—public power—for private purposes. It has eroded society's disciplining sense of the true costs of things.

The era of the servile state began during the last American experience of real scarcity. It began in 1933, when the governor of the then largest state became President and altered the relationship between the citizen and the state. Today citizens receive more than ever from government and government receives less respect than ever.

And today the governor of the largest state—California, cornucopia in the national imagination—is saying that government has been too responsive to people's appetites. He has become spectacularly popular while saying that "an austere, leaner" life is probably inevitable, and will be good for our souls. It is appropriate that the foremost prophet of governmental Niebuhrism sees a connection between statecraft and soulcraft.

[November 10, 1975]

Woody Hayes: Kulturkampf in Columbus

Philistines think football is just a game for chunky boys wrapped in plastic and pads, going thunk and crunch. Professor Woodrow W. Hayes thinks it is Kulturkampf pitting the religion of sacrifice against the slothfulness of our lax age.

This Saturday brings the climactic event in this year's Hayes crusade for the stern virtues of pain and discipline, virtues that Hayes says produce winning football teams and civilizations. In Columbus, Ohio, Hayes' Ohio State University Buckeyes will play a game of tackle with the young scholars from the University of Michigan. The U. of M. ranks near the top of Hayes' long list of things (like good losers, and people who take vacations) that the modern world would be better off without.

Columbus, Ohio, home of OSU, is a no-nonsense community where a TV commercial offers piano buyers a free shotgun as a bonus. Ann Arbor, home of the U. of M., sometimes calls itself the "Athens of the Midwest," which suggests a certain hauteur. Buckeye partisans think Ann Arbor is a nest of tea sippers, sunk in the cultural distractions and mad luxury that led to the fall of Babylon.

Football brings out the sociologist that lurks in some otherwise respectable citizens. They say football is a metaphor for America's sinfulness. You know: the violent seizure of real estate, sublimated Manifest Destiny, oh! bury my heart in the end zone. They say football imitates life. Hayes thinks it's the other way around.

Hayes can explain any historic event in football terms: he says the Greeks had a "home field advantage" over the Persians at the Battle of Salamis. (It is the license of genius to talk like that about a naval battle.) His heroes are Emerson and General Patton. He is a voracious reader of military history, probably because war is OSU football carried on by other, less violent, means.

He quotes Sir Joshua Reynolds ("There is no expedient to which men will not resort to avoid the real labor of thinking"), Persian proverbs ("Luck is infatuated with the efficient"), and Louis Pasteur ("Chance favors the prepared mind") to illustrate his belief that civilization is built by winners. Winners are people who scorn delights, live laborious days, and keep the ball on the ground.

His football style is a philosophical statement—"three yards and a cloud of dust," or, less delicately, "grinding meat." He favors the running game because it favors those who can administer and absorb pain, two abilities that mark life's winners. "If it comes easy," Hayes says, "it isn't worth a damn." He thinks the forward pass is a modernist heresy, worse than gun control and almost as bad as deficit spending.

He thinks the purpose of football is to let spirited lads behave on Astroturf the way another Ohioan (and another Hayes pin-up), General William Tecumseh Sherman, behaved on southern turf.

Like General Patton, Hayes has public relations problems. He periodically cuffs gentlemen of the working press. When Hayes is not angry he is a puma of extremely uncertain temperament, gifted with words of molten passion that would not be tolerated beneath deck on a troop ship. When he is angry he is like those creatures that lurk in hollow trees. His glare freezes the marrow, and causes brave men to run like scalded cats.

In a 100-megaton rage he will bite the palm of his hand until it bleeds. Once, after four hours of looking at game films, a bleary eyed assistant asked Hayes for a rest. "Don't anyone—ever—tell me we've had enough," Hayes roared. Then, according to a witness, "Woody took those clenched fists and smashed himself high alongside both cheeks, obviously as hard as he could swing his hands. . . . When he's beside himself with rage and has no other way to express himself, he'll punch himself silly." The next day Hayes had two black eyes.

Hayes, like bouillabaisse, is an acquired taste. But in an age of plastic politicians, professors professing the day's flaccid consensus, and tradesmen who can't do their jobs, Hayes is a glorious anachronism.

He seems to have wandered off a Grant Wood canvas and into a Wagnerian opera. He is American Gothic with intimations of Götterdämmerung, Bayreuth in Columbus. There is an ugly rumor that Hayes is mellowing. Don't believe it. As he paces the sidelines Saturday, hollering "Blow, winds, and crack your cheeks!" or words to that effect, he will be backed by his legions, nationwide, the closet Woody freaks. They are not quite sure what he is, but they know he is no imitation.

[November 19, 1974]

Lord Shaftesbury's Conservatism

In her splendid biography *Shaftesbury: The Great Reformer, 1801–1885* Georgina Battiscombe says of the seventh Earl of Shaftesbury, "No man has in fact ever done more to lessen the extent of human misery or to add to the sum total of human happiness." That is a defensible claim for the foremost champion of the poor in nineteenth-century Britain.

That Shaftesbury was a conservative gives his luminous career a special interest to those of us who are interested in broadening and deepening the somewhat narrow and negative social prescriptions of American conservatism. His life reminds us that a determined assault on poverty is not only compatible with conservatism, but should be one of its imperatives in an urban, industrialized society.

Ms. Battiscombe makes it abundantly clear that Shaftesbury's reforming passion derived from his idiosyncratic personality, as well as from his social philosophy: "He had the misfortune to be born with, as it were, a mental skin too few; all his feelings were abnormally intense. . . . The sight of a hungry, ill-treated child would reduce him to tears; he was literally 'tortured,' to use his own word, by the thought of suffering."

Shaftesbury spent his life exposing himself to the sight of suffering. Substitution of steam for water power had caused factories to move from the banks of rural streams to towns containing ample cheap labor. The factories often were supplied with wagonloads of children from poorhouses.

Shaftesbury's most testing struggle was for the Ten Hours Bill to guarantee children a ten-hour workday, with an eight-hour Saturday. He also worked to change society's belief that children became adults at age thirteen.

He aided "climbing-boys," often the illegitimate children of prostitutes—some as young as four—who were forced shrieking into narrow chimneys to sweep. Theirs was an occupation (from which Oliver Twist narrowly escaped) which often inflicted an excruciating death from the occupational disease, cancer of the scrotum.

He aided children in coal mines, children whose exploitation was memorably defended by a Shropshire witness: "There are very few under six or seven who are employed to draw weights with a girdle 'round the body, and only when the roof is so low as to prevent the smallest size of horses or asses being employed." Some children were born where they were to work, as a woman miner recalled: "I had a child born in the pit, and I brought it up the pit-shaft in my skirt; it was born the day after I were married—that makes me to know."

There were 30,000 unattended children, some as young as three, living off their wits in London streets. For them, Shaftesbury founded Ragged Schools.

He (like Dickens in *Bleak House*) forced public attention to common lodging houses, where a missionary who managed not to faint from the stench said: "I have felt the vermin dropping on my hat like peas." With the help of cholera epidemics, Shaftesbury launched sanitary reform, which established the principle that public health is a public responsibility—this at a time when a London water company was pumping from the Thames directly opposite a huge sewer outlet.

Shaftesbury was one of those Tories—the young Disraeli was another—who were early and vigorous in fighting, for impeccable, conservative reasons, the rawest abuses of the early industrial system. He understood that conservatism is about the conservation of certain values and these values have social prerequisites. He believed that nothing is more subversive of a nation than the existence of a permanent underclass, blocked from upward mobility, and brutalized by conditions destructive of institutions like the family and values like gentleness and lawfulness that are, conservatives know, mainstays of civilization.

Today in the United States it is only too apparent that the existing quiltwork of welfare systems, far from preventing the emergence of such a class, may be helping to create it, and certainly the welfare system does not prevent conditions that would inflame a modern Shaftesbury.

Attention paid to Shaftesbury's career might give some American conservatives a quickened interest in the problem of poverty, and in improving—which does not necessarily mean just pruning—the welfare system. If conservatives do not interest themselves in this, it will be fair to assume that they have at least a mental skin too many, and have inadequate mental material beneath that skin.

[October 24, 1975]

Hugh Hefner's Faithfulness

Hugh Hefner was born in Chicago, hog butcher of the world, city of the broad shoulders. He built a $200 million empire on breasts, and then took up California dreaming.

Oh, westward the course of empire. Hefner, the tuning fork of American fantasies, knows that if F. Scott Fitzgerald were writing today he would say that California, not Long Island, is "the fresh, green breast of the new world."

". . . I thought of Gatsby's wonder when he first picked out the green light at the end of Daisy's dock. He had come a long way to this blue lawn, and his dream must have seemed so close that he could hardly fail to grasp it. . . . Gatsby believed in the green light, the orgiastic future that year by year recedes before us."

Hefner, guiding genius of *Playboy* magazine, has grasped his dream. He savors it here in his 30-room mock-Tudor mansion set on five acres where apes gambol and peacocks strut and movie stars materialize at odd hours in Hefner's private copse of redwoods, all God's creatures having fun being rich together.

". . . for a transitory enchanted moment man must have held his breath in the presence of this continent, compelled into an aesthetic contemplation he neither understood nor desired, face to face for the last time in history with something commensurate to his capacity for wonder."

Step now out of history, into Hefner's house at the edge of the conquered continent. The house is a monument to wondrous, conquering technology. It is a wired cocoon staffed by platoons of servants, including five full-time electricians who nurse Hefner's toys—pinball machines, stereos, television cassette machines, video tape cameras, hidden panels concealing movie projectors. The house is Hefner's triumphant gesture as the quintessential modern man: he suspends time, imposing his will upon his world with electronic will.

"If personality is an unbroken series of successful gestures, then there was something gorgeous about [Gatsby], some heightened sensitivity to the promises of life. . . ."

The center of Hefner's controlled universe is The Bed. Its dials and buttons summon servants and operate a movable canopy of mirrors and various audio and visual entertainments, including a low-light video tape camera pointed at The Bed.

"[Gatsby] knew women early, and since they spoiled him he became contemptuous of them, of young virgins because they were ignorant, of the others because they were hysterical about things which in his overwhelming self-absorption he took for granted. . . . The most grotesque and fantastic conceits haunted him in his bed at night. . . . For a while these reveries provided an outlet for his imagination; they were a satisfactory hint of the unreality of reality, a promise that the rock of the world was founded securely on a fairy's wing."

Hefner meets the world not much, and always on his terms. To mingle with his flamingos and macaws and other birds of paradise, your name must be on the guard's admittance list, and Infirmity is never on the list.

". . . and Gatsby was overwhelmingly aware of the youth and mystery that wealth imprisons and preserves. . . ."

One of those dials or buttons in the house—he's not sure which one but he's sure there is one—is capturing time itself on low-light video tape, ready for instant replay, world without end, amen.

"As I went over to say good-by I saw that the expression of bewilderedness had come back into Gatsby's face, as though a faint doubt had occurred to him as to the quality of his present happiness."

Hefner entertains many things, but never doubts. He knows that push-pin is as good as poetry, and backgammon is better. He plays games in Shangri-la, frantically, and he plays, he says, "for escape." They are a part of living out "the adolescent fantasies I've never really lost."

"The truth was that Jay Gatsby of West Egg, Long Island, sprang from his Platonic conception of himself. He was a son of God—a phrase which, if it means anything, means just that—and he must be about His Father's business, the service of a vast, vulgar and meretricious beauty. So he invented just the sort of Jay Gatsby that a seventeen-year-old boy would be likely to invent, and to this conception he was faithful to the end."

[March 22, 1975]

Calvin Coolidge: Pickles and Ice Cream

A partially scrutable Providence has used Independence Days to evince Its special interest in this Republic.

On Independence Day, 1863, Lee retreated from Gettysburg and, more important, Vicksburg fell, dooming the rebellion and giving Lincoln a chance to say: "The Father of Waters again goes unvexed to the sea." On Independence Days, Presidents John Adams, Thomas Jefferson (both in 1826) and James Monroe (in 1831) were called to Judgment. And on Independence Day, 1872, in a Vermont village with the resonant name of Plymouth Notch, the thirtieth President was born.

Calvin Coolidge lacked the modern politician's sparkle, but not because, as H. L. Mencken charged, he was weaned on a pickle. A more plausible explanation of Coolidge's distinctive charisma is that he first glimpsed the majesty of the presidency when, as an impressionable youth, he heard a public address by Benjamin Harrison.

Coolidge had the divine gift of concision concerning pastoral theology (explaining a minister's sermon on sin: "He said he was against it"), abstract theology ("Inflation is repudiation"), and applied theology ("The business of America is business"). That's Calvinism.

On August 2, 1923, President Harding was called to a Glory exceeding even that of the presidency to which Vice President Coolidge ascended. That day, Coolidge's youngest son, Calvin Jr., 14, was working in a Connecticut tobacco field.

The next day another boy said: "If my father was President, I wouldn't be working here." Young Calvin replied: "If my father were your father you would."

A chip off the old block of flint, Calvin Jr. was the apple of his father's eye. He developed blood poisoning from a blister he got playing tennis on the White House grounds. He suffered a lingering death in the summer of 1924.

One day the distraught President, knowing his son's affection for animals, took some lettuce and, crawling on hands and knees in the White House garden, captured a rabbit to show his dying son. Later Coolidge would only say: "When he was suffering he begged me to help him. I could not." That was the eloquent reticence not of a passionless man, but of a man with a sense of life's limits, who knew that life's most important tragedies involve personal rather than public affairs, and who would not spread around himself the turmoil he felt.

Harry Truman once said that Coolidge didn't do much but there wasn't much to be done then. Coolidge said: "If you see ten troubles coming down the road you can be sure that nine will run into the ditch before they reach you."

Recently a *New York Times* commentator reached deep into his duffle bag of historical parallels and extracted this: "President Ford is probably the least receptive chief executive toward government intervention in the market place since Calvin Coolidge." Perhaps Ford knows that under Coolidge's benign neglect the federal budget shrank, the national debt declined, unemployment was only 3.6 percent, and consumer prices actually fell 2.3 percent.

During the Coolidge years the number of automobiles on American roads rose from eight million to twenty-three million and radio sales rose from $60 million to $842 million. These two great technological revolutions of mobility and communication, which helped Americans acquire an expanded sense of nationhood, began transforming American society during the Coolidge years when, we are told, nothing much happened. Actually, Americans were moving around, Minnesotans meeting Mississippians, all humming the same broadcast songs, like "Barney Google" and "It Ain't Gonna Rain No More."

A cold rain fell when Coolidge was laid to rest in the rocky New England countryside in January, 1933. He died in Northampton, Massachusetts, and the Mayor announced that, as a tribute, the stores would not close during the funeral: "Every nickle counts. If the business places close they might lose some sales and that is exactly what Calvin would not want."

Coolidge had died unexpectedly, alone at home. His wife was out shopping on . . . Main Street.

Coolidge is a faded memory, as unromantic as old commerce statistics, relegated to that remote corner of the national consciousness reserved for quaint and faintly ridiculous characters. This does no credit to the nation which under his stewardship enjoyed a 45 percent increase in the production of ice cream.

[July 2, 1975]

Alf Landon's Little House on the Prairie

Seven decades ago, while at the University of Kansas, Alf Landon successfully agitated for the elimination of the ice cream course from his fraternity house menu.

I know what you are thinking: Republicans always want to take the ice cream out of life. But Landon never has been an anti-hedonist. As a college blade he introduced the tuxedo on campus. He always has been frugal, and he thought his fraternity was living beyond its means. But his life has been an ice cream life.

Today, at eighty-seven, he lives with zest, going horseback riding before 7 A.M. and attending to business interests (radio stations and oil wells). Longevity makes some people melancholy as they survive their contemporaries. Landon's desk is piled high with books (like a recent biography of Huey Long) about contemporaries long gone. But because of his lively spirit and curiosity, today's Americans also are his contemporaries. The sheer sweetness of his temperament is apparent in his inability to say anything harsh about anyone.

In 1936 he was the Republican candidate against President Franklin Roosevelt. He lost forty-six of the forty-eight states, carrying only Maine and Vermont. He didn't expect to win and knew what to do when he lost. He went duck hunting. There is a duck decoy on his kitchen table, where he sits sipping coffee. He is wearing a bright canary yellow cardigan, and he gently thanks his pet canary for "a very nice song."

The 1930s were passionate years, filled with passionate public figures. Landon was not one of them.

True, his candidacy was backed by passionate people. As the election approached, the Chicago *Tribune* telephone operator greeted callers, "Good morning. Did you know you have five days to save your country?" A typical *Tribune* news story began, "Governor Alfred M. Landon tonight brought his great crusade to save America to Los Angeles." The 1936 Republican platform breathed fire: "America is in peril. . . . [New Deal] actions are insufferable. . . . This election transcends all previous political divisions. We invite all Americans . . . to join us in defense of American political institutions."

But Landon never breathed fire. And he never burned with ambition.

Of course, it used to be normal for politicians (for example Adlai Stevenson in 1952) to make an elaborate show of praying that the cup of power would pass from their lips. But while their parched lips were praying, their eyes were fastened on the cup. Today's politicians do not even pretend, and campaigning never ends as candidates work like dray horses for four years to win the White House. These dray horses must seem strange to Landon, which is to his credit.

After two successful terms as Kansas' governor (he was the only Republican governor re-elected in 1934) he made a brief, decorous appearance on the stage of national politics. Then he turned away from ambition, and walked off the stage. He could have been elected to the Senate, but chose not to run.

Like little Dorothy in *The Wizard of Oz,* who did not enjoy Oz, Landon did not like our political Oz, Washington, with all its political wizards. Like Dorothy, he just wanted to come home to Kansas, to build the huge white house he lives in today, and raise his children. "We preferred the comparatively simple but more intelligent life of Kansas to Washington," he says. "There are some intelligent people in Washington. More of 'em in Kansas."

Besides, he says, he thought the Republican party needed one person who was free of ambition for elective office. Landon did not know that in 1974 the Republican party would be the ideal place for people without ambition for elective office.

[November 7, 1974]

Hitler: Radically Free

The study of medicine begins, in a sense, with the study of death. Similarly, the study of modern politics begins with the study of Hitler. In him and his enterprises are visible the worst pathologies to which modern societies are prey. It is therefore good that many people seem to have Hitler on their minds. These people are the reason John Toland's 1,035-page *Hitler* is on the best-seller list.

The continuing market for books about Hitler is due in part to a vicarious fascination with evil. Biographies of Al Capone sell better than biographies of Dwight Eisenhower. And what might be called the "Hitler nostalgia" also is part of the nostalgia for World War II. That was a great simplifying event, the last occasion (some people think) when the United States was unquestionably in the right.

But more serious interest in Hitler is related to this fact: his regime was founded at least in part in the heart as well as on the neck of a great civilized nation. The regime was run, as all large states are run, by civil servants whose principal attributes were what normally are called virtues—patriotism, a sense of duty, regularity. Men who would never cheat at cards or condone adultery by a fellow civil servant condoned unrationalizable evils.

Hitler was decorated for bravery in World War I but never became a warrant officer because, a regimental officer said, "We could discover no leadership qualities in him." Until he was 30 he had not joined a political party or given a political speech. But shortly after his fifty-sixth birthday he killed himself in the ashes he had made of much of Europe.

Since then Hitlerology has been a growth industry. The first phase of its growth, in the early postwar days, satisfied the public's thirst for details of the private life of a man whose public life was all too familiar. These early Hitler books usually were of what one critic calls the "I Was Hitler's Toothbrush" genre, and were generally rubbish. They were of two kinds, one portraying Hitler as a demon, simply not human, and the other portraying him as a *Teppichfresser,* a carpet-chewer, mad as a hatter.

The second phase of Hitlerology featured serious books burdened by their authors' implacable loathing. Usually it was loathing for the man. Golo Mann, a distinguished historian, would only refer to Hitler by the letter "H." Sometimes it was loathing for the nation of which the man was, allegedly, a "natural" product. William Shirer believes there is a straight line of German fanaticism connecting Luther to Hitler.

To say Hitler was a demon, to deny his humanity, is comforting to the rest of the human race. Similarly, to reduce him to a mere manifestation of a dark continuity in German history is to limit his disturbing relevance. Both misdescriptions of Hitler obscure the extent to which his career calls into question comfortable assumptions about the durability of civilization.

The Thousand-Year Reich fell 988 years short, but it took an unprecedentedly ferocious war to make it fall. And that it existed at all is evidence of the speed with which re-barbarization can occur in even a nation of deep and admirable culture.

Hitler was the founder of a secular religion, and was tireless in performing priestly functions at events like Nuremberg rallies, with holy relics like the "Blood Flag" from the attempted putsch of 1923. His rise to power was a meeting of the man and the moment, but his perverse genius—it was nothing less—was in seeing an aching emptiness in people that his passion could fill.

Finally, part of the fascination of Hitler is this: if you accept the modern notion that freedom is just the absence of restraints, then Hitler was a radically free man, a man operating on society from outside, unrestrained by any scruples or ties of affection. Toland calls him "perhaps the greatest mover and shaker of the twentieth century." There is only one thing that can be said for Hitler's vicious life: it refutes the theory that only vast, impersonal forces, and not individuals, can shake the world.

[November 4, 1976]

Simone Weil: The Politics of Self-absorption

Simone Weil died in 1943, a burnt-out case at age thirty-four, but her soul lives on in many echoes. Today she is better known than when she lived, and the quickening interest in her among intellectuals says something about the temper of our time.

She is the subject of a thick (577-page) new biography by her friend Simone Petrement. It has been said that a well-written life is as rare as a well-lived one. This is, unquestionably, a poorly written biography that does not deserve reviews of the prominence and warmth it has received. And Simone Weil's was not a well-lived life.

Her life ended in an English nursing home. The coroner called it suicide—voluntary starvation. She wouldn't, and then couldn't, eat. She was showing solidarity with people suffering privation in occupied Europe. It was her final futility, the last episode of a body overtaxed by an overheated mind.

Born in Paris to secular Jewish parents, she was precocious and sickly. Even as an adult, as teacher and writer, she was childlike. Her parents had to follow her about, begging her to eat and rest. She saw the world as a painting in snow and ink, a moral drama lit by lightning, the underprivileged against the powerful.

She wept about famine in China, but frequently was inexcusably rude to persons near at hand. By ruthless suppression of her femininity, including flamboyantly masculine dress, she placed herself in that class of intellectuals who (in Tom Stoppard's words about James Joyce) want "their indifference to public notice to be universally recognized."

Through most of her life she was a fanatic in search of a fanaticism. She sought it first in left-wing politics. She believed, or tried to believe, many things, moving from the fringe of communism to the threshold of Catholicism, always at full throttle. She compressed into a short span most of the disillusionments of the century.

Having concluded that there is no such thing as a "good" state, she decided that socialism, under which the state is everyone's employer, was the most dangerous social arrangement. But she soon lost hope for anything better. Her fierce disappointment with politics epitomized a mood—a generalized hostility to the responsibilities of power—that is a constant temptation to intellectuals.

Weil was manually clumsy, and weak, except in her will, which was too strong for her own good. So she put herself through experiences—factory and agricultural work—which she thought would "purify" her. She became a nuisance and a worry to those around her.

When visiting friends she would sleep on the floor next to an empty bed. While working for peasants who tolerated her, she would insultingly accuse them of insensitivity because they ate cheese while the

Indochinese were hungry. She was, in short, a parody of the self-absorbed intellectual.

She shattered her health with empty gestures, like not eating to show solidarity with this or that group. The baffling questions are: how could someone so learned, and so determined to be virtuous, behave with such futility? And why is her life as well as her thought so attractive to many intellectuals?

Her life of elaborately, not to say ostentatiously, cultivated self-denial reflected a theme of her later writings, a peculiar idea of the morally responsible life. It is the idea that the goodness of an act consists solely in the goodness of the motive, not of the consequences. This is a disastrous approach to social affairs, where policies have complex consequences, and consequences are more important than motives. Her obsession with her own motives allowed her to offer pacifism as a response to Hitler.

The politics of this and other nations have been shaped by the idea, especially popular among intellectuals, that well-motivated policies of social engineering are justified by their motives. That is why the widening gap between intentions and results is less disturbing to many intellectuals than you would reasonably expect.

At the end, Simone Weil's life was a tragic model of self-absorbed rightmindedness. And at the end, her thought was a recipe for irresponsibility.

[February 3, 1977]

Anne Frank: The Triumph of Faith over Experience

AMSTERDAM, THE NETHERLANDS—This coastal city of placid canals, a city hard won from a turbulent sea, this most bourgeois city, has a distinctive and agreeable architecture that is evidence of the inventiveness of the human spirit when challenged by the tax collector. Many houses are very deep but very narrow, so built because, for years, real estate taxes were based on the width of a building's canal frontage.

To solve the lighting problems created by this deep backward extension of row houses, some houses were built almost as two houses, with a small passageway connecting the front house with a back annex, and a small, light courtyard between them. Not all such houses have back annexes, and in those that do the annex can be hidden by concealing (with, for example, a bookcase) the entrance to the annex.

The four-story house on the canal at 263 Prinsengracht is typical. But because history touched it, hard, it is a shrine of sorts, a symbol of the triumph of faith over experience.

"8–9–10 July 1942. We put on heaps of clothes as if we were going to the North Pole, the sole reason being to take clothes with us. No Jew in our situation would have dreamed of going out with a suitcase full of clothing. . . . So we walked in the pouring rain . . . each with a school satchel and shopping bag filled to the brim. . . . We got sympathetic looks from people on the way to work. You could see by their faces how sorry they were they couldn't offer us a lift: the gaudy yellow star spoke for itself. . . . When we arrived at the Prinsengracht, Miep took us quickly upstairs and into the 'Secret Annexe.' She closed the door behind us and we were alone."

They lived behind the bookcase-that-was-a-door until August 4, 1944, when someone told the Gestapo that Dr. Otto Frank's family, and some others, eight in all, were hiding there. The Gestapo deported the inhabitants, one of whom, Dr. Frank's fifteen-year old daughter, left behind three notebooks.

Great men's war memoirs tell the story of making history. But the most widely read book of the Second World War, Anne Frank's diary, is a view of history from the receiving end.

"6 June 1944 . . . the best part of the invasion is that I have the feeling that friends are approaching. . . . I may yet be able to go back to school in September. . . . 15 July 1944 . . . in spite of everything I still believe that people are really good at heart."

After the August 4 arrest, the Franks were sent to Auschwitz. A survivor says Anne, unlike most prisoners, retained her capacity for tears: there was, for example, the time the Hungarian children had to stand naked in the rain for half a day because they were brought too soon to the gas chamber.

At Auschwitz her head was shaved: the Reich needed women's hair for (among other things) packing around pipe joints in U-boats. Then for no knowable reason the Thousand-Year Reich shipped the fifteen-year-old girl to the Bergen-Belsen concentration camp. She died there in March 1945, shortly before the camp was liberated. She probably died of typhus, but no one really knows, because she was just another wraith-like person in the mud behind the wire.

A year earlier she had written: "23 February 1944 . . . Nearly every morning I go to the attic. . . . From my favorite spot on the floor I look up at the blue sky. . . . 'As long as this exists,' I thought, 'and I may live to see it, this sunshine, the cloudless skies, while this lasts, I cannot be unhappy.' . . . Riches can all be lost, but that happiness in your own heart can only be veiled, and it will still bring you happiness as long as you live. As long as you can look up fearlessly into the heavens."

Today visitors climb the ladder-like steps to see the attic where, from that window, Anne Frank looked up into the Dutch sky. Recently someone climbed to the top of those sad steps, pulled a pencil from his pocket and signed the woodwork: "D. E. Gomez, Leadville, Colo., January, 1976." Let the record show that the war made Western Europe safe for Gomez and his pencil.

[February 26, 1976]

Jean-Paul Sartre Discovers Democracy

Connoisseurs of modern casuistry glowed with anticipation when Jean-Paul Sartre visited Portugal. The hour had produced its man, and he did not disappoint us.

Sartre swallowed his distaste for military things and pronounced a benediction on Portugal's military regime. The Portuguese army, he said, "is not like any other," because it represents all classes of society. No one ever accused Sartre of sharing John Stuart Mill's understanding of representative government, or logic or liberty or anything else. Sartre, you see, is an existentialist.

For those who have not read the classics of existentialism (for example, Sartre's *Being and Nothingness*), let me explain: existentialism often seems to be the belief that because life is absurd, philosophy should be, too. Sartre is the patron saint of those people in black turtlenecks and black moods who used to frequent the Boulevard St. Germain, comparing notes on the emptiness of life. Sartre himself looks like a man who received some bad news in 1947 and hasn't gotten over it.

Actually, he received some disagreeable news then, but he quickly recovered. In about 1947—about thirty years late—Sartre received the news that the Soviet government had millions of slave laborers. Sartre's newspaper, *Les Temps Modernes,* promptly declared that the Soviet government should not be condemned for this because the Soviet government is against slave labor "grosso modo"—an Italian phrase meaning, roughly, "in the big picture."

Sartre recently took his finely honed moral senses to Lisbon, where he dined at the Red Barracks and talked socialism with the Light Artillery Regiment. Regarding political philosophy, this regiment is the All Souls College of the new Portugal.

Imagine, if you can, Sartre clutching a tattered copy of Marx's *Eighteenth Brumaire* and lecturing artillery lieutenants on "the reification of alienated production" and the "historically progressive consequences of primitive accumulation." To help you imagine that, let me remind you of the book Sartre wrote in 1960, a hymn of praise for Castro's Cuba.

Sartre had seedlings of doubts about Castro's commitment to democracy. But Castro nipped these doubts over a glass of warm lemonade.

Trujillo's Dominican Republic regime called itself "neo-democracy." Mussolini called his regime "ennobled democracy." There is an African government that describes itself as "directed democracy." In Cuba, Sartre discovered "direct democracy."

At a roadside stand, Sartre and Castro were served warm lemonade, and Castro got hot. The warm lemonade, he said, "reveals a lack of revolutionary consciousness." The waitress shrugged, saying the refrigerator was broken. Castro "growled" (Sartre's approving description): "Tell your people in charge that if they don't take care of their problems, they will have problems with me." Sartre was deeply stirred:

"This was the first time I understood—still quite vaguely—what I called 'direct democracy.' Between the waitress and Castro, an immediate secret understanding was established. She let it be seen by her tone, by her smiles, by a shrug of the shoulders, that she was without illusion.

"And the prime minister—who was also the rebel leader—in expressing himself before her without circumlocution, calmly invited her to join the rebellion."

Fifteen years later Cubans still enjoy "direct democracy": they have no elections, but they can shrug their shoulders. Sartre still chooses to live in Paris, enduring the elections and other tediousness of bourgeois democracy, but he has discovered yet another improved form of democracy, this time in Lisbon's Red Barracks. As any existentialist above the rank of colonel knows, life is absurd and the art of politics should imitate life.

[April 22, 1975]

Gus Hall's Vigorous Obedience

Another precinct has been heard from. Comrade Gus Hall, general secretary of the Communist party, USA, and the party's presidential candidate for the second time (he got 25,343 votes in 1972—.03 percent of the vote), has launched his campaign with a stirring defense of détente.

Notice of this event appeared deep in the *New York Times,* next to a story with an overshadowing headline:

Hair Stylist Tops Field
in Annual Fiddle Contest

Is that any way to report on the vanguard of the proletariat, cheek-by-jowl with a fiddle contest? Well, actually, yes. When I called Manhattan information for the number of the Communist party, USA, the operator, who surely is one of the toiling masses, responded with a question: "Is that 'Communist' with one 'm' or two?" Her question illustrated the futility of both New York's educational system and the CPUSA.

The CPUSA has been relegated to what Communists call "the dustbin of history." But as recently as the early 1950s it received public attention out of all proportion to its importance, just as the John Birch Society did in the early 1960s. Both were examples of the spectres a bored society invents for its own titillation.

The CPUSA's last chance to become consequential was in 1948 when many of its agents and sympathizers penetrated the Progressive party presidential candidacy of Henry Wallace who, poor dear, was the last to know. If Communists had been successful, through him, in denying Truman the presidency, then it might have been possible to take them seriously. But with Communist help Wallace managed to finish fourth, behind Strom Thurmond's Dixiecrat candidacy. It is not the least of the ironies of the CPUSA's history that it became something of a national obsession for a few years after 1948, after it had conclusively demonstrated its impotence.

American Communists have always been demoralized by the knowledge that they lack courage proportional to desire, and that not even their desire matches their imported rhetoric. What Madame de Staël said of Germans always has been true of American Communists: they are "vigorously obedient" to every exigency of Soviet foreign policy. Thus the *New York Times* noted with nice dryness that Hall's remarks about détente "parallelled" recent Soviet press attacks on U.S. critics of détente.

Actually, one can almost admire the energy the CPUSA has invested over the years in rising above self-respect in the name of obedience. The Moscow trials (which prompted Alger Hiss' admiring statement: "Joe Stalin certainly plays for keeps"), Soviet duplicity during the Spanish Civil War, the Soviet alliance with Hitler, the partition of Poland, the attack on Finland—all of these drew CPUSA applause.

Lillian Hellman's anti-Nazi drama *Watch on the Rhine* was attacked by party critics in 1940 and praised by them in 1942. This revolution in aesthetic standards was brought about at dawn, June 22, 1941, when Germany attacked Russia.

There are scores of reasons why the CPUSA is today and always has been a Potemkin village, an empty, cardboard party. But not the least of the reasons is the CPUSA's marvelous ability to evoke laughter, as when a prominent member announced that he had "unmasked the open Trotskyism" of a rival faction.

General Secretary Hall, who presumably will rule Soviet America after the revolution, is an Ohioan, a bureaucrat lacking only a state. He and fellow operatives are gray reminders of what Thomas the Cynic says in Ignazio Silone's *School for Dictators:* "No dictator has ever had trouble finding civil servants."

But were it not already as petrified as a dinosaur's skeleton, we would want to embalm the CPUSA, the better to preserve it as a monument to the leftism of irrelevance, and to that wit who coined a slogan appropriate for a CPUSA factional fight involving Jay Lovestone: "Lovestone is a Lovestonite." It runs in the family. When the French Communist party recently decided to drop references to the "dictatorship of the proletariat," a disillusioned former member declared that the party "is guilty of Right-wing Trotskyist petty bourgeois deviationism of an opportunist, social-Fascist character."

[June 20, 1976]

Shirley MacLaine: Hyperventilating in China

Shirley MacLaine, showgirl and McGovern campaigner, visited China and found a society worthy of her admiration. She vents her admiration in *You Can Get There from Here*, a wonderfully funny book.

"What we saw in action in China so disturbed us that we got physically sick." No, tyranny (she calls it "totalitarian benevolence") doesn't sicken her. She got sick (clogged sinuses) because her "basic values," which must have the strength of rice pudding, crumbled at the sight of China's "harmonious society." People react to tyranny—I mean, totalitarian benevolence—in odd ways: Ms. MacLaine calls for antihistamines.

China also affected the MacLaine lungs and tear ducts. When young Chinese applauded her passing car, "my eyes were wet and I was breathing hard and sighing deeply."

At a nursery school the scales fell from the MacLaine eyes and she saw the superiority of Chinese education. The scale-remover was a children's skit about a tot who hogs the only toy airplane until her peers recite to her the relevant thoughts of Chairman Mao. Ms. MacLaine nearly hyperventilated: "I took a deep breath and tried to absorb what I was seeing: the glorification of selflessness, the refusal to ostracize her, the endorsement of group action, the reverence for the wisdom of Chairman Mao. . . ."

After visiting a day-care center, Ms. MacLaine wept again, free at last from "the strain of guilt that accompanies the realization that there are profoundly positive aspects to the bogeyman called Communism." It was quite a day: "That night, I had one of two erotic dreams I had in China."

She could not suppress her suspicion that China isn't, well, *sexy* enough. But her anxiety was assuaged by her guide, who explained: "Our young people are not attracted by looks or sex appeal, but by political ideology. . . ."

When Lady Astor met Stalin, she asked: "When are you going to stop killing people?" When Ms. MacLaine met Ms. Chou En-lai, she asked about "the role of the artist in China." Ms. Chou explained that the state must control art because there is "no essential difference between art and politics in serving a new nation and that the *people* were the highest priority." Ms. MacLaine was going to grill Ms. Chou more, "but her gentility and the clarity of her explanation made me hesitate."

The MacLaine mind "tossed with ideas and insights, some as flinty and hard as the landscape." For example:

"The notion of freedom was suddenly suspect; in American terms, there were no freedoms here of the variety we cherished: no freedom to publish, no opposition political parties, no freedom to write books or create works of art. But there were other freedoms: freedom from starvation, discrimination, exploitation, slavery, and early death."

And that idea wasn't the flintiest. When her guide reported that Western male visitors are interested in machines that make money, and female visitors are interested in jewelry, Ms. MacLaine was dazzled: "It seemed a pretty fair summing up of two thousand years of Western civilization."

Ms. MacLaine, who spent 1972 instructing the Democratic party in the nuances of reform, likes Chinese reform. China has proven that it is "possible somehow to reform human beings and here they were being educated toward a living communal spirit through a kind of totalitarian benevolence."

As Ms. MacLaine "sloshed around in the tub" in Yenan, "I thought of America's climate of anger, violence, crime, and corruption, of her selfishness and deception, and her freewheeling abuse of freedom." That is the authentic voice of the American left: come home, despicable America, to Yenan, where there is no freewheeling abuse of freedom.

Back home, her consciousness raised, Ms. MacLaine made the long march to Las Vegas to put her art in the service of the masses. Ms. MacLaine, we've met you before, in a Saki short story, "The Byzantine Omelette":

"When she inveighed eloquently against the evils of capitalism at drawing-room meetings and Fabian conferences she was conscious of a comfortable feeling that the system, with all its inequalities and inequities, would probably last her time. It is one of the consolations of middle-aged reformers that the good they inculcate must live after them if it is to live at all."

[April 3, 1975]

Tania Becomes Conscious

Appropriately, it began in Berkeley where, ten years earlier, the politics of extended adolescence had become fashionable campus radicalism.

On the evening of February 4, 1974, American radicalism took a quantum leap from unintelligent operetta to unintelligible theater of the absurd. Armed members of something called the Symbionese Liberation Army abducted Patricia Hearst, daughter of a "corporate enemy of the people."

Eight days later, the SLA released a tape of Ms. Hearst sounding frightened, and of "General Field Marshall Cinque" demanding a multi-million-dollar food giveaway. On April 3 a photograph accompanying a tape showed Ms. Hearst posing in front of the seven-headed cobra symbol of the SLA. She was wearing a beret, brandishing a carbine.

On the tape she said she had joined the SLA and taken a new name for her new self: Tania. She reviled her father as a "corporate liar," denounced nuclear power plants (the energy crisis was competing with the SLA as a news story) and told her boyfriend: "I've changed—grown. I've become conscious."

Of course, "conscious." Previously Ms. Hearst saw as through a glass, darkly. Now Tania sees face to face. Re-born, converted, baptized in political chiliasm, nestled in a political church-militant, wrestling with an unconverted world. "One thing I know, that, whereas I was blind, now I see." (John 9:25).

People can rise on stepping-stones of their dead selves to higher things. But they also can descend on such stepping-stones to lower things. And surely no "self" should be as fragile, as easily discarded, as Ms. Hearst's was.

The fragility of the individual's sense of self is a persistent theme of modern literature, and the malleability of persons is the assumption of modern radical movements. Arthur Koestler's *Darkness at Noon*, a classic of political literature, features the sinister Gletkin, who is gifted at mind bending—or, as it is known today, "consciousness raising."

The disturbing thought is not that the SLA had some cunning Gletkin who destroyed Tania's sense of her former self. The disturbing thought is that no Gletkin was needed.

There is nothing new about radicalism inflaming the comfortably born. Henry Hyndman, a major figure in early British socialism (the model for the role of John Tanner in Shaw's *Man and Superman*) used to go to Hyde Park Corner, dressed to the nines, to berate working-class listeners for permitting people like him to live in splendid houses.

But Ms. Hearst's radicalism is not redeemed by such puckish self-knowledge. Indeed, what self remains to be known? She evidently has only some consuming need that is soothed by her new political catechism that teaches her to love the "people" and despise her father.

When exasperated by the question of why he became a socialist, John Strachey (whose father was a leading conservative journalist of his generation) would answer that he did it to spite the world because he had failed to make the cricket eleven at Eton. There are worse reasons for becoming a socialist; but Strachey was joking. Nevertheless, he may have been closer to the truth than he realized. Bertrand Russell once asked Strachey why he had become a socialist: "Did you hate your father, your childhood, or your public school?" Strachey, finding nothing odd about the assumption planted in the question, replied: "A bit of all three."

Political beliefs about the external world can, and radical beliefs often do, reflect the believer's inner turmoils, and satisfy his emotional needs. If Tania ever comes in from the cold, perhaps she will explain what she was driven to look for out in the cold. Until then, this episode of fanaticism, like all such, calls to mind Max Beerbohm's inspired caricature featuring Dante Gabriel Rossetti and Benjamin Jowett.

Rossetti, a suitably exotic Pre-Raphaelite figure, his hair tumbling in lush profusion, stands on a high scaffold in the Oxford Union, where he is painting murals of the Arthurian legends. The diminutive but formidable Jowett, great master of Balliol College, dressed in Victorian black, peers up at Rossetti's Arthurian murals and inquires: "And what were they going to do with the Grail when they found it, Mr. Rossetti?"

[February 1, 1975]

Patty Hearst Comes Home

According to her parents, whose fortitude has been stirring throughout, Patty Hearst, confined to a tacky jail cell and thus prevented from arriving at her Finland Station, would rather like to go home now, thank you. And you can appreciate her point of view. The revolution has come a cropper.

Only an incorrigible cynic would conclude that she has lost her desire to grind the faces of the jackbooted ruling classes that grind the faces of widows and orphans. But she, like Trotsky, has learned to her sorrow that being Trotsky is no bed of roses. War to the knife is all very well, but it is time for an armistice. Aside from Ms. Hearst and a friend, the other soldiers in the Symbionese Liberation Army, both of them, were arrested while jogging—training, no doubt, for a tactical retreat to some mountain redoubt.

You might think that completion of the nineteen-month hunt for Ms. Hearst is a coda to the decade of political infantilism that began where the hunt began, in Berkeley, across the Bay from where it ended, in San Francisco, Ms. Hearst's home town. But infantilism knows no season or home town, and shortly after Ms. Hearst was captured a bomb exploded at a Safeway grocery store in Seattle. Someone called the police to explain, helpfully, that the bomb was in retaliation for Ms. Hearst's arrest.

It is a good thing the caller called, because without this exegesis it might have been difficult for the rest of us, in the fog of our bourgeois Weltanschauung, to understand the connection between Ms. Hearst and the Seattle Safeway. Today's radicals subscribe to principles of social change that are not apparent to the untutored eye.

In a sense, Ms. Hearst never left home. Certainly her "rebellion" —if that, rather than "tantrum," is the right word—was directed at her home, at her parents. The rebellion never changed the world beyond the confines of her large-hearted parents' large house. Aside from her parents, whose sturdy hearts not even she could break, the only people she injured were her adopted friends in the SLA. Emboldened by her conversion, some of them got killed in a Los Angeles shoot-out.

When booked by the San Francisco police, she pluckily listed her occupation as "urban guerrilla." The San Francisco police, who recently brought their city to its knees with an illegal strike, could teach Ms. Hearst a thing or six about urban guerrilla action. But for the moment they were on the side of the law, and she, too, was playing her assigned role, as render of the social fabric.

Playing your role is not easy when you cannot even keep your name straight. When the judge asked her if she was Patricia, she said yes, thereby rendering obsolete all those wall posters celebrating "Tania" clutching her carbine. That is what novelists have been trying to tell us about modern man's sense of self—easy come, easy go.

Of course her "army" always was to a real army what "Woodstock Nation," that rock festival in the mud, was to a real nation. It was the sort of invention that one would expect from not very bright children who, looking down the vista of the years, see no farther than tomorrow noon.

The revolution was a classic expression of the use children make of dimly understood adult words and forms of action. Like most radical politics of the recent American past, the "revolution" was private psychotherapy masquerading as political action.

The "revolution" launched by the "army" ended as it began: a smashingly successful media event. It was the headline of the hour, even unto an NBC special that delayed the Johnny Carson "Tonight" show. That delay discomfited more Americans than any revolutionary group could reasonably have hoped to discomfit at one fell swoop.

The mills of justice grind slowly, but the presses of publicity roll swiftly. More than Ms. Hearst needs a lawyer she needs an agent to fend off publishers, and to arrange for serialization of her memoirs in women's magazines. It is the genius of America that "revolutions" launched by "armies" spill more ink than blood.

[September 22, 1975]

P. G. Wodehouse: Soufflé Chef

British school boys relished his short stories during the Boer War and today, nearly seventy-five years and more than seventy-five books later, he is recognized as one of the century's masters of English prose. Indeed, *Punch* magazine once said that criticizing him would be like taking a spade to a soufflé. Next Tuesday at his Long Island home P. G. Wodehouse, who has just published a new novel and is working on his next, will celebrate his ninety-third birthday.

In an age when craftsmanship seems to be a depleted cultural resource, he has worked with words the way a silversmith works with his metal, and has brought pleasure to scores of millions. His most familiar characters are Bertram Wooster, a well-born, well-intentioned nitwit who goes through life in the protective custody of his erudite gentleman's gentleman, Jeeves, whose imperturbable grace is conveyed in Wodehouse's description of him as "a procession of one." Since 1916, when the estimable Jeeves appeared, Wooster has been tumbling into and Jeeves has been pulling him out of hilarious imbroglios at Blandings Castle, the Drones Club, and Brinkley Court, Market Snodsbury.

The purity of Wodehouse fun is never spoiled by the rude intrusion of serious ideas, and this purity offends some somber moderns. The bleak utilitarianism of the modern age leads to the disparagement of Wodehouse's works as "escapist." The strange thing is that anyone would be so fond of the cares and conditions of the modern world as to deplore literature that helps people escape to the Blandings Castles of their minds.

Wodehouse, creator of an innocently idyllic England, has been in a sort of self-imposed exile from England for four decades, because of an innocent mistake in a time when innocents were casualties. In 1940 he probably forgot there was a war on. In any case, the Germans captured him at his villa in France. Released, he passed through Berlin, where a CBS radio correspondent (America was not yet in the war) asked him to broadcast to his American readers.

He foolishly made a few utterly nonpolitical broadcasts, and the Germans made propaganda use of the fact that a distinguished British writer was broadcasting from Berlin during Britain's darkest hour. He is so gloriously out of place in this century that he had a childlike incomprehension of the century's principal product, total war. That is not an excuse, but it is a fact. And it caused him to blemish his relations with his country.

In any case, Malcolm Muggeridge, in his recent autobiography, reports that Wodehouse contributed, in an appropriately inadvertent way, to the war effort: "The Germans, in their literal way, took his works as a guide to English manners, and used them when briefing their agents for a mission across the Channel. Thus, it happened that an agent they dropped into Fen country was wearing spats—an unaccustomed article of attire which led to his speedy apprehension."

Wodehouse's full political philosophy is contained in the passage where Wooster wonders why the titled Spode, who likes politics, does not run for Parliament. Wooster's Aunt Dahlia explains:

" 'He can't, you poor chump. He's a lord.'

'Don't they allow lords in?'

'No, they don't.'

'I see,' I said, rather impressed by this proof that the House of Commons drew the line somewhere."

Wodehouse's philosophy of life is apparent (to the tutored eye) in the passage where Wooster, as usual, is trying to solve someone else's romantic problems and is making a hash of things. While dressing for dinner, and failing (as usual) to tie his tie properly, Wooster tries to convince the unflappable Jeeves that a shattering crisis is at hand.

Jeeves, refusing to be distracted from the important things, says: "We can but wait and see, sir. The tie, if I might suggest it, sir, a shade more tightly knotted. One aims at the perfect butterfly effect. If you will permit me—"

Wooster, exasperated: "What do ties matter at a time like this?"

Jeeves, serene: "There is no time, sir, at which ties do not matter."

[October 10, 1974]

Winston Churchill: In the Region of Mass Effects

In 1895 Sir William Harcourt said to him: "My dear Winston, the experiences of a long life have convinced me that nothing ever happens." But some things did happen in the stimulating next five decades, in which Winston Churchill was the largest figure.

After the Great War a critic (perhaps remembering Mr. Dooley's suggestion that Teddy Roosevelt should title his war memoirs "Alone in Cuba") dryly noted that "Winston has written his autobiography and called it *The World Crisis.*" And, in truth, Churchill, an adventurer, wanted to live life at a fever pitch, in the thick of the world crises that exhilarated him.

Some exhilarated him too much. Recall Margot Asquith's chilling description of the evening of August 4, 1914, at No. 10 Downing Street. The cabinet meeting ends, the ministers depart, grim and sorrowful, except for Churchill, who descends the stairs, smiling. This exhilaration distinguishes Churchill from the only greater democratic leader. Lincoln hated every minute of war.

Like Burke and Disraeli, Churchill was a literary man. Like his predecessors among great democratic leaders—Lincoln, Lloyd George —and like none since, Churchill's rhetoric was the essence of his statecraft. Its distinctive, ponderous cadences invited parody, defied imitation.

Many people found the archaisms of his rhetoric offputting. Such people—who, I wager, prefer Le Corbusier to Wren, Hemingway to Flaubert—dismiss his rhetoric as the unfortunate consequence of exposing an impressionable youth to Gibbon and Macaulay. But it was much more than the residue of others' eloquence.

People who look upon history as a story of unreasonably slow progress from darkness to our current enlightenment, and hence as just a tiresome prelude to modernity, should dislike Churchill's rhetoric. In it style and substance are fused. As Isaiah Berlin wrote, Churchill's rhetoric was "an inspired, if unconscious, attempt at revival. It went against the stream of contemporary thought and feeling . . . a composite weapon created by Mr. Churchill to convey his particular vision. In the bleak and deflationary Twenties it was too bright, too big, too vivid, too unsubtle. . . . Revivals are not false as such: the Gothic Revival, for example, represented a passionate, if nostalgic, attitude towards life. . . . It sprang from a deeper sentiment and had a good deal more to say than some of the thin and 'realistic' styles which followed."

The greatest leader of the "century of the common man" was born in a palace, and was sustained throughout his life by a romantic sense of continuity with aristocratic traditions of government. In 1937 he wrote: "Nowadays when 'one man is as good as another—or better,' as Morley once ironically observed, anything will do. The leadership of the privileged has passed away; but it has not been succeeded by that of the eminent. We have entered the region of mass effects."

Unfortunately, just "anything" will not do. Democracy confers privileges, not eminence. It confers the privileges of power upon people who must lead by persuasion, or not at all. When the powerful are not persuasive, they retain power but lack authority, and public opinion itself is inchoate—a mass effect.

But perhaps the leadership problem Churchill delineated is more than a problem. Perhaps it is an insoluble dilemma of democracy, and the region of mass effects is everywhere.

It may be that, as Plato and most serious political philosophers since him have argued, democracies always contain the bacillus which produces decay and turns democracies into regions of mass effects. This bacillus is a corrosive suspicion of figures like Churchill, who are too vivid—too excellent—by half.

Nature is not democratic. The virtues of a Churchill—independence, imagination, spiritedness, character, eloquence—are not distributed with democratic evenness across the population, one allotment, one voter. A Churchill threatens the premise most cherished in a decayed democracy, the idea that "one man is as good as another—or better."

We probably shall not see his like again. If, as the philosophers teach, decay is the condition toward which all democracies tend, then certainly we shall not see again the like of Winston Leonard Spencer Churchill, born 100 years ago, November 30, 1874.

[November 26, 1974]

Charles Lindbergh, Craftsman

In 1926 a young poet expressed her, and the day's, yearning for something magical:

> Everything today has been heavy and brown,
> Bring me a unicorn to ride about the town.

The next year the poet and the world met Charles Lindbergh. The poet soon would be Anne Morrow Lindbergh.

It is fifty years since the day—May 21, 1927—the silvery Spirit of St. Louis landed in Paris, thirty-three hours from Long Island's muddy Roosevelt Field. Lindbergh's flight is still vivid in the memory of a passing generation.

Nostalgia is a distorting lens, and the 1920s are bathed in the soft glow of remembered radiance. But the decade has a hard, unlovely dimension. Frederick Lewis Allen called the period "the ballyhoo years" given to "crazes" for the Charleston, Mah-Jongg, crossword puzzles, Florida real estate, flagpole sitting, and marathon dances. And news became entertainment.

In 1925 Floyd Collins, trapped in a Kentucky cave, took eighteen days to die. By the time he did, there was a tent city of journalists at the mouth of the cave, churning out copy for a mesmerized world. That summer millions of words poured through the transatlantic cable from Dayton, Tennessee, scene of the Scopes trial.

Lindbergh's flight was the first elevating spectacle of the "wired world." Most fame is an artifact, a perishable consumer product. Lindbergh's fame was different because he was different. In a day when a senator (Charles Curtis, R-Kan.) could endorse Lucky Strike cigarettes, Lindbergh did not commit testimonials or movies. He was a craftsman whose craft—aviation—was an end in itself. It was a "vocation" in a semi-religious sense.

Antoine de Saint-Exupéry, author of *Wind, Sand and Stars* and *Night Flight,* disappeared during a flight in 1944. He has been called "the Joseph Conrad of the air." Conrad used ships and the sea for voyages through the inner man. Saint-Exupéry, like another brilliant writer on aviation, Charles Lindbergh, was moved to mysticism by access to the sky.

But one striking thing about the literature of the air is that there is so little of it, compared with the literature of the sea. It was over in a blink of an eye, that moment when aviation stirred the modern imagination. Aviation was transformed from recklessness to routine in Lindbergh's lifetime. Today the riskiest part of air travel is the drive to the airport, and airlines use a barrage of stimuli to protect passengers from ennui.

The marvels of science have made science seem less marvelous. And as mankind's capacity for awe contracts, mankind is diminished. Consider one of the most remarkable features of the public mind today —the vanishing interest in space exploration. The average American is so jaded that he is unmoved by the idea of exploring beyond the thin atmosphere of this tiny cinder that orbits in a corner of a solar system in the whirl of the universe.

Lindbergh was born in 1902, almost two years before Kitty Hawk. As a child in Minnesota, automobiles were less familiar to him than were his grandmother's recollections of the Sioux uprising of 1862. Having experienced the compression of time, and having come to question the pace and direction of change, Lindbergh became a student of man's pre-history, flying to primitive regions in advanced planes.

A schoolmate asked one of Lindbergh's children: "Didn't your father discover America?" Well, no, he didn't, exactly. But he was a catalyst of the nation's rediscovery, in a frivolous decade, of old virtues, including bravery, solitary discipline, and applied knowledge.

Lindbergh died in 1974. He is buried in a grave cut from rock in Maui, Hawaii. His epitaph reads, "If I take the wings of the morning and dwell in the uttermost parts of the sea. . . ."

Maui is about as far away as America gets from Roosevelt Field, which is now a shopping center.

In Brendan Gill's new illustrated essay, *Lindbergh Alone,* the last photograph is a stunning snapshot of young Charles, alone on a raft, probably on the Minnesota headwaters of the Mississippi. Several hundred miles south, the Father of Waters flows past Hannibal, Missouri. There, another boy, and the American imagination, set out on a raft. That image of American adventure is recurrent, and thus imperishable.

[May 15, 1977]

Huey Long: "I Bet Earl Bit Him, Didn't He?"

He was "Tom Sawyer in a toga," a rustic Caesar more absolute in his domain than any American politician before or since. Then on September 8, 1935—forty years ago—an assassin shot the Kingfish.

Huey Long's bodyguards blasted sixty-one holes in his assassin. Excess was the rule in Long's Louisiana.

Because Long was the raucous voice of the rural poor, the people he called "one-gallus boys," he is remembered as a regional anachronism. But he seized the opportunity modern society provides for the politics of class antagonism.

Some say that Long was the first southerner since the great Virginia Founding Fathers to have an original idea about government. But that assessment underestimates John Calhoun and overestimates Long. Long's idea about government was just to use it, everywhere at once, with exuberant ruthlessness, to succor friends and harry enemies. Regarding everything from the placement of roads to admission to Louisiana State University, Governor Long's discretion expanded, driving out law.

Henry Adams' description of Theodore Roosevelt fits Long: "He was pure act." Imagine with what disdain Long would have regarded George Wallace, who is pure pose.

Wallace is a campaigner, not a ruler. He is entirely self-absorbed, consumed by status anxieties, uninterested in changing the world, interested only in the warmth of resentments shared with friendly audiences.

Long knew he was just as good as anyone else, and he was cold-blooded enough to understand how bad that was. Unlike Wallace, who resents his enemies for what he thinks they think about him, Long hated his enemies for what they obviously were. They possessed power that he wanted to use to force a new distribution of the rewards of life.

Long knew where he was going, and cut a swath through institutions to get there. He had an egalitarian's sense of what would be equitable in social results. But he had no sense of fairness in social procedures. Destination was everything, due process was nothing, as he turned the state into a crude engine of redistribution.

State policemen were used to escort balky legislators to their desks. Long had signed, undated letters of resignation from supposedly independent state officials. He had damaging dossiers about people on his enemies list. When asked if patronage workers were compelled to pay 10 percent of their salaries to the Long machine, a Long operative indignantly said, gracious, no, they "had to pay that 10 percent voluntarily."

Long always had a killer instinct for driving opponents beyond defeat, into the dust, out of politics. This instinct grew after the attempt to impeach him in 1929 for unspeakable acts and unnatural practices, like wanting to tax Standard Oil. The proceedings lacked even what passed for decorum in Long's Louisiana.

When brother Earl brawled with a pro-impeachment legislator, Huey chortled: "I bet Earl bit him, didn't he? Earl always bites." (Earl had bitten.) Huey had, as they say, enough money to burn a wet mule, so the impeachment was settled by the price mechanism. As a Long supporter explained one man's helpfulness: "They bought him, and we bought him back."

By 1935 Long was a U.S. senator, dictating a book titled *My First Days in the White House,* and dictating Louisiana policy through his chosen governor, a cipher. That policy included final consolidation of power: election officials and poll watchers would be picked by the governor's men.

An opponent warned: "I can see blood on the polished marble of this capitol." Like the Louisiana sky before a thunderstorm, the atmosphere was thick with latent violence.

Long's dying words on September 10 were, "God, don't let me die. I have so much to do." But by the time his assassin stepped from behind the pillar in the capitol, Long had done enough to make more than a few Louisianans think of assassination as tyrannicide.

After Long, as before him, the rich were rich and the poor were poor. But the poor were left with sustaining memories of a time when the rich had known fear. The memories were better than beans and gravy, and were more than the poor usually got from government.

[September 3, 1975]

Russell Long: The Sovereignty of Chromosomes

May is the month of buses in Washington, buses bringing boys and girls to see senators and other monuments. Bumper-to-bumper buses form a yellow and silver snake at the foot of Capitol Hill. On a gray morning after a dawn downpour last week the buses were disgorging their tumbling cargo while Senator Russell Long, a Louisiana Democrat, was nearby, tucking into bacon and eggs in the senators' dining room. For the boys and girls a glimpse of him would be a glimpse of a large chunk of government, and a slice of the century.

First there is the shock of recognition. You are struck by the sovereignty of chromosomes. The cherubic jowliness, the shrewdness, the quickness, intelligence and humor, the darting eyes and restless energy: an assassin shot Huey Long forty-two years ago but his genetic code goes marching on.

Russell Long's early memories of Washington are of getting lost in the lower corridors of the Capitol, while Daddy was upstairs raising Cain. He became a senator when the Constitution let him, when he was thirty, in 1948. It will be *at least* 1992 before he removes himself from his office in the Richard Brevard Russell Office Building. Then he will have spent forty-four of his seventy-four years in the Senate (Huey Long lived only forty-two years) and will have passed Georgia's late Senator Russell as the longest-serving man of first-rate powers in our time.

Politics often is a family affair, and not just for Adamses, Rockefellers, and Kennedys. In 1931 Richard Russell was sworn in as governor by his father, Georgia's chief justice. Byrds were in seventeenth-century Virginia politics and Harry Byrd Jr. is Long's colleague on the Finance Committee, which Harry Byrd Sr. chaired before Long.

Russell Long's Uncle Earl was the most *novel* governor in postwar America. He had Flaubert's flair for *bons mots* ("Jimmie Davis loves money like a hog loves slop") and a gift for concision, as when he described an uncle's death: "He got drunk and pulled a man out of bed and got into bed with the man's wife, and the man got mad and shot my poor uncle, and he died." Earl's downfall was loose living and the premature declaration that "niggers is people too."

Earl was made possible by Huey, bless him. Huey dominated Louisiana more thoroughly than anyone has dominated any other state. He was a master of the politics of class antagonism. He broke the state to his saddle, yanked the cinch tight, slammed his spurs into its flanks, and set about the business of government, as he understood it: helping friends and hurting enemies.

Russell is the least Technicolor Long. When aroused he resembles a hurricane coming ashore among the pines of Plaquemines Parish. But power means not needing to raise your voice. Anyway, Long need not outshout people he can outthink. He is one of those legislators whose eminence does not carry to the country. He is less well known than most members of the Foreign Relations Committee, the most publicized and least consequential Senate committee. He may be the cleverest senator and his Finance Committee has more than its share of Senate talent, including Democrats like Lloyd Bentsen and Daniel Moynihan, and Republicans like Robert Packwood and John Danforth. The Finance Committee deals with the arcane essence of the modern state, the tax code. More than any other committee, it requires both mastery of complex detail and a comprehensive view of social processes.

Long's heroes were populists of various stripes, Andrew Jackson and Senators Robert La Follette of Wisconsin and George Norris of Nebraska. And Huey Long, whose dying words were, "God, don't let me die. I have so much to do." Huey's son says he is doing Father's work, not roughly redistributing wealth but sculpting the revenue system that shapes production and, hence, the distribution of wealth. He says that changes in the revenue system often are better than appropriations at accomplishing government's ends. That is, government is better at providing incentives than services. For example, he would rather give employers tax breaks for providing employees with health insurance than have government provide comprehensive health insurance. Father wanted to make "every man a king" by squeezing the rich. Son wants to make "every man a capitalist" with tax incentives for employee stock-ownership plans.

After the new boys in town tried to kill some Louisiana water projects, Long introduced himself to a White House meeting: "I'm Russell Long and I'm chairman of the Senate Finance Committee." Asked why he did this, he replied solemnly: "I saw a lot of strangers and I wanted to make sure they knew who I was." Now they know that his committee is the eye of the needle through which Carter's camels must pass. Does that help explain Carter's tendency toward fiscal conservatism? Golly no, says Long, radiating innocence: "Nobody should blame me if the guy has some sense."

Many of Carter's energy proposals will suffer the scrutiny of Long's committee. When Carter takes up the nasty task of restoring the soundness of the Social Security system, Long's committee will take charge of Carter's suggestions. When, in the fullness of time, Carter launches his frail craft of welfare reform, the port it will sail toward will be Long's committee. Long's committee was an unsafe haven for Nixon's "Family Assistance Plan," the special project of White House aide Daniel Moynihan. Later, when Professor Moynihan did an autopsy on the lifeless remains of FAP, he decided it died from, among other things, a collision with Russell Long.

Now Senator Moynihan is the most junior Democrat on Long's committee, and he is impatient. Long says, sweetly: why, shucks, "I guess Senator Moynihan knows just what he wants." But, the chairman did not need to say, Moynihan will do a load of compromisin' on the way to the horizon.

When Carter tackles the tax system that he says is a "disgrace to the human race" he will be dealing with Long's committee, which has done rather a lot to make the tax system what it is. But Long does not expect the administration's "reforms" to be offensive. With the calm confidence Mark Twain attributes to a Christian with four aces, he says: "I have a *vague* idea of what the *general* direction will be." And well he should. Carter's assistant secretary of the treasury for tax policy is a gentleman who for many years was one of Long's most trusted and talented aides.

That is how a prudent chief executive accommodates a powerful legislator. It may not quite square with a purist's theory of the proper separation of powers. But such theories are little brooks that flow with the landscape. They are turned aside by boulders like Russell Long, who was part of the Washington landscape long before Carter was, and will be when Carter has gone.

[May 16, 1977]

Franco: Last from the Rock Pool

A recent joke in Madrid was that Franco refused a gift of a baby elephant because he said he could not bear to watch it grow old and die. That is political humor suited to a regime in which the only constitution that matters is that of the leader.

Franco was the last inhabitant of Europe's rock pool of rightist dictators. His was the longest caretaker government in modern European history. He was in his peculiar and unattractive way a "great man": his nation's experience in his lifetime was, in large measure, his signature.

After the fifty-eight-year rule of Emperor Haile Selassie ended last year, Franco's became the longest term of personal power in the world. He ruled half of Spain six months after the civil war broke out in July 1936, and ruled all of it after April 1939.

Widespread fear and exhaustion are the despot's friends. They were Spain's afflictions in 1939, when Franco began his ferocious retribution against the defeated loyalists. No one knows how many scores of thousands of executions occurred after the daily trials between 1939 and 1941. No one knows how many people were sent to jail as a result of the retroactive Law on Political Responsibilities, under which any Republican recruit who had won promotion was assumed to have committed a crime.

Spanish history is a long line of precedents for ruthlessness, to which the Republican side added 60,000 executions during the war. Then, and during the savage years immediately before and after, Spain resembled Goya's masterpiece of Spanish art, the *Third of May, 1808*, which depicts the slaughter of Spaniards by Napoleon's firing squads.

But in Spain's twentieth-century agony, Spaniards were on both ends of the guns.

Franco did not destroy the Republic. By July 1936—the military rising—the Republic was in an advanced and irreversible process of disintegration, the result of political, regional, and anti-clerical fanaticism. Spain was doomed to a bloody passage toward some sort of dictatorship.

The victor, Franco, was a European eccentric. Ruler of a notoriously emotional nation, he did not emote. Never has a man been so enigmatic while having so little to be enigmatic about. Beyond cold vengeance, he desired nothing but order.

Like another and immeasurably greater European eccentric, De Gaulle, Franco found the twentieth century strange and distasteful. Whereas De Gaulle tried to transcend his time, tapping what he considered the eternal wellspring of his nation's greatness, Franco was content to sit on the bent back of his nation, hoping that the century would mend its ways or go away.

Last spring, when Spaniards moved their clocks ahead an hour to daylight saving time, a magazine published a cartoon in which a man says to another: "So now our clocks will be the same as the rest of Europe." The other man replies: "Clocks, yes, but calendars, no."

The fact that political humor has been published in Spain is evidence that the Spanish calendar is not just a yellowed and blood-stained page from 1939. But the question now to be answered is this: have the passions of 1939 vanished forever, or have they merely gone underground, like a stream, to emerge later, full force?

It has been well said (by British Prime Minister Stanley Baldwin) that a durable despotism is like a great tree, majestic in its way, like any extravagance of nature, but like everything in nature, mortal. And when it blows down you see that nothing has grown beneath it to replace it, the saplings have died or been stunted for want of air.

Where the public landscape has been kept flat, producing only untested leaders and conspiratorial organizations, gusts of passion encounter nothing to deflect and temper their sweep. And there is no reason to believe that Franco's deep freeze of political life did anything to alter the fact that excess is the distinguishing mark of the Spanish spirit. So there is reason to believe that Franco's caretaker regime will prove to have been only a long parenthesis between turmoils.

[October 31, 1975]

Franco: The Art of Ingratitude

Franco never smiled easily or often. But he must have permitted himself a thin trace of cynical amusement when, as a result of some recent executions, he became the object of protest demonstrations in European capitals, like Stockholm. The demonstrators denounced him as a fascist, and as Hitler's helper. Similar judgments have figured prominently in recent writings about Franco.

In 1939, at a victory mass in Madrid, Franco asked God to be pleased with the efforts of his army which "in Thy name has vanquished with heroism the enemy of the truth in this century." Like most people who advise God that their only interest is restoring Truth to her throne, Franco found a throne for himself. Then he just sat there.

Hitler, the risen corporal, and Mussolini, the playtime Roman pro-consul, used military trappings to stir their mass movements. Franco, a soldier of almost pathological bravery (at thirty-three one of the youngest European generals since Napoleon), trusted no "movement," only the army. Hitler and Mussolini were men of intoxicating words. Franco was perhaps the worst public speaker of the 1930s, the decade of demagogues. Having no ideology, he had nothing to say.

Fascism is a revolutionary ideology, a mishmash of romantic nationalism and anticapitalism. Like Portugal's Salazar, also inaccurately called a fascist, Franco was from the start a reactionary, not a revolutionary. He was a throwback to the pre-democratic age when despotism felt little need to justify itself with ideology.

A Latin American dictator recently expressed a political philosophy as simple as Franco's: "I will die and my successor will die and you will not have elections." In 1936, and ever after, Franco believed, without malice or philosophic pretense, that Spaniards could not cope with democracy.

Franco was pleased to call himself "Caudillo of the Crusade," meaning by "crusade" the civil war against the Republic, which he called "anti-Spain." He never was a crusader of the European right; he never was a leader of the fascist movement. That movement was never crucial to his purposes; when he no longer needed it, he caused it to atrophy.

During the Second World War, he was a stumbling block to various Axis plans. And Spain was less helpful to Hitler than was Sweden, which did well selling him iron ore.

When Spain's civil war erupted, both sides appealed almost immediately for foreign help. Franco received men and material from Hitler and Mussolini, as the Republican side did from Stalin and foreign volunteers.

Hitler and Mussolini (who claimed that Italy contributed 1,000 planes, 6,000 dead, and 14 billion lire to Franco's victory) expected more than mere thank-you's in return when the Second World War began. But all they got from Franco was some volunteers, a few mining concessions, and some warm, turgid, rhetorical support.

Hitler asked repeatedly and in vain for Franco to declare war; for Franco to seize Gibraltar; for Franco to permit German forces to cross Spain to seize Gibraltar. General Jodl, head of the German High Command, told the Nuremberg Tribunal that Franco's obstinacy was a major cause of Germany's defeat. Jodl was wrong about that, but was understandably bitter about Franco's unhelpfulness.

Franco behaved like a seventeenth-century Spanish statesman for whom the central fact of life was Spain's weakness, and her need to accommodate the great powers. As long as German divisions were at the Pyrenees, Franco was merely unresponsive to Hitler. When the tide of war shifted, so did Franco, accommodating the Allies in various ways. He allowed the evacuation of Allied casualties through Barcelona, and the passage through Barcelona of Allied supplies destined for southern France.

Regarding his fascist benefactors, Franco's policy was an echo of a nineteenth-century nationalist from another Mediterranean peninsula. When Cavour, architect of Italian unity, succeeded with the help of northern liberals, he was asked what his policy would be regarding those who had helped him. He replied that he would "astonish the world with our ingratitude."

[November 3, 1975]

Hubert Humphrey: The Art of Exhaustion

For fifteen minutes the Humphrey tendency to talk like a Sten gun was held in precarious check. But now he is explaining what Presidents should be, and colliding images are bending each other's fenders.

Presidents should be like ol' Doc Sherwood of prairie memory, "someone who can deliver a baby in the middle of the night with the lights off and comfort the mother and make sure the baby lives," and Presidents should see that the "orders go out to the troops," and Presidents should make the White House into what Woodrow Wilson said it should be, "the nation's classroom."

Hubert Humphrey says he is not a candidate for the office of baby-delivering, orders-barking schoolmarm. He has his pride, and remembers an earlier Minnesotan, Harold Stassen. But buried among his many reasons for not actively seeking the nomination is this reason: uncharacteristic restraint might be the best tactic for getting nominated. Unquestionably he still wants to end his career by turning the White House into a little red schoolhouse.

Like Moses, Humphrey was discovered early in his career, at the 1948 Democratic convention, when he was mayor of Minneapolis. Like Moses, Humphrey was a bundle of opinions, including those about civil rights which, forcefully put to the convention, were part of the pretext for the foreordained Dixiecrat walkout. Since then he has been near the center of the national stage.

Nevertheless, some critics wrongly persist in seeing Humphrey's career as vindication of the Biblical warning, "Unstable as water, thou shalt not excel." Actually, he is not unstable, only ebullient in his upper-Midwest liberalism which is, if anything, too stable by half.

He has excelled at everything that senators do, except shimmying up the greasy pole from the Senate to the presidency. But because his disappointments have come so near the summit of public life, his public image is of a man toward whom Destiny has had the warmest of intentions without getting around to implementing them. John Kennedy rode to the presidency over him, and Lyndon Johnson's presidency collapsed around Humphrey, nearly burying him under the debris.

Kennedy's rocket ignited with a victory over Humphrey in the

1960 West Virginia primary. Lyndon Johnson wanted Humphrey not only as Vice President but as a hostage against a hostile future in which, Johnson suspected, the liberal coalition would crack and turn on him. In 1968 it turned on Humphrey, too, and did not consider the alternative until election eve, when voters began moving Humphrey's way in unprecedented waves. If the election had been eight hours later, today Humphrey might be completing his second term.

If ever a man had an excuse for smoldering resentment about life's close calls, he is Humphrey. But the closest thing to roughness in his temper today is an ever-so-slight waspishness—beeishness, really—about what he considers an uninspiring crop of presidential pretenders. But even when goaded he will not say mean things about rivals.

Most successful politicians, and all modern Presidents from Franklin Roosevelt on, emphatically including Dwight Eisenhower, believed that the quality of life was improved by a sprinkling of a few enemies around the landscape. Humphrey has, Lord knows, an abundance of the animal spirits that fuel most politicians, but he does not have, and never has had, blood in his eye. That is one reason why he can hope that next year's Democratic convention will turn its glazed eyes to him.

Humphrey's hope must be that all announced candidates will limp into Madison Square Garden next July with torn ligaments and many contusions, and nearly prostrate from their exertions in the spring scrimmages, too weak to seize the prize and too irritated with one another to collaborate in conferring the prize on one of their number. Then perhaps they will settle on the least disliked alternative, perhaps Humphrey, who will be waiting to ride the wave of other people's exhaustion toward a green shore.

In his long public life of obtrusive goodwill and usefulness, Humphrey has exhausted more people than he has persuaded. But in politics the effect sometimes is the same.

[September 19, 1975]

Gordon Allott: Not a Malleable Man

Some of the friends of Gordon and Welda Allott gathered recently to wish them well on their return to Colorado, which Gordon Allott represented in the Senate for eighteen years. It was a cheerful affair, not only because Washington is fun to leave these days, or because Colorado is a nice place to go at any time. The Allotts are just too sensible to allow a lost election to dampen their spirits.

It was my privilege to serve on Senator Allott's staff from January 1970, when I first came to Washington, until January 1973, when he left the Senate, having lost his bid to become Colorado's first four-term senator.

A number of factors contributed to his defeat a year ago this week. We made a number of campaign mistakes. But the most important factor was that Colorado has changed a lot since 1954, and Gordon Allott is not the sort of malleable man who finds it easy to pour himself into new molds.

He came of age in the Depression, living in the unglamorous parts of Colorado, around the smoking steel mills of Pueblo, and the barren wheat growing flatlands around Lamar in eastern Colorado. As senator he worked hard to promote the economic growth of Colorado, his attachment to the traditional Western growth ethic reinforced by his Depression experience. His constituents approved.

But he did his job so well that he changed his constituency, and brought about his defeat. He helped bring Colorado light technological industry that does not use Colorado's most precious resource, water. And this brought a large influx of new voters, young, educated, affluent, mobile, and quite unresponsive to the stolid Westernness of Allott's policies and personality.

The new Coloradeans are environmentalists of the sort who want to slam the door behind them when they cross the Kansas border into Colorado. They—and the industry that employs them, and the expensive ski facilities that accommodate them—have made Colorado more like Connecticut or New Jersey than any state contiguous to it. And now Colorado has senators who grew up in Connecticut and New Jersey.

In spite of the changes in the state, Allott only lost by approximately one-half of one percentage point. And he might not have lost at all if the White House had granted his urgent request for an end-of-campaign visit by the President to Colorado.

Allott made the request about ten days before the election, when a poll showed him below 50 percent and a substantial number of voters undecided. The political experts in the White House assured us that they had secret information that we would win by 15 percentage points. Besides, they said, snow on the ground might hold down the size of a crowd for a presidential visit, and that might hurt his image.

But the plain truth is this.

For twenty years Allott had been a supporter of Nixon. For four years he had been an especially loyal supporter of even President Nixon's dumbest causes—the SST, Judge Carswell. But when it came time for Nixon to do Allott a favor, all Allott got was an evasive, dishonest and, in the end, contemptuous refusal.

The day before the election, Nixon flew across country to vote in California. He made several stops, but none in Colorado.

In the days after the election the White House, flushed with victory, made clear its lack of sympathy for Allott. According to the hard-boiled Darwinism in vogue at the White House in November and December 1972, those politicians who need the help of friends in order to survive deserve neither friends nor survival.

A year ago the reigning philosophy was survival of the fittest, and Mr. Nixon and his agents were feeling remarkably fit. Today Mr. Nixon has all the friends he has earned and deserves.

Now Mr. Nixon may not survive. He certainly won't be saved by Allott's vote in the Senate. But Allott has more important things to worry about, like what the trout are hitting up at Electra Lake, north of Durango.

[November 8, 1973]

Nixon: The Way He Was

On August 6, 1974, the 781st day of the 784-day Watergate debacle, Barry Goldwater became fastidious. He told White House chief of staff Alexander Haig: "We can't support this any longer. We can be lied to only so many times." Richard Nixon had exceeded the generous quota of permissible lies.

This snapshot of a politician putting his foot down is from *The Final Days*. To read it is to plunge again into the dark stream of Watergate, to smell the acrid ozone of baseness that hung over Washington two years ago. Authors Bob Woodward and Carl Bernstein are scrupulously reportorial, and their narrative should subvert some Watergate myths.

It pleases many people to regard Watergate as a brush with an emerging police state, the Republic rescued by an intrepid Congress.

The conventional wisdom is that Watergate was a manifestation of the "imperial presidency."

But when, precisely, was Nixon's presidency in its imperial phase? Nixon could not get his way with the school-lunch program, and he felt threatened by a "Jewish cabal" in the Bureau of Labor Statistics. Some emperor.

The plug-uglies behind the White House crime wave were not a bit like those Mohicans who crept through dry forests without rustling a leaf. From the burglars who couldn't burgle to the biggest enchilada of them all, they floundered, like insects in yogurt. As for Congress, it insisted that the central question was: "What did he know and when did he know it?" In retrospect the most interesting question is: "What did we all know, and when did we know it?"

By April 30, 1973, after Nixon's first television speech about Watergate, we knew he was stonewalling about several burglaries. By October 20, 1973, the Saturday Night Massacre, we knew there were grounds for believing that he and some of his closest aides had been bribed by ITT, the dairy industry, and an international rogue (Vesco). We knew he had cheated on his income taxes and lavished public funds on his private estates. He had hired rather more thugs than necessary and had (in Madison's words describing a ground for impeachment) "neglected to check the excesses" of his hirelings. He fired a special prosecutor and was fighting like a wounded puma to hide tapes, which obviously did not contain evidence of innocence. By April 30, 1974, when the transcripts were released, we knew he had plotted obstruction of justice, urging blackmail payments and perjury. Yet most congressmen still were talking about the need to find a "smoking gun."

The unadorned truth is that if Nixon's lawyers had not sent the Senate Watergate committee a memo containing an exact quote from a Nixon-Dean meeting, the committee staff probably would not almost inadvertently have caused Alexander Butterfield to volunteer information about the taping system. And Nixon would still be President.

Woodward and Bernstein have a nice sense of such contingencies, and they have a gift for details that illuminate a narrative like lightning flashes. For example: Kissinger despised Secretary of State William Rogers and delighted in humiliating him. Haig relished recounting how, in 1971, when he was Kissinger's obedient servant, Kissinger dispatched him to Rogers's office to make sure the secretary's television was on when Nixon made the surprise announcement that he would visit China. As Haig would tell the story, Rogers, who had no knowledge that Kissinger's negotiations with the Chinese had succeeded, was mortified. And Kissinger was tickled by Rogers's suffering. (Rogers disputes Haig's story.)

Haig joked about a homosexual relationship between Nixon ("our drunken friend") and Bebe Rebozo, and imitated what he called Nixon's limp-wrist manner. Kissinger sneered at "our meatball President" who couldn't fathom "anything more complicated than a *Reader's Digest* article" and who would cause "a nuclear war every week" were it not for Kissinger. Kissinger cultivated movie stars, returning their calls before returning Nixon's calls. Kissinger reviled a colleague as a "psychopathic homosexual," and squabbled with Haig for the suite next to Nixon's in the Kremlin.

Ken Clawson, the White House's foremost public-relations expert, said the day before the transcripts were released: "Watergate is going to go away tomorrow." Nixon's inept attorney, James St. Clair, was so ignorant of Watergate details that another lawyer had to explain references in the "smoking gun" tape of June 23, 1972. Columnist Joseph Alsop used breakfast, lunch and intimations of Armageddon to lobby tormented Representative Tom Railsback of Illinois, a key Republican on the impeachment panel: Railsback must support Nixon for the nation's sake. (Railsback didn't.)

Senator Hugh Scott fidgeted while J. Fred Buzhardt, a Nixon lawyer, administered a bit of preventive blackmail: support Nixon or the world may learn about Scott's abuses of power in influencing awards of government jobs and contracts. John Ehrlichman, who learned something about blackmail at the hands of Howard Hunt, called Julie Eisenhower the night before the resignation: her father should pardon him or he might embarrass her father.

Pounding the rug sobbing "What have I done? What has happened?" her father had been pounded shapeless on the anvil of Watergate. A hollow man accustomed to leaning on stuffed men, he had watched his brave daughter Julie crisscross the nation defending him against charges he knew were true. By 1974, Nixon was no stranger to abasement. In the pathetic 1952 "Checkers speech," he invoked his wife's cloth coat. Then he was photographed weeping on Senator William Knowland's shoulder. At his "last press conference" in California in 1962 he whined that he would not be kicked around anymore. But by August 1974, he had been dragged along "the hang-out route" and millions of Americans had savored the distinctive tang of his private conversation. The transcripts were a best seller, and he was a laughing-stock.

The acid of resentment had ulcerated his personality until self-pity was his only unimpaired faculty. He had measured out his life in forkfuls of chicken à la king at banquets given by strangers. His long trek through the Valley of Humiliation had brought him to the White House East Room on August 9, 1974, to tell his staff, and a national television audience, that his father had owned "the poorest lemon ranch in California." "Nobody will ever write a book, probably, about my mother." "I am not educated, but I do read books." Wearing glasses for the first time in public, he read Teddy Roosevelt's words about the death of his first wife: "And when my heart's dearest died, the light went from my life forever." Nixon was equating the loss of a political office with the death of a wife. Today, as one sifts the bleached bones of the Nixon presidency, that last grotesquery still startles.

[April 5, 1976]

Alexander Bickel, Public Philosopher

In 1969 Professor Alexander Bickel of the Yale Law School was invited to address a gathering of Yale alumni on the subject "What is happening to morality today?" He said: "It threatens to engulf us."

He meant that we are living in "an age of assaultive politics." The legal order is battered by "a prodigality of moral clauses," each of which is immoderately righteous, and gifted at rationalizing disobedience of the law and disregard of the traditions of civility. Bickel returned to this theme in the January 1974 issue of *Commentary* magazine, in the most brilliant political essay of the year, "Watergate and the Legal Order."

He argued "that much of what happened to the legal and social order in the fifteen years or so before Watergate was prologue." In those years three distinct groups—white southern militants, the civil rights movement, the white middle class anti-war movement—preached disobedience to law, and practiced what they preached.

Watergate, Bickel said, was not a radical departure from the course of our recent history. Rather, it was a manifestation of the radicalism that, for fifteen years, in various guises, had been challenging "the premise of our legal order," the idea that the complicated arrangements of the legal order are themselves "more important than any momentary objective."

Bickel's most acute—and acutely resented—perception was this: the impatient, righteous, anti-institutional impulse that helped produce Watergate had been much in evidence—and much applauded—in government itself, since the mid-Fifties.

"The assault upon the legal order by moral imperatives wasn't only, or perhaps even the most effectively, an assault from the outside. It came as well from within, in the Supreme Court headed for fifteen years by Earl Warren. . . . More than once, and in some of its most important actions, the Warren Court got over doctrinal difficulties or issues of the allocation of competences among various institutions by asking what it viewed as a decisive practical question: if the Court did not take a certain action which was *right* and *good,* would other institutions do so, given political realities? The Warren Court took the greatest pride in cutting through legal technicalities, in piercing through procedure to substance. But legal technicalities are the stuff of law, and piercing through a particular substance to get to procedures suitable to many substances is in fact what the task of law most often is."

The "derogators of procedure and of technicalities, and other anti-institutional forces, rode high, on the bench as well as off." In a democracy the derogation of concrete institutions inevitably becomes a populist celebration of an abstraction—"the people." Thus, as Bickel noted, it was "utterly inevitable that such a populist fixation should tend toward the concentration of power in that single institution which has the most immediate link to the largest constituency"—the U.S. presidency.

So, Bickel said, we wound up with "a Gaullist presidency . . . needing no excuse for aggregating power to itself beside the excuse that it could do more effectively what other institutions, particularly Congress, did not do very rapidly or very well, or under particular political circumstances would not do at all. This was a leaf from the Warren Court's book. . . . I don't know when Mr. Nixon caught the liberals bathing, but he did walk off with their clothes, and stood forth wearing the plebiscitary presidency. . . ."

The truth that Bickel wanted us to see, before it is too late, is that Watergate was an episode in what is becoming a tradition. It was an eruption, in a new form, of a familiar anti-institutional righteousness, the assaultive politics of the populist impulse. Thus Watergate, although past, is prologue, part of the engulfing stream of moral righteousness.

But the truth Bickel wanted us to see is an unwelcome, and hence an unheeded, truth in this year-end atmosphere of national self-congratulation about "surviving" Watergate.

Hell, Hobbes said, is truth seen too late. Republics—at least fortunate republics—can be saved from damnation by a few constitutionalists like Bickel. But threats to republics are many and constant. Great constitutionalists are few and mortal. Alexander Bickel, the keenest public philosopher of our time, died of cancer late in this, his forty-ninth, year.

[December 26, 1974]

PART 2

Issues

How Far out of the Closet?

On June 7 Miamians will vote on repeal of an ordinance banning discrimination in housing, jobs or public accommodations based on "affectional or sexual preferences." There is a lot of ideological freight packed into the idea that homosexuality is merely a "sexual preference," and echoes of the argument about Miami's ordinance are traveling through the nation like cantering horses.

Homosexuals have come out of their closets and into society. The civil service and many corporations have announced that they do not discriminate against homosexuals. There are homosexual bars (many thousands of them), movies, campus organizations, hotels, resorts, apartments, churches, newspapers, clothing stores, neighborhoods, bookstores, synagogues, beaches, dances, bathhouses, Alcoholics Anonymous groups, softball teams and even a credit union. Avowed homosexuals have been elected to state legislatures. Thus Professor Ernest van den Haag, a philosopher, notes that "by no stretch of the imagination is the situation of homosexuals analogous to that of blacks. . . . There is little actual employment discrimination against them, and it is rapidly diminishing. . . . The attempt to restrict employer freedom in their favor appears to be motivated by a wish for equal recognition, for equal prestige, rather than by a grievance about material disadvantage."

Homosexual militants correctly regard ordinances like Miami's as a form of ideological militancy. The ordinances are weapons in a battle to force society formally to indicate that homosexuality is a matter of indifference. They are declaratory acts announcing that homosexuals' differences are not things about which society can care. Sexual appetites are as morally irrelevant as skin color, so people must be coerced into disregarding homosexuality when, say, hiring teachers.

Ordinances like Miami's are part of the moral disarmament of society. Once they establish society's official indifference to homosexuality, society will be hard put to find grounds for denying homosexuals the right to marry. (If ratified, the Equal Rights Amendment might easily be construed to invalidate laws prohibiting homosexual marriages. Homosexuals complain that such laws constitute discrimination on the basis of sex, discrimination that denies them tax, social security and inheritance advantages, "family" health insurance and other benefits.) Next will come the right of homosexuals to adopt children, to have homosexuality "fairly represented" as an "alternative life-style" in every child's sex-education classes, and in literature in public libraries.

If you think that is a caricature of possibilities, consider the case of the Minneapolis man who applied to be a Big Brothers companion to a fatherless boy. He listed as a reference a renowned homosexual who had "married" his male roommate in college, so officials asked the applicant if he was a homosexual. He said his "serious affections" were for males, so the officials said they would tell any mother before he could be Big Brother to her son. He cried "discrimination!" and city civil-rights officials suggested Big Brothers should stop inquiring about "sexual preferences"; that Big Brothers staffers should attend seminars given by homosexuals; and suggested that Big Brothers should undertake "affirmative action" by advertising for volunteers in homosexual periodicals.

Now, the idea that homosexuals are bent on recruiting children is a canard. Further, the causes of homosexuality are a mystery. There is no evidence that it is congenital. There do seem to be some correlations between homosexuality and particular kinds of parental relations. A divided American Psychiatric Association has removed homosexuality from its list of "mental disorders," conceding only that unhappy homosexuals suffer "a sexual-orientation disturbance." This brought labels into line with the modern theory that all notions of moral normality are "mere" conventions, or utterly idiosyncratic.

But surely homosexuality is an injury to healthy functioning, a distortion of personality. And the grounds for believing that it is a socially acquired inclination are reasons for prudence. To the extent that homosexuality is, in some sense, a "choice" of character, as many homosexuals insist, then that choice may be influenced by various things, including a social atmosphere of indifference, or sustained exposure to homosexual role models, such as teachers.

Much advocacy of the homosexuals' cause involves what van den Haag calls "the Kinsey syndrome." The fallacious idea is that if behavior, such as homosexuality, is statistically "normal," meaning only that it is frequent, then it is or ought to be approved by moral norms. Van den Haag notes: "Kinsey originally was an entomologist, and in insect societies there are no moral norms separate from biological reflexes or statistical frequencies."

The Kinsey syndrome is in tune with the times. To the extent that there is today something that can be called "the public philosophy" it is: "Different strokes for different folks." The notion that no form of sexuality is more natural, more *right* than any other is part of something larger. It is a facet of the repudiation of the doctrine of natural right on which Western society rests. According to that doctrine, we can know and should encourage *some* ways of living that are right because of the nature of man.

A traditional function of law is to point people toward more human ways of living, to shore up what the community considers essential values. That a particular value needs no support from the law is an empirical claim, perhaps wrong, but arguable. That there are *no* essential values, or none that is any of the law's business, is as absurd as the idea of a polity with no notion of "the public good." True, a liberal society concerns itself with a minimum of essentials, but surely healthy sexuality is one; the family, and hence much else, depends on it.

It will be said that a society like this one, which seems indifferent to so much, *should* be indifferent to homosexuality. A society in which pornography is a constitutionally protected growth industry cannot convincingly condemn homosexuality merely because homosexuality often reduces sex to the physical. The homosexual subculture based on brief, barren assignations is, in part, a dark mirror of the sex-obsessed majority culture.

But a society swept by the trendy thought that "liberation" from "mere" conventions is an inherent good soon finds that its values have been reduced to desiccated concepts like "change" and "free choice of life-styles." But *especially* in such a society many people want a few rocks to cling to in the riptide that washes away old moral moorings. Opposition to Miami's ordinance is a way of saying "Enough!" And it is eminently defensible.

[May 30, 1977]

In Cold Blood

Much opposition to capital punishment is, like mine, a strong emotion searching uneasily for satisfactory reasons to justify it. Such reasons cannot be found in the Constitution.

The founders did not consider capital punishment "cruel and unusual" and neither does today's nation: since 1972, thirty-five states have enacted death penalties. The Supreme Court says the ban on "cruel and unusual" punishment must draw its meaning from "evolving standards of decency." Those standards do evolve: the First Congress passed a statute prescribing thirty-nine lashes for larceny, and one hour in the pillory for perjury. And one prescient congressman opposed the "cruel and unusual" clause because it might someday be construed to ban such "necessary" punishments as ear-cropping. Someday capital punishment may offend the evolving consensus.

But, then, someday mutilation may again be acceptable. Mutilation (castration of sex offenders; removal of brain segments from the unmanageably deranged) is still practiced in some Western societies. Aggressive "behavior modification" techniques result from the "progressive" theory that sin is sickness, so crime ("deviant behavior") is disease. That theory prescribes therapy instead of punishment, and assigns the "curing" of criminals to persons C. S. Lewis called "official straighteners."

Today "progressives" oppose capital punishment, but it is not invariably a conservative policy. It was used aggressively in the United States, Britain and France when these liberal societies were in their most rationalist phases. They then had extravagant confidence in carefully calibrated punishments as means of social control. On the other hand, one can imagine a conservative like Dostoevski (who knew something about crime and punishment) disdaining capital punishment as a deterrent. Such a conservative would argue that people in this fallen world cannot transcend the impulse to sin; and it is impious to believe that even savage punishment can overcome that impulse, and thus do the work of God's grace. Dostoevski did say:

". . . that evil is buried more deeply in humanity than the cure-all socialists think, that evil cannot be avoided in any organization of society, that a man's soul will remain the same, that it is from a man's soul alone that abnormality and sin arise, and that the laws that govern man's spirit are still so unknown, so uncertain and so mysterious that there cannot be any physicians, or even judges, to give a definitive cure or decision."

Today, "cure-all progressives" oppose capital punishment on the ground that crime is a product of individual or social pathology and therefore rehabilitation is the only just purpose of punishment. The logic of this theory also discounts the deterrent value of punishment: no sickness—crime no more than the common cold—can be deterred by threats. But as Robert Bork, Solicitor General of the United States, says, "The assertion that punishment does not deter runs contrary to the common sense of the community and is, perhaps for that reason, a tenet fiercely held by a number of social scientists."

In fact, there is ample evidence that the rates of many specific crimes are related negatively to the likelihood of punishment, and its severity. Capital punishment surely would deter double parking. And one elaborate statistical study suggests that capital punishment deters murder, that each execution may save as many as eight lives. But because such studies must grapple with many variables, and because murder is frequently a crime of passion, not calculation, the most that can be said confidently is that it is not clear that capital punishment does not deter murder.

The silliest argument against capital punishment is that, deterrent or not, it is wrong because it is "retributive." *All* punishment is retribution. The Marquis of Halifax was delightfully quotable but wrong when he said: "Men are not hanged for stealing horses, but that horses may not be stolen." The point to be made against Halifax is a logical, not an ideological, point. It is that the word "punishment" is only properly used to describe suffering inflicted by authority in response to an offense. Punishment is always (to use the archaic verb) "to retribute," to "pay back for" guilt.

Although concern for rehabilitation or deterrence *may* influence how a crime is punished, punishment *must* have a retributive dimension. Thus "the object so sublime . . . is to make the punishment fit the crime." And a "fit" is a rough proportionality between what the criminal suffers and what his victim suffered.

The complaint that capital punishment is "retributive" is a muddled way of charging that it is disproportionate. It occasionally has been grotesquely so: British laws prescribed death for damaging Westminster Bridge or impersonating a Chelsea pensioner. But the nub of the matter still is: is death a disproportionate response to murder?

Many thoughtful persons argue that categorical opposition to capital punishment even for murder depreciates life. They say, rightly, that one function of law is to affirm and thereby reinforce values, and that one way law should do this is by making punishments "fit" crimes, not criminals. They say, rightly, that society's justified anger should be tamed and shaped by law into just retribution. And they also insist that to punish the taking of life with less than death reinforces the modern devaluing of life. Society, they say, must take lives to demonstrate that it properly reverses life.

These people understand the problem, if not necessarily the solution. The problem is the cheapening of life in our time. But other persons of sobriety reasonably regard capital punishment as part of the problem, not of the solution.

These are interesting times. The nation's security rests on its ability to deter aggression by treating its principal enemy's civil population as hostages; that is, national security rests on the credible threat of a form of warfare universally condemned since the Dark Ages, the wholesale slaughter of noncombatants. The nation's capital has just become the nation's first city in which abortions exceed births. The growing torrent of violence and other pornography in popular entertainment is both cause and effect of the desensitization of the nation.

Now a Texas convict is demanding prompt execution and asking that his death be televised. He says the spectacle would shatter support for capital punishment. Perhaps. But I detect no evident standards that would prevent, say, NBC from making an execution part of "The Big Event" and finding ample sponsors from among the companies subsidizing today's television violence.

Pending more powerful evidence that capital punishment is a powerful deterrent saving innocent lives, the burden of proof is still on those who say that today the valuation of life can be enhanced by violent deaths inflicted by the state, in private, in cold blood.

[November 29, 1976]

Discretionary Killing

It is neither surprising nor regrettable that the abortion epidemic alarms many thoughtful people. Last year there were a million legal abortions in the United States and fifty million worldwide. The killing of fetuses on this scale is a revolution against the judgment of generations. And this revolution in favor of discretionary killing has not run its course.

That life begins at conception is not disputable. The dispute concerns when, if ever, abortion is a *victimless* act. A nine-week-old fetus has a brain, organs, palm creases, fingerprints. But when, if ever, does a fetus acquire another human attribute, the right to life?

The Supreme Court has decreed that *at no point* are fetuses "persons in the whole sense." The constitutional status of fetuses is different in the third trimester of pregnancy. States constitutionally can, but need not, prohibit the killing of fetuses after "viability" (twenty-four to twenty-eight weeks), which the Court says is when a fetus can lead a "meaningful" life outside the womb. (The Court has not revealed its criterion of "meaningfulness.") But states cannot ban the killing of a viable fetus when that is necessary to protect a woman's health from harm, which can be construed broadly to include "distress." The essence of the Court's position is that the "right to privacy" means a mother (interestingly, that is how the Court refers to a woman carrying a fetus) may deny a fetus life in order that she may lead the life she prefers.

Most abortions kill fetuses that were accidentally conceived. Abortion also is used by couples who want a child, but not the one gestating. Chromosome studies of fetal cells taken from amniotic fluid enable prenatal diagnosis of genetic defects and diseases that produce physical and mental handicaps. Some couples, especially those who already have handicapped children, use such diagnosis to screen pregnancies.

New diagnostic techniques should give pause to persons who would use a constitutional amendment to codify their blanket opposition to abortion. About fourteen weeks after conception expectant parents can know with virtual certainty that their child, if born, will die by age four of Tay-Sachs disease, having become deaf, blind and paralyzed. Other comparably dreadful afflictions can be detected near the end of the first trimester or early in the second. When such suffering is the alternative to abortion, abortion is not obviously the greater evil.

Unfortunately, morals often follow technologies, and new diagnostic and manipulative skills will stimulate some diseased dreams. Geneticist Bentley Glass, in a presidential address to the American Association for the Advancement of Science, looked forward to the day when government may require what science makes possible: "No parents will in that future time have a right to burden society with a malformed or a mentally incompetent child."

At a 1972 conference some eminent scientists argued that infants with Down's syndrome are a social burden and should be killed, when possible, by "negative euthanasia," the denial of aid needed for survival. It was the morally deformed condemning the genetically defective. Who will they condemn next? Old people, although easier to abandon, can be more inconvenient than unwanted children. Scientific advances against degenerative diseases will enable old people to (as will be said) "exist" longer. The argument for the discretionary killing of these burdensome folks will be that "mere" existence, not "meaningful" life, would be ended by euthanasia.

The day is coming when an infertile woman will be able to have a laboratory-grown embryo implanted in her uterus. Then there will be the "surplus embryo problem." Dr. Donald Gould, a British science writer, wonders: "What happens to the embryos which are discarded at the end of the day—washed down the sink?" Dr. Leon R. Kass, a University of Chicago biologist, wonders: "Who decides what are the grounds for discard? What if there is another recipient available who wishes to have the otherwise unwanted embryo? Whose embryos are they? The woman's? The couple's? The geneticist's? The obstetrician's? The Ford Foundation's? . . . Shall we say that discarding laboratory-grown embryos is a matter solely between a doctor and his plumber?"

But for now the issue is abortion, and it is being trivialized by cant about "a woman's right to control her body." Dr. Kass notes that "the fetus simply is not a mere part of a woman's body. One need only consider whether a woman can ethically take thalidomide while pregnant to see that this is so." Dr. Kass is especially impatient with the argument that a fetus with a heartbeat and brain activity "is indistinguishable from a tumor in the uterus, a wart on the nose, or a hamburger in the stomach." But that argument is necessary to justify discretionary killing of fetuses on the current scale, and some of the experiments that some scientists want to perform on live fetuses.

Abortion advocates have speech quirks that may betray qualms. Homeowners kill crabgrass. Abortionists kill fetuses. Homeowners do not speak of "terminating" crabgrass. But Planned Parenthood of New York City, which evidently regards abortion as just another form of birth control, has published an abortion guide that uses the word "kill" only twice, once to say what some women did to themselves before legalized abortion, and once to describe what some contraceptives do to sperm. But when referring to the killing of fetuses, the book, like abortion advocates generally, uses only euphemisms, like "termination of potential life."

Abortion advocates become interestingly indignant when opponents display photographs of the well-formed feet and hands of a nine-week-old fetus. People avoid correct words and object to accurate photographs because they are uneasy about saying and seeing what abortion is. It is *not* the "termination" of a hamburger in the stomach.

And the casual manipulation of life is not harmless. As Dr. Kass says: "We have paid some high prices for the technological conquest of nature, but none so high as the intellectual and spiritual costs of seeing nature as mere material for our manipulation, exploitation and transformation. With the powers for biological engineering now gathering, there will be splendid new opportunities for a similar degradation of our view of man. Indeed, we are already witnessing the erosion of our idea of man as something splendid or divine, as a creature with freedom and dignity. And clearly, if we come to see ourselves as meat, then meat we shall become."

Politics has paved the way for this degradation. Meat we already have become, at Ypres and Verdun, Dresden and Hiroshima, Auschwitz and the Gulag. Is it a coincidence that this century, which is distinguished for science and war and totalitarianism, also is the dawn of the abortion age?

[September 20, 1976]

Freedom and the Busing Quagmire

President Ford is being criticized for "raising" the busing issue, as though the issue has been dormant, and as though it is unseemly for elected officials to interfere with the judiciary's formulation of the nation's school and racial policies.

Ford seems to understand that the only reason we cannot say for sure that busing is a failure is that it is not clear what busing is supposed to achieve. Ford's remedy is to stipulate precisely the wrong that busing can be used to right, and to limit the duration of any busing program.

Ford's bill would authorize busing only to correct *de jure* segregation, racial separation deliberately caused by local decisions, as in drawing school zones or locating school construction. That is distinguished from *de facto* "segregation," the result of living patterns. The bill also would require courts to "make specific findings concerning the degree to which the [minority] concentration . . . in the student population of particular schools affected by unlawful acts of discrimination presently varies from what it would have been in normal course had no such acts occurred." If that determination is impossible, courts would be limited to busing to reproduce only the degree of integration that would have existed in the schools had there been no de jure segregation. They could not aim, as Boston's Judge Garrity has aimed, at sweeping district-wide integration that would not have occurred "in normal course." And even the limited busing programs that could be based on such precise findings could continue only five years.

The remarkable thing is not Ford's proposal, but that the meandering path of the law since 1954 has made such a common-sense proposal necessary.

In 1954, in *Brown v. Board of Education,* the Supreme Court seemed to say that states cannot classify pupils racially. But today many school districts are under court orders to do precisely that. In Boston and elsewhere, busing for desegregation is being superseded by busing for integration, understood as "racial balance." *Brown* seemed to mean that no child would be barred from a school because of race. Today court orders exclude many white children from their neighborhood schools solely because they are white.

The government has wandered waist-deep into the busing quagmire because no coherent reasoning has shaped policy since 1954. *Brown* held that "separate" school *systems* are "inherently" unequal. Obviously dual school systems had to be dismantled. But what *is* a dual school system, and *when* is it dismantled? *Brown* did not explicitly answer these questions, in part because the answers seemed obvious. Dual systems were those where segregation had been explicit policy; dismantling meant making the law "color blind."

But since then, persons ardently in favor of coerced integration as distinct from coerced desegregation have accomplished a semantic sleight of hand. In 1954 "segregation" meant compulsory separation of races. Today it is a synonym for "absence of integration," which has its own synonym: "racial isolation." Having tailored the terminology to blur distinctions, busing enthusiasts argue as follows:

If "separate" *school systems* are "inherently" unequal, surely "separate" *schooling* is, too. So "racial isolation" in any school violates black children's right to "equal protection of the law."

Ample scholarship disputes, and none unambiguously supports, the tattered theory that compulsory mingling of white and black children improves academic attainment or racial harmony. Sociology is catching up with common sense. But some sociological abstractions survive long enough to produce constitutional absurdities. Today there are judges and others who seem convinced that once a scintilla of de jure segregation has been found in a system, the Constitution mandates that no school shall be all white or more than half black.

If, as they believe, all that matters is the allegedly harmful consequences of "racial isolation" or the allegedly beneficial consequences of "racial balance," then it is pointless to distinguish between de jure and de facto segregation. All that matters is the presumed effect of "racial isolation," not whether it was produced deliberately by state action or accidentally by uncoerced living patterns. Thus the "equal protection" clause licenses federal judges to shuffle children in buses, in perpetuity.

But, a new Supreme Court decision indicates reason is staging a comeback.

Having found de jure segregation in Pasadena, California, in 1970, an especially exuberant federal judge declared that "at least during my lifetime there would be no majority of any minority in any school in Pasadena," and ordered local officials to remedy "racial imbalance." Pasadena satisfied the "no majority" standard in 1970. But by 1974 the normal movement of people into, away from, and within the district had produced black majorities in five of thirty-two schools, and officials were ordered to comply again with the "no majority of a minority" standard. Last week the Supreme Court ruled that Pasadena need not rezone its school system every year. The Court said school authorities need not alter racial patterns that they do not cause. That is Ford's position.

The complicated shiftings of urban populations, as in Pasadena, account for the special importance of Ford's proposal that courts should determine the *specific* extent to which minority concentrations in *particular schools* are the result of deliberate segregative acts. That is a stiffer standard than the one Judge Garrity used to convulse Boston. Drawing on the Supreme Court's language in a 1973 Denver case, Garrity ruled that de jure segregation anywhere in a school system justifies the assumption that "racial isolation" anywhere in the system is deliberate. That assumption justifies "remedies" inflicted on the entire system.

Ford's proposal that judges be limited to rectifying just the specific effects of intentionally segregative acts would prevent them from using busing as a perpetual churning device to blend society. It would prevent judges from altering enrollments when those enrollments reflect living patterns displeasing to judges like Detroit's Roth. Roth has sniffily scolded blacks because they, "like ethnic groups in the past, have tended to separate themselves and associate together."

Ford's critics say that his proposal would make it harder to begin and perpetuate busing. I say: Good. Busing is a kind of conscription. Any conscription is a significant excision from American freedom and can be justified only rarely, and for clear, urgent goals. Ford's critics say that his proposal will block judicial attempts to enforce a constitutional mandate for an integrated society. Again: Good. The Constitution mandates a free, not a "racially balanced," society.

[July 12, 1976]

Common Sense on Race

The Supreme Court has just handed down a decision about the perennial American problem, race. The Court said nothing astonishing. Indeed, what is astonishing is that confusion about the nation's premises is so substantial and widespread that the court was compelled to say what should be obvious. The Court, in effect, said this: the fact that a suburb is predominantly white does not impose upon it a constitutional duty to alter its racial composition.

Arlington Heights is a predominantly white suburb of Chicago. In 1971, a Chicago development corporation decided to build subsidized housing there for low- and moderate-income tenants. But the decision was contingent upon Arlington Heights' rezoning a particular tract of land to permit multiple-family dwellings. The rezoning request was debated at several well-attended hearings, and then refused.

The developer and several individuals sued. The district court upheld the refusal, noting that the record of the hearings indicated that it was based on two legitimate reasons. First, construction of the housing project would depress property values in a neighborhood where people had purchased property in the expectation that such construction would not occur. And this expectation was reasonable because Arlington Heights' long-standing policy was not to permit multi-family dwellings in such areas.

Next, the court of appeals agreed that the rezoning refusal was not racially motivated, but said the "ultimate effect" of the refusal was "racially discriminatory" because it had a disproportionate impact on blacks. Based on family income, blacks constituted 40 percent of the Chicago-area residents eligible for such subsidized housing, although blacks constitute a much lower percentage of the Chicago-area population.

The court of appeals even declared that Arlington Heights was "exploiting" the growth of its area by allowing itself to *remain* so predominantly white. The court found intolerable the discriminatory *effect* of the community's refusal to take affirmative action (revision of zoning policy) likely to increase the number of black residents. So the court ruled that the community's adherence to its traditional zoning policy violates the equal-protection clause of the Fourteenth Amendment.

The Supreme Court's rejection of this tortured argument was refreshingly crisp. The Court held that even when an official policy bears more heavily on one race than another, proof of a discriminatory *purpose* is required to prove a violation of the equal-protection clause.

The Court cited its own reasoning in a 1976 case involving two blacks whose applications to become District of Columbia policemen had been rejected. They challenged the constitutionality of a standard government literacy test on the ground that a higher percentage of blacks than whites failed the test. This meant, they said, that the test violated their right to "due process." But the court noted that the two blacks "could no more successfully claim that the test denied them equal protection than could white applicants who also failed." And the Court stated the principle that was to govern the Arlington Heights case: the *racially disproportionate impact* of a policy is without constitutional significance unless there is proof of a *racially discriminatory purpose.*

The Court could hardly have ruled otherwise. If it had, it would have imposed on government a bizarre and paralyzing new constitutional duty: before government could take any significant action, it would have to establish that the action probably would not have a disproportionate "racial impact." And in the immortal words of the late Ernest Bevin, "If you open up that Pandora's box you'll find a lot of Trojan horses inside."

The principle the Supreme Court affirmed in the Arlington Heights case is no more than common sense. But common sense is more than can confidently be expected from government today. After all, in the Arlington Heights case the court of appeals tried to impose on a *town* the sort of "affirmative action" that is odious when imposed on employers whose payrolls do not have a government-approved racial or sexual composition. That is, the court said, in effect, that Arlington Heights' traditional zoning standards must be changed because they help produce a community with the "wrong" racial statistics.

Today the government is pressuring many employers, such as universities, to end "underutilization" of this or that racial or ethnic or sexual group, and to bring their faculties and student bodies into line with this or that federal agency's statistical notion of justice. Such "affirmative action" decrees are the most serious violations of academic freedom in U.S. history. They abridge the academy's most fundamental freedom, the freedom to select its professorship solely in accordance with standards of scholarly excellence. And in attempting to regulate the racial and ethnic composition of student bodies, philistines from the government have tried to abridge the freedom of the university to control its curriculum. Sidney Hook has reported this example:

"At one Ivy League university, representatives of the Health, Education and Welfare Department demanded an explanation of why there were no women or minority students in the graduate department of religious studies. They were told that a reading knowledge of Hebrew and Greek was presupposed. Whereupon the representatives of HEW advised orally: 'Then end those old-fashioned programs that require irrelevant languages. And start up programs on relevant things which minority group students can study without learning languages.'"

Such know-nothingism is a stench in the nostrils of reasonable people, and not just because it repudiates the principle that government should be color-blind. It also repudiates the premise of our legal system and political order. That tenet is this: justice can only result from considering the rights and interests of *individuals*. The idea behind "affirmative action" programs is this: justice results from government's assigning rights and benefits to racial and ethnic *groups*.

Government is telling universities that they must take "affirmative action" to alter the racial and ethnic composition of their faculties and student bodies in order to approximate what the government considers the "right" balance of government-approved groups. In the Arlington Heights case, the court of appeals ruled, in effect, that that suburb has a constitutional duty to take "affirmative action" (to revise its zoning policy) in order to acquire a racial composition pleasing to that court. That the Supreme Court thwarted the court of appeals is gratifying. That the Supreme Court had to thwart it is depressing.

[January 24, 1977]

Marijuana and Moral Taxidermy

Mississippi, not known for restless modernity, has joined the growing list of states that have liberalized marijuana laws. As a legislator explains: "When your friends get caught with this stuff, or your relatives or their children, well . . ."

Most lawmakers are middle-class parents, and most marijuana users are young members of the middle class. Entrepreneurs supplying America's thirteen million regular users are engaged in organized crime on behalf of the middle class. Laws are no match for powerful appetites, and laws inconvenient to the middle class are doomed. Marijuana is ubiquitous and eventually will be legal.

"Decriminalization" (legalizing possession for use but not for sale) obviously will be just a step toward full legalization, then full commercialization through the American genius for mass production, advertising and marketing. And the attitudes and arguments bringing this about could also bring about the commercialization of cocaine.

The advance of such recreational drugs is irresistible, meaning that society lacks strong convictions that would cause it to resist. The "counter-culture" is no longer so "counter."

Marijuana's advocates who assert its harmlessness speak with more certainty that certitude. (It took many decades to establish the indictment of tobacco, and a decade ago next to nothing was known about marijuana.) Still, marijuana may indeed be less physiologically damaging than alcohol and tobacco.

On the other hand, the inadequacy of the ethic of individualism is apparent in the fact that people tend to worry about recreational drugs only in terms of damage to users' physiologies rather than damage to the community's character. The assumption is that, unless marijuana use involves dramatic pathologies, then alcohol and marijuana are comparable ways of satisfying comparable desires.

But Samuel McCracken, a professor of humanities, has argued that alcohol (like tobacco and coffee) is not comparable to marijuana in one crucial respect: use of marijuana is apt to be part of a strong self-definition by "true believers."

The difference is between "drugs of dependency" and "drugs of belief." Drugs of dependency are supposed to restore normal human attributes. Drugs of belief are supposed to confer extraordinary attributes.

Users of alcohol (or tobacco or coffee) depend on it for help in acquiring attributes (such as tranquility, poise, alertness) that, ideally, people possess without chemical assistance. Users of marijuana believe in its ability to confer capacities (special sensitivity, creativity, understanding) impossible for non-users.

The marijuana user's distinctive phrase—"turning on"—expresses the assumption that human beings are mechanical devices which need chemical additives in order to attain full life. For them, recreational "drugs of belief" are instruments of moral taxidermy, filling users with sensitivity.

Such drugs involve the conjunction of simple chemistry and simple-minded ideology. They generate sensations for persons who believe there is special merit in the passive absorption of such sensations.

Marijuana is the recreational drug of a generation born into postwar abundance and raised by parents who believed vaguely in "self-realization." The marijuana generation arrived at universities in the 1960s, and embraced the idea that standards external to the individual (grades, tests, required curricula) were "repressive" impediments to "self-realization."

Marijuana was an emblem of the anti-institutional sensibility of the 1960s, part of the cult of the expansive ego. The 1960s preferred music loud (The Rolling Stones), lyrics more elemental than grammatical ("I can't get no satisfaction"), theater adolescent *(Hair),* cinema shocking *(MASH, Bonnie and Clyde).* The 1960s celebrated senses over mind, believing in "sensitivity training" and "consciousness raising." The big book was *The Greening of America,* a hymn to Consciousness III.

It was, as historian Richard Hofstadter said, an age of intellectual rubbish, including the beliefs that came with recreational "drugs of belief."

Commercialization of such drugs will express society's indifference to the drug culture's baggage of beliefs. Inevitably, the "progressive" person will paraphrase Jefferson to justify indifference. He will say that no belief or drug "picks my pocket or breaks my leg." Such words express indifference to all but physical and property damage.

But self-government presupposes certain character traits, including moderation, reasonableness, discipline and other attributes of mind.

Any democratic government that neglects to nurture such traits, and is indifferent to the weakening of them, fails to maintain its cultural prerequisites. In the end, it will be no more durable than smoke.

[May 1, 1977]

God Went Too Far

It is a tale of two cities. Detroit, says Thomas Murphy, provides "what people want," but Washington is telling people, in effect, "you're not smart enough to know what's best for the country, so we've got to decide for you."

Murphy, chairman of General Motors, is understandably unhappy that henceforth the kinds of cars marketed will be determined more by government than by consumers. He is worried that Washington may impose stiff taxes on large cars. And he is alarmed by the law which stipulates that by 1985 each manufacturer's "fleet average" must be 27.5 miles per gallon.

The effect of that law will be to establish a small quota for the large cars the public emphatically prefers. A manufacturer might have to sell four subcompacts in order to keep its "fleet average" in balance and earn the right to sell another large car.

The only American-built car that currently gets at least 27.5 m.p.g. is Chevrolet's Chevette. It is a cute roller skate, but not what you would choose to drive from Baltimore to Yosemite, and not what a family of five or more (23 percent of families) can drive anywhere.

If Americans become unhappy about the size of new cars, they may start maintaining their 100 million old cars rather than trading them. Cut new car sales by even 10 percent and you will have an earthquake in the automotive and related industries such as steel, rubber, glass and rugs. Last year, manufacturers made fifty million square yards of rugs for automobiles.

For decades, the evolution of automobiles was at the expense of efficiency. Automatic transmissions decreased fuel efficiency 10 percent; air conditioning took another 10, as did emission standards and every additional 500 pounds. Each additional 100 horsepower took 7 percent.

Since 1974, GM's fuel efficiency has improved nearly 50 percent, and this year its big cars shrank a foot without sacrificing interior space. But there are limits to what engineering ingenuity can do in preserving dimensions while increasing efficiency. And the government believes that much more fuel efficiency is needed.

It is needed because no politically possible increase in the price of gasoline will cut demand substantially. Even at 60 cents a gallon, the average driver of a big car would save only $100 a year in gasoline costs by switching to a compact. So the key to substantial conservation is not trying to get people to drive less. It is getting them into more efficient cars.

But Murphy's question echoes across the land: what entitles government to censor consumer preferences? It is a serious question from a serious man, and the answer in part, is this:

Government exists not merely to serve individuals' immediate preferences, but to achieve collective purposes for an ongoing nation. Government, unlike the free market, has a duty to look far down the road and consider the interests of citizens yet unborn. The market has a remarkable ability to satisfy the desires of the day. But government has other, graver responsibilities, which include planning for the energy needs, military and economic, of the future.

Unfortunately, many citizens today think of themselves primarily as consumers, and think government's primary duty is to facilitate enjoyable consumption. So liberals, rallied by "consumer advocate" Ralph Nader, advocate a "consumer protection agency." Conservatives respond by championing "consumer sovereignty." Both sides seem to agree about one thing: citizens should be regarded, primarily, as consumers, and policy should serve consumption.

Advertising is a barometer of this climate, and ads for Johnnie Walker Scotch enjoin, biblically: "Honor Thy Self." The woman in the ad for L'Oréal hair coloring proclaims, defiantly: "I deserve it!" McDonald's declares, humbly: "You deserve a break today." The recurring message is: ever better consumption is no more than Americans deserve.

God (according to Jefferson) endowed mankind with an inalienable right to pursue happiness. But 201 years after Jefferson explained God's will, many Americans think happiness is a Buick Electra, and the government is hinting that God went too far.

Neither Detroit nor Washington is recognizably the City of God. But Washington is the city of government and is responsible for stipulating the national interest. So, increasingly, it will discourage some ways of pursuing happiness.

[April 17, 1977]

Risk-Taking on the Road

Thinking he heard thunder, my neighbor went to close his car windows. Actually, he had heard a commonplace tragedy, the making of a statistic. A woman died and a man nearly did in an occurrence shocking but routine: an automobile accident.

The car veered out of control on Connecticut Avenue, hit trees, fragmented, and broke in half. Three of us arrived immediately. Emergency equipment arrived quickly. Cleaning up took hours.

In 1900, this "village," six miles from the White House, was where Washingtonians came for country breezes. Today, it is a small incorporated area near the center of a sprawling metropolis, adjacent to Washington's city line. It is divided by Connecticut Avenue, which passes around a traffic circle as it enters Maryland. Trees on the circle are heavily scarred from crashes. Crumpling steel and crying sirens are common sounds here as on many urban thoroughfares.

Increasingly, American driving reflects, I think, the sublimated fury of persons heading for infuriating jobs, the animal spirits of persons whose lives allow little scope for such spirits. As Daniel P. Moynihan wrote years ago, the automobile is "both a symbol of aggression and a vehicle thereof. . . . It is a prime agent of risk-taking in a society that still values risk-taking, but does not provide many outlets."

The endless epidemic of accidents is one of the nation's gravest public health problems. Automobile deaths and injuries have costs beyond counting, and are a special plague to the young. Of every 100,000 males age fifteen, about 1,100 will die in accidents, most involving automobiles, before age twenty-five—a death rate twenty times worse than polio inflicted at its worst.

As Moynihan notes, the social life of most Americans "now primarily takes the form of driving to a place where alcohol is consumed." And because traffic laws are widely ignored, almost everyone is a lawbreaker, and the incidence of arrest in America may be the highest of any nation in history. Repairing and replacing wrecked cars may provide 20 percent of business for the automobile industry, the nation's most important.

Such statistics are as lifeless as the woman who lay beneath blankets on the Connecticut Avenue median strip. But they describe a river of sorrow flowing from monstrously irrational behavior.

Most drivers frequently exceed speed limits; only 25 percent use seatbelts; only 4 percent use harnesses. Because slaughter behind the wheel is deeply rooted in aggression and other irrationality, it is very difficult to substantially reduce accidents by reforming drivers. So government has tried to reduce the severity of injuries received in accidents.

The public disliked, and the government quickly disconnected, the ignition "interlock" system that prevented cars from starting when safety belts were unfastened. Today, new cars just make a brief buzz of disapproval.

Government may yet require "passive restraints"—air bags that instantly inflate to cushion passengers in collisions. There is evidence that they would save many thousands of lives annually and may be one answer to what Moynihan has called "the seeming incompatibility of safe driving and mass driving." That is a considerable problem in a nation where more people drive than pay taxes or vote.

Air bags require no forethought by drivers, so they are suited to the American driving public. The air bags would probably cost manufacturers less than $100, a fraction of what car buyers exuberantly spend when loading their cars with snappy wheel covers and other options.

Long before the most recent Connecticut Avenue death, I regretted having once argued that government has no business requiring drivers to buy and use inexpensive devices, like seat belts, that might save them from self-destruction. There is a pitiless abstractness, and disrespect for life, in such dogmatic respect for the right of consenting adults to behave in ways disastrous to themselves. Besides, too many children passengers are sacrificed on that altar. And a large part of the bill for the irrationality of individual drivers is paid by society.

Most important, society desensitizes itself by passively accepting so much carnage.

On Connecticut Avenue that evening, the police operated with the weary patience normal to those who are paid to look unblinkingly at what people do to themselves. "Go home," a policeman finally said, with barely noticeable disgust, to people milling around the debris. "Go home and watch television." After a while, we did.

[April 14, 1977]

Ennui, Decadence and the *Times*

Times Square, rotting core of the Big Apple, is home of this city's last growth industry, pornography. This industry has flourished in spite of a naive hope. And it represents the failure of a public policy which, like so many liberal policies, was based on nothing more than naive hope.

Some aspects of American life have become so vulgar that they can hardly be discussed without contributing to the coarsening of American life. Elbow your way through the dirty men in dirty raincoats in the crowded peepshows, and read for yourself the vivid descriptions of current fare.

These are no longer just seedy storefront operations. On Eighth Avenue there is a pornography supermarket, a Piggly Wiggly for perverts. It is located in a modern building owned and shared by a major bank. Pornography in this area has achieved new sliminess involving children and animals. And business is booming.

The pornographic impulse has found a grotesque new outlet in depictions of violence, as you can see at the large Broadway cinema presenting *Snuff*. This is an unspeakably crude, and successful, attempt to cash in on rumors that some South American films show real murders.

Until the final minutes *Snuff*, a Latin import, is a mishmash of shootings and knifings and general mayhem. This part is not much worse than what erupts on U.S. television at 9:01 each night (after the "family" hour), and it is no worse than such American films as *The Texas Chainsaw Massacre*. But the final minutes of *Snuff* show a man dismembering and disemboweling a live woman, using a knife, wireclippers, and an electric saw.

Feminist groups, with their wondrous knack for missing the larger point, denounce *Snuff* as an insult to women. *Variety*, the show business newspaper with a curious sense of fun, calls *Snuff* the "spoof of the year."

Closed-circuit television on the street in front of the theater coyly says *Snuff* could only have been made in South America, "where life is cheap." Assuming that the final murder is a fake (and it almost certainly is), the advertising is patently misleading. Think about this. The path of American law has been steadily downhill to this point: today pornographers need fear no law, except, perhaps, consumer protection laws which conceivably could insist that films promising a real murder must show a real murder.

The final murder, whether real or fake, is repulsive enough to convince the dismal viewers who want to believe it is real. That includes most of the people who shell out $4 to see it. When I saw it, the final butchery left the viewers tittering merrily, asking one another "It was real, wasn't it?"

The audience would have been surly if the murder had been unconvincing. Remember a decade ago, the Connecticut woman who went to see *I Am Curious, Yellow*? That Swedish film was a great shocker in the quaint 1960s, even though it only showed simulated sexual intercourse. That wasn't enough for the Connecticut woman, who indignantly declared: "I paid for filth and didn't get filth!"

The *New York Times* is located about a block from the *Snuff* theater. The *Times* has not been hostile to the court decisions that have virtually obliterated all legally enforceable standards of public decency. But the *Times* is unhappy about being in the middle of the open sewer of pornography that midtown Manhattan has become. In 1969, when pornography was comparatively tame, the *Times* voiced a familiar hope. It was the empty hope that is the usual rationalization offered by people who are appalled by censorship and by the predictable consequences of no censorship:

"The insensate pursuit of the urge to shock, carried from one excess to a more abysmal one, is bound to achieve its own antidote in total boredom. When there is no lower depth to descend to, ennui will erase the problem."

What rubbish. *Snuff* is the lowest depth so far, but human ingenuity will find lower depths to conquer. Perhaps pornographers will reach the "lowest possible" depth that the *Times* anticipates with equanimity. But if that happens, and if the *Times*' hope is realized, our society will be moved only to boredom by (say) films of real murders. Another name for such ennui is decadence.

[March 28, 1976]

The Desensitization of America

Mayor Richard Daley is "asking" Chicago's City Council (that is like Napoleon asking a drummer boy to drum) for an ordinance that would prohibit anyone under eighteen from seeing films consisting substantially of "assaults, cuttings, stabbings, shootings, beatings, sluggings, floggings, eye gougings, brutal kickings, burnings, dismemberments," among other things.

Supporters of the ordinance say such films are "training films" for potentially violent children. Libertarians respond, reflexively, that an ordinance keeping children from seeing *The Texas Chainsaw Massacre* is the thin end of the totalitarian wedge. Both arguments are weak.

Libertarians should wonder whether a desensitized society will be sensitive to any civilized values, including those that sustain the First Amendment. And those advocating the ordinance should note that two ambitious investigations—those for the Commission on Causes and Prevention of Violence and the Commission on Obscenity and Pornography—failed to find evidence that depictions of violence "teach" children violent behavior, except the innocuous behavior of play.

Occasionally, depictions of violence lead directly to emulative violence, as in Boston in 1973, when a woman was set afire with gasoline the night after a local television station showed a movie featuring a similar atrocity. But such episodes, however horrible, do not define the problem of depicted violence.

Such isolated correlations *are* important. So, too, is the point made by Professor James Q. Wilson:

"It would have been as difficult for a social scientist in Nero's Rome to prove that a person who himself did not read erotic stone tablets was thereby benefited as it would have been for a social scientist in Cromwell's London to prove that a person who did read dirty books was thereby harmed; yet most historians would agree that the different attitudes toward public obscenity in Nero's Rome and Cromwell's London are not irrelevant to understanding those very different societies."

For an understanding of our society, view the stunning presentation prepared by J. Walter Thompson, the huge advertising agency. *The Desensitization of America* uses film, slides, music and narration to sledgehammer the viewer with the "sensory overload" of increasing violence and vulgarity in music, magazines, television and movies.

The presentation spares nothing, not television game shows ("The qualifications of a contestant include greed, exhibitionism, and a gift for going bananas at a moment's notice"), not the stridencies of "family comedies" like "Maude" and "All in the Family," or the splashy violence on "Starsky and Hutch." The presentation argues (among other things) that television is the cause of even worse effects:

"As long as television gives entertainment away free, Hollywood is in a special kind of trouble—it has to try to do something that television can't or won't." And: "What seems to be at work is the basic law of pop culture in America in the 1970s. You can't survive unless you top yourself. Each sensation has to be replaced by a greater sensation. Inflation is the law of the land."

Thus, Al Goldstein (a pornographer who has been associated with a newspaper the name of which does not belong in newspapers) declares: "We can no longer make money in pornography alone. I am now selling tastelessness." That may strike you as a distinction without a difference until you open *Hustler,* a magazine founded by a dreadful young entrepreneur who had the audacity to think that hordes of Americans will enrich someone who will fill a magazine with material too vulgar for *Playboy.* The wages of such sin are stupendous. *Hustler* is expected to turn a $13 million profit this year.

But don't look in *Hustler,* look across your living room. Recently, at 9 P.M. Eastern time, the three networks offered the movie *Slaughterhouse Five;* a dramatization of the Charles Manson murders; and *Streets of San Francisco* featuring a double rape (featuring it twice, first in the "teaser" at the opening of the show).

The message of the J. Walter Thompson presentation is that companies should ask more of a program they sponsor than that it attract a large audience: they should care how it attracts the audience. The presentation is a heartening example of an important business rising above the morals of the marketplace.

[May 27, 1976]

Pornographic Minds

Today's pornographers do not slight the life of the mind. Hugh Hefner of *Playboy* dabbles in metaphysics. And Larry Flynt, publisher of *Hustler,* has a sociological flair. He explains his magazine's success in terms of the American psyche, as revealed in the public's enthusiasm for the film *Jaws:*

"The shark wasn't repressed. The American people don't like repression."

It is impossible to describe *Hustler*—even to report what is printed on the cover—without becoming a collaborator in its assault on sensibility. Suffice it to say that Flynt has been a pathbreaker in the accelerating movement to make human sexuality resemble that of sharks and other unrepressed creatures free from "hangups."

But recently Flynt came a cropper in Cincinnati, a community now being roundly despised by libertarians. Flynt was fined and sentenced to seven to twenty-five years in prison for participating in "organized crime" (the crime of distributing obscenity), and was fined and sentenced to six months for "pandering obscenity."

The Supreme Court has ruled that there is no inherent incompatability between the First Amendment and statutes regulating obscenity. And it has held that juries can judge obscenity by "community" standards which, the Court's language suggests, can be statewide standards. There is little reason to doubt that *Hustler* would be declared obscene by community standards applied by any representative jury in any of the fifty states.

If Flynt and his lawyers did not know that, by Ohio standards, he was manufacturing and distributing obscenity, then he has foolish lawyers and his lawyers have a fool for a client. But, of course, he not only knew but boasted that he was producing obscenity unprecedented in a slick, mass-circulation magazine.

Much criticism of the *Hustler* ruling involves general objections to the "organized crime" charge. Critics also questioned why the distributor of the magazine was not the appropriate target, if there was to be prosecution, and emphasized the fact that Flynt operates out of Columbus, Ohio, not Cincinnati. But these arguments are not the essence of the familiar libertarian complaint, which is that "obscenity is a subjective matter."

The idea is that "obscenity" (unlike, presumably, "unfair labor practices" or "fraudulent advertising") is too subjective a category for other than capricious use in law. So censorship of obscenity is opposed by those who cite the "slippery slope" argument: "Once it starts, how will society know when to stop?"

The "slippery slope" argument also can be made against taxation and police: taxation might become expropriation; police forces might become gestapos. But self-government rests on the confidence that communities can, generally and within a tolerable margin of error, make reasonable distinctions.

The "slippery slope" argument against censoring *Hustler* is that the magazine, although loathsome, must be protected or even James Joyce's *Ulysses* cannot be protected. The logic of this argument is that censorship of anything endangers everything because all standards are equally arbitrary and idiosyncratic.

Former Supreme Court Justice William Douglas, the archetypal libertarian, argued with equally vehemence and thoughtfulness two propositions. One was that, "The idea of a Free Society written into our Constitution . . . is that people are mature enough . . . to recognize trash when they see it." His second proposition was that the distinction between trash and art is merely a matter of taste, no more rationally defensible than a taste for anchovies or ripe olives.

But if there are no critical standards to identify trash, then there are no critical standards to identify art. The authentic voice of this school of thought is the New York Civil Liberties Union, which has argued that tattooing is an "art form" enjoying First Amendment protection.

A few years ago, the Court wisely modified (by broadening) the principle that a community could only regulate material that is "utterly without redeeming social value." There is, alas, no community so far off the beaten path of the march of intellect that some professor cannot be flown there to testify to the cathartic, or otherwise redemptive, value of anything.

Indeed, a professor, Walter Bennett, declares that even laws prohibiting incest are unconstitutionally repressive: "It seems clear that the incest taboo is not instinctive but the product of cultural conditioning, because no aversion to sexual intercourse between relatives exists in animals other than man."

So goes the libertarian argument for ending the "repression" that prevents human sexuality from ascending to the level of shark sexuality —the level of "animals other than man."

[February 17, 1977]

The Not-So-Mighty Tube

In simpler days it was said that the hand that rocked the cradle ruled the world. Today, says Professor Michael J. Robinson of Catholic University (in *The Public Interest*), the rule of television rocks the world: "In the 1950s television was a *reflection* of our social and political opinions, but by the 1960s it was an important *cause* of them." He insists that television journalism did "engender" fundamental changes, "moving us" toward conservatism, and entertainment programing is a "fomenter" of social liberalism, "fostering" and "pushing us toward" change.

"Mary Tyler Moore and 'Mary Tyler Mooreism' seem to have been unusually effective in 'consciousness raising.' Between 1958 and 1969, the percentage of women accepting the idea that a woman could serve effectively as President actually *declined* by 3 percent. But between 1969 and 1972, the proportion of women who came to accept the idea of a female President *increased* by 19 percent. . . . During those first two seasons in which Mary Richards and Rhoda Morgenstern came to television, the level of public support among women for a female President increased more than among any other two-year—or ten-year— period since the 1930s."

The *post hoc, ergo propter hoc* fallacy involves mistaking mere antecedents for causes: the cock crows and then the sun rises, so the crowing caused the sunrise. Did prim Mary cause consciousness to rise? Does the water wheel move the river? Television conforms entertainment to market research, struggling to paddle as fast as the current. Robinson finds it ironic that entertainment programing, the servant of commerce, is supportive of "social liberalism," which he identifies with "hedonism and libertarianism" (and "Maude"). But commerce, which profits from the sovereignty of appetites, has never been a conservative force.

Television is not always benign or even innocuous. When vacuous or violent it is enervating and desensitizing; and it has influenced, often unfortunately, the way Americans campaign for office and for change. But it is more mirror than lever.

Robinson believes the "audio-visual orgy of the 1960s" shifted "power" upward toward the President and downward toward "have-nots" such as the civil-rights movement, and other "groups wretched or angry or clever enough to do what was needed to become photogenic." But Kennedy, constantly on television and consistently stymied by Congress, learned that conspicuousness is not power. Jimmy Carter, who uses television even more assiduously than Kennedy did, is learning that television does not make governing easier. Americans have developed fine filters for what they consider static, commercial and political, so Carter's media blitz about the energy crisis was like water thrown on sand: it left little trace. Thanks in part to broadcasting, political rhetoric has become like advertising, audible wallpaper, always there but rarely noticed.

Robinson notes that the 1963 "March on Washington" ("the greatest public-relations gambit ever staged") capped five months of intense civil-rights coverage, during which the percentage of Americans regarding civil rights as "the most important problem facing America" soared from 4 to 52. But it is unhistorical to say that this means the networks had begun "to define our political agenda."

Television did not give civil-rights leaders the idea of a March on Washington or make the idea effective. In 1941 the mere threat (by A. Philip Randolph) of a march frightened FDR into important policy changes. The civil-rights movement did not start with television, but with the moral and social changes wrought by the Second World War. The movement's first great victory was the Supreme Court's 1954 desegregation decision, when television was in its infancy. (During the two television decades the least "photogenic" branch of government, the judiciary, has grown in importance relative to the other branches.) The movement had on its side great leaders, centuries of grievances, the Constitution, and justice. It benefited from television, but did not depend upon it. Television hastened change a bit, but probably did not determine the direction or extent of change. What television did on its own (for example, manufacturing Stokely Carmichael as a "black leader") was as evanescent as most shoddy fiction.

When Robinson says "Nixon would have lost in 1968 had it not been for network news coverage of politics between 1964 and his election," he must mean either that LBJ would have been re-elected but for disintegration at home and defeat abroad; or that without television Americans would not have minded disintegration and defeat; or that without television there would not have been disintegration and defeat. The first idea is true but trivial; the last two are false.

The United States has never had national newspapers, so the focus of news was local. But network news is "national news." So, Robinson says, television has shifted frustrations toward the national government. But the centralization of power in Washington began well before television and would have "nationalized" news, and frustrations, with no help from television. Robinson believes that television journalism, although accused of liberal bias, has recently stimulated political conservatism. But the limitations of government would have become apparent, and the conservative impulse would have had its day, even if television had developed only as an entertainment industry.

To represent situation-comedy shows as shapers of the nation's consciousness is to portray the public as more passive and plastic than it is. To represent television journalism as a fundamentally transforming force is to make the nation's politics seem less purposeful, more mindless, more a matter of random causes than is the case. The contours of history are not determined by communications technology, however much it pleases people to think that history is what, and only what, can be seen at home. To see the rise of blacks, or the fall of LBJ, as primarily a consequence of television is to hollow out history. It discounts the noble and ignoble ideas and passions, heroes and villains and common people who make history.

In the silly movie *Network,* millions of Americans are prompted by a deranged anchor man to sprint to their windows to shout, "We're mad as hell and we won't take it any more." Modern man, proudly sovereign beneath a blank heaven, is prone to believe that "they" (evil persons, irresistible impulses, impersonal forces) control the world. Astrology, vulgar Marxism and Freudianism, and other doctrines nourish this need. So does the exaggeration of media influence. Journalists and perhaps even serious scholars, such as Robinson, who study television, are prone to believe that it turns the world. But the world is not that easy to turn.

[August 8, 1977]

Prisoners of TV

Disparagement of television is second only to watching television as an American pastime. And most disparagement of television is a series of footnotes to Fred Allen, who called television "bubble gum for the eyes." He meant that television is not nourishing.

Most of it is unnourishing. But so is most criticism of it. And recently Eric Sevareid, who has brought a touch of class to print and broadcast journalism, rounded on the critics.

"For TV," he said, "the demand-supply equation is monstrously distorted . . . TV programing consumes eighteen to twenty-four hours a day, 365 days of the year. No other medium of information or entertainment ever tried anything like that. How many good new plays appear in the theaters of this country each year? How many fine new motion pictures? Add it all together and perhaps you could fill twenty evenings out of the 365."

As a station manager says, "Hell, there isn't even enough mediocrity to go around." Certainly, it is sentimental to believe that "Our Miss Brooks" on radio was superior to "Mary Tyler Moore" on television. And it is nutty to suppose that people would read more Virginia Woolf if they watched less "Laverne & Shirley." It may even be true that, as Sevareid insists, television stimulates more conversation than it suppresses: "Nonconversing families were always that way." And the theory that television kills reading is a theory killed by a fact: since television, book sales have grown much faster than the population has grown.

It is not true that sponsors control the content of television entertainment. True, the gas utility that sponsored the drama *Judgment at Nuremberg* bleeped out the words "gas ovens" from references to Nazi crimes. But television would be better if sponsors concerned themselves with content by boycotting gratuitously violent shows like "Starsky and Hutch." To its great credit J. Walter Thompson, the largest advertising agency, is warning sponsors about their complicity in what the agency calls "the desensitization of America."

As the networks scramble for audiences, they do contribute to the coarsening of American life by edging ever closer to the soft-core pornography of violence and sex. But without excusing the networks, it must be said that they are pulled along, downward, by movies. Television increases the dosages of shocking material in order to grab the attention of audiences that have become blasé about the sort of mayhem in movies like *Taxi Driver* and *Marathon Man.* The most violent *entertainment* on television recently was in movies like *The Godfather* and *The Wild Bunch.* Of course worse violence was in news stories from Beirut.

Television news is, it seems, a burr under the nation's saddle. It is watched, voluntarily, by scores of millions, and it is criticized, incessantly, by (it sometimes seems) as many.

Part of the problem is that television news is so brief. Subtract commercials and there are twenty-two minutes in the news portion. A two-minute story is longer than most. It has been noted that a transcript of the network news would not fill half the front page of the *New York Times.* But such quantitative comparisons miss this point: a page of print cannot have the unique impact of, say, thirty seconds of film showing Joe McCarthy bullying a witness, or retreating Vietnamese soldiers clinging to helicopters.

Taking twenty-two minutes to cover the world *is* a bit like taking a teacup to empty the ocean. And the compression of television journalism magnifies the importance of editorial judgments, and hence magnifies the suspicion of "bias." The viewer is a volunteer, but he also is, in a sense, as Sevareid says, a prisoner: "A newspaper or magazine reader can be his or her own editor in a vital sense. He can glance over it and decide what to read, what to pass by. The TV viewer is a restless prisoner, obliged to sit through what does not interest him to get to what may interest him."

Recently David Brinkley wondered why NBC had routinely run a two-minute story of indecisive, unremarkable fighting in Beirut. It was, he believes, a story of interest to only a tiny fraction of NBC's viewers. Brinkley thinks the problem is that television has adopted newspapers' standards of news, standards that are inappropriate for television because viewers, unlike readers, cannot "skip around." But viewers *can* skip around, to competing programs. And they may skip unless a program provides a steady dosage of what a camera provides best, *entertaining action.*

When wondering why NBC aired the story of inconclusive Beirut violence, Brinkley concluded: "We couldn't even use the excuse that the story was easy to get. It wasn't. It was hard, dangerous work for a correspondent and a camera crew and it was sent to the United States by satellite, which is expensive." But the difficulty of getting a story, far from being an excuse for not getting it, can be a "reason" for getting it. The Beirut story *was* hard to get. But *only* television could get the sight and sound of battle.

"And in the end," Brinkley asks, "after all the work, danger, time and money, who really wanted to see it? In my opinion, almost nobody." I disagree. Perhaps the Beirut war scenes are *precisely* the sort of thing viewers want to see.

Brinkley, a superb professional, assumes people watch news in order to see newsworthy things. So, regarding the scenes of meaningless Beirut violence, he asks, "Why bore the audience any more than necessary?" *Bore* the audience? With *war?* Not likely. Brinkley's audience does not consist of Brinkleys. His news show is a brief information program, sandwiched between an afternoon of entertainment and an evening of entertainment. A lot of people turn on news shows in search of . . . entertainment.

Television's raison d'être is the camera. Television is not always "bubble gum for the eyes," but it always is *for the eyes.* People do not stare at their refrigerators. They stare at their television sets, expecting remarkable sights to appear there. And even unnewsworthy fighting is a riveting *sight.* As a newsgathering instrument a camera is at once powerful and limited. It can never produce a picture of an idea. It always can produce vivid pictures of action. Such pictures *can* be invaluable journalism. They can hardly fail to be entertaining.

While Brinkley and Sevareid and other good people in television journalism have been working to make their powerful technology serve the public good, some bad people in television entertainment have found their own uses for Beirut. NBC's "Saturday Night," which fancies itself satire, saw comic potential in a ghastly picture of a body being dragged behind a car through Beirut. The "Saturday Night" satirists superimposed a "Just Married" sign on the car. That's entertainment, at least for "Saturday Night's" desensitized young audience, television's children.

[January 10, 1977]

Government by Adding Machine

President Carter has joined Senator Birch Bayh's crusade to abolish the electoral college, the world's most tested and vindicated mechanism for choosing a chief executive. For years the Indiana Democrat has been advocating direct election so "the people" can choose presidents, and because the electoral college is "undemocratic" and dangerous.

One of Bayh's terrors is the "faithless elector" who does not vote for the candidate who carries his state. Actually, of the 17,000 electors since 1789, about ten have been "faithless," none has altered an election. If this specter haunts Bayh, it can be exorcised by abolishing the *office* of elector, and leaving the electoral vote in peace.

Bayh also says the electoral college must go because in three elections (1824, 1876, 1888) the electoral-vote winner was not the popular-vote winner. Actually, even if in "only" forty-five of forty-eight elections the same person won both, that would not justify Bayh's calling the electoral college "electoral roulette."

In 1876 and 1888, exuberant fraud on both sides probably involved more votes than the narrow victory margins. In 1824 all four candidates were together on ballots in only five of twenty-four states. Six states (including New York) had no elections: the legislatures selected the electors. Only about 350,000 of the 4 million eligible white males voted. Andrew Jackson won 38,149 more votes than John Quincy Adams, but neither had a majority of electoral votes. So the House of Representatives decided, picking Adams.

This was before the emergence of the two-party system. But Bayh says the events of 1824 (and 1876 and 1888) justify fundamental constitutional revision.

Actually, an electoral-vote victory by a candidate who loses the popular vote by a substantial margin is improbable and has never happened. And only extremely dogmatic majoritarians think democracy would be "subverted" (Bayh's word) if the electoral college gave the presidency to a candidate who lost the popular vote by a wafer-thin margin. It is odd to say that the "nation's will" could be "frustrated" in a standoff.

Bayh is fond of the somewhat feverish thought that under the electoral college a candidate "could" win with just 25 percent of the vote by narrowly winning in the eleven largest states, even if he did not get a single vote in any other state.

But under direct election a candidate "could" sweep Alaska's 231,000 eligible voters, lose forty-nine states by an average of 4,700 votes, and win. This "possibility" is about as probable as the one that Bayh is fond of imagining.

Bayh is not apt to produce what Madison was too sober to attempt, a constitutional arrangement under which no unwanted outcome is even theoretically possible. Serious people consider probabilities, not possibilities. And direct election would make probable a grave difficulty.

The electoral-vote system, combined with the winner-take-all rule (a custom, not a constitutional requirement), discourages ideological third parties: such parties are unlikely to win pluralities in many states, so they are effectively shut out of the decisive electoral-vote competition. But direct elections would *incite* such parties. They could hope to prevent any candidate from receiving a national majority, or even an impressive plurality of popular votes.

Bayh's remedy for this defect in direct elections poses a substantial danger. He proposes a *second* election, a runoff between the two leaders, if neither gets 40 percent the first time. But a runoff would be an *incentive* to minor parties. They would try to force a second vote so they could sell their support.

Bayh evidently is undisturbed by the fact that direct election might frequently produce "41 percent" presidents. The electoral college has only produced three presidents with such low pluralities, in 1824, before the two-party system developed, and in 1860 and 1912, when the two-party system was in disarray. But Alexander Bickel of Yale warned that direct elections might make disarray permanent:

"The monopoly of power enjoyed by the two major parties would not likely survive the demise of the electoral college. Now, the dominance of two major parties enables us to achieve a politics of coalition and accommodation rather than of ideological and charismatic fragmentation, governments that are moderate, and a regime that is stable."

The genius of the Constitution is the effect it has on the *character* of majorities. The electoral college promotes unity and legitimacy by helping to generate majorities that are not narrow, geographically or ideologically, and by magnifying (as in 1960, 1968, 1976) narrow margins of victories in the popular vote.

Such considerations are of no interest to single-minded majoritarians, who consider democracy a matter of mere numbers. They note that in 1976, 123,545 Alaskans determined three electoral votes, one for each 41,181 voters, but in California (7,867,043 voters, forty-five electoral votes) there was only one electoral vote for each 174,823 voters. Is an Alaskan four times more powerful than a Californian? Is a Californian more powerful because he helps to determine a larger bloc of electoral votes? Bayh says that in any case, the system is "undemocratic." His understanding of democracy has the charm of simplicity: "Every vote should count the same." That, he says, is constitutional propriety, as stated by the Supreme Court in its "one man, one vote" reapportionment ruling. But Bickel revealed the foolishness of this argument by expressing it this way: "It is time for the system to be ideologically pure. The Court has said that the Constitution commands equal apportionment. We should, therefore, reapportion the presidency. In effect, we must now amend the Constitution to make it mean what the Supreme Court says it means."

As Irving Kristol and Paul Weaver have written: "In recent decades, the democratic idea has been vulgarized and trivialized. From being a complex idea, implying a complex mode of government, appropriate to a large and complex society, the idea of democracy has been debased into a simple-minded arithmetical majoritarianism—government by adding machine."

Defenders of the electoral college are defending not an eighteenth-century artifact, but a system that has evolved, shaping and shaped by all the instruments of politics, especially the two-party system. It is an integral part of a constitutional system with premises too subtle and purposes too varied to be summed up in slogans like "one man, one vote." Bayh insists that the electoral college "is, by simple definition, undemocratic." But this constitutional democracy was not devised by, and should not be revised by, persons addicted to simple definitions of democracy.

[April 4, 1977]

The Hot Seat: From Expectations to Entitlements

Joseph Califano, 45, was minding his own business as a Washington lawyer and doing nicely, thank you. A dutiful Democrat, he aided Jimmy Carter's campaign. Then without provocation or warning, Carter rounded on Califano and made him Secretary of Health, Education and Welfare. So now most of the smoldering issues of the day ("equal opportunity," desegregation, the welfare "mess," the "crisis" of rising medical costs) crowd around him like an armed host.

He is armed with sacks of money: HEW's budget is 35 percent of the federal budget. But of HEW's $161.7 billion, all but $16 billion is for mandated payments. Social security and medicare account for 78 percent of Califano's funds.

He has scads of helpers, 142,000 employees. But some of them are problems, not problem solvers. One of an HEW secretary's more testing chores is to be mau-maued by the department's "civil rights" firebrands. They include folks who argue with scary sincerity that federal retribution should be visited upon any high school guilty of premeditated sexism in the form of a father-son dinner. But department morale depends, in part, on the secretary seeming to have his mental windows open to such breezes of progressive thought, so he must be able to look perfectly grave while listening to perfect nonsense.

If the waves of such contention can be smoothed by the oil of Califano's considerable charm, he still will be the focus of great expectations. So it is fortunate that Califano is acquainted with disappointment, having been present at the creation of the Great Society. He became a White House adviser in 1965 when Lyndon Johnson decided that a merely good society wouldn't do. In those heady days the White House became a Vesuvius of proposals; Congress became a Niagara of legislation. And many of the programs, problems and expectations that now confront Secretary Califano were shaped by the child Califano in the White House. Consider, for example, the problem of rising medical costs.

Over the past decade the cost of medical care has risen twice as fast as the general cost of living. Today nearly 10 percent of GNP is spent on medical care, and costs are rising $1 billion a month. By 1980 a day in the average hospital may cost well over $200. Today the costs of medicine surge through the economy.

In 1965, the last year before Medicaid and medicare, government paid just 12.8 percent of the nation's health bill. Today it pays 28.6 percent. This redistributes but also disguises the costs of medicine. According to a survey, the average family of four thinks it pays about $1,000 a year for medical care. Actually, that family's doctor bills, insurance premiums and taxes used for medicare and Medicaid amount to about $2,500 a year.

This is the price of progress, social and technological: more sophisticated medicine is becoming more accessible to more people. But when the government subsidizes demand for a service, the cost of the service will rise. Today government is subsidizing excessive provision and excessive utilization of care.

William Lilley, former director of the Council on Wage and Price Stability, notes that when you buy a car you do comparison shopping and pay for your choice directly, but when you get sick, your doctor's choice is decisive: "the provider of the service decides the level of services, the cost of the services, the place where they are provided." Understandably, both doctor and patient want "the best." After all, as the saying goes, health is "priceless." Anyway, the patient has only a diluted interest in the price of the service he is receiving. Less than a third of personal health costs are paid by individuals, and patients who have paid their insurance premiums figure they have already paid for their care.

Reasonable people disagree about what constitutes "reasonable" health costs. But the trend of costs is *politically* intolerable for the Carter administration. If costs continue to rise at the current rate, the bill for medicare and Medicaid will rise 75 percent, from $38 billion to $66 billion, in four years. If they do, a balanced budget and many other promised blessings will become even more remote.

Califano is being stern about hospital costs. He forbids them to rise more than 9 percent annually. The pity is that no one thought to do that earlier, but once this strategy is perfected perhaps the administration will apply it to Château Lafite-Rothschild. But the cost of medicine is just a sliver of one many-faceted problem (providing better and more equitably distributed care) that Califano is now supposed to "do something" about. His fundamental problem is that a revolution is in full swing. Sociologist Daniel Bell writes:

"What is clear is that the revolution of rising expectations, which has been one of the chief features of Western society in the past twenty-five years, is being transformed into a *revolution of rising entitlements* for the next twenty-five . . . the particular demands will vary with time and place. They are, however, not just the claims of the minorities, the poor or the disadvantaged; they are the claims of *all* groups in the society, claims for protections and rights —in short, for *entitlements.*"

Bell notes that the fastest growing sectors of Western societies are health, education and government. He cites evidence that on any particular day in 1974 more than a third of California's population (a total about as large as the state's civilian labor force) was receiving institutional care, in schools (not counting colleges), hospitals, prisons, daycare centers, nursing homes. This is an example of the extraordinary expansion of what Bell calls the "public household," an expansion that involves two problems:

"One is the increasing 'overload' of issues which the political system may simply be unable to manage. The virtue of the market is that it disperses responsibility for decisions and effects. The public household concentrates decisions and makes the consequences visible. The second problem is that, because of the pressure of the rising entitlements, there is a constant tendency for state expenditures to increase, requiring more taxes to pay for services, and stimulating more inflation. . . . Both are, simply, prescriptions for increased political instability and discontent."

The focus of discontent often is HEW, that great gray growth at the foot of Capitol Hill. There is nothing gray about Califano, whose mind is a rainbow of redistributionist plans. And before long he will be, figuratively speaking, black and blue. As the focus of the scramble for entitlements, he is the lead tenpin in the bowling alley of modern government.

[March 7, 1977]

Society As a Field of Forces

In William Dean Howells' novel, *A Traveler from Altruria* (1894), an American expresses the optimistic assumption underlying the theory of social Darwinism:

"You know we are sort of fatalists here in America. We are great believers in the doctrine that it will all come out right in the end."

"Ah, I don't wonder at that," said the Altrurian, "if the process of natural selection works so perfectly among you as you say."

Natural selection as social policy—survival of the fittest—is for the hairychested. It is not for the person whom Theodore Roosevelt in *The Strenuous Life* (1899) called "the overcivilized man who has lost the great fighting, masterful virtues."

America has always had an abundance of the undercivilized. Certainly San Francisco has its share, including a sizable majority of its police and firemen. Recently police, quickly joined by firemen, staged an illegal strike to enforce a demand for double the 6.5 percent pay increase offered by the city board of supervisors. These strikers ignored court orders and carried pistols while picketing.

In social Darwinism, "best" means "strongest." In San Francisco, where social Darwinism prevails, the strongest are prospering. Mayor Joseph Alioto declared that he would never surrender to such illegal coercion, then declared an "emergency," and surrendered, granting everything the strikers demanded, including amnesty.

As redundant evidence that Alioto, like most politicians, doesn't mean what he says, the episode was uninteresting. But it was interesting as evidence of the extent to which "public safety unions" have adopted the social Darwinist ethic that might makes—no, *is*—right.

Actually, there is no reason to expect San Francisco police to be more law-abiding than their New York counterparts, who recently rioted on the Brooklyn Bridge to make an economic point. And there is no reason to expect New York police to be more law-abiding than New York lawmakers like Congressperson Bella Abzug who, in a recent oration at a Manhattan protest rally, urged subway and bus riders to break the new law increasing transit fares by fifteen cents.

Ms. Abzug's opinion that a fifteen-cent fare increase is a grievance fit for redress by mob force reflects the distemper of the times.

In the nineteenth century, social Darwinism was adored by entrepreneurs who thought it conferred respectability on rapaciousness. Today, for the same reason, it is the civic religion of other untamed forces, like municipal trade unions and, if Ms. Abzug gets her wish, urban mobs.

Our society is becoming a field of forces in which politicians pick their favorite forces and back them against rival forces, including the law, which, increasingly, is regarded as just another "force." Law once was considered the codification of community interests and shared values. But there is no community, only private interests contending in the struggle for survival of the fittest.

It is not surprising that Darwin, suitably bowdlerized, should have seized the American imagination. Americans confidently expected that struggle—called "competition," social nature red in tooth and claw—would produce an ever-upward evolution of the gross national product, and hence of society.

The American assumption was that self-interest could substitute for such virtues as public spiritedness in binding a community together. Unfettered passion for aggrandizement would overcome scarcity, and thus all serious social tensions would evaporate.

But it is still not clear that a society so constituted can even endure. It has special tensions.

The rationale for obeying its laws is self-interest. So when people —like San Francisco policemen—decide that obedience to law is interfering with their aggrandizement, obedience stops, abruptly, and without qualms of conscience.

Such a society, held together in lawfulness by momentary calculations, is inherently unstable. It becomes more so as struggling interest groups become more organized and fierce, and as politicians like Alioto and Ms. Abzug reward, and even incite, lawlessness.

The rule of law is the way society, acting reflectively, imposes mind on the world. Today, as force replaces law in San Francisco and New York and elsewhere, our society is becoming plastic to the appetites of its "fittest"—most ruthless—factions.

Americans always have had a soft spot in their hearts for hardhearted doctrines like survival of the fittest. Indeed, nineteenth-century social Darwinists expressed their social philosophy in the slogan "root, hog, or die." Now as then it is a sound rule, for hogs.

[August 27, 1975]

The Dispersal of Cities

Tulane University, a gracious host and a glutton for punishment, imports legions of speakers each year for a renowned series of forums on current discontents and trends. This year I moderated, sort of, a panel on urban problems. The melee involved, among others, Bella Abzug, who soon may offer New Yorkers the gift of herself as mayor.

The Abzugian style of public speaking is a throwback to the days before electronically amplified sound. She has complete confidence in her opinions and none at all in the clip-on microphone bobbing like a cork on the heaving ocean of her bosom.

In voice fortissimo, she instructed the panel, the audience, and nearby neighborhoods that the "urban crisis" proves that Washington is "not interested in the majority of people and where they live." Actually, the most important fact about "urban crisis" is that few people are directly afflicted by it.

It is frequently said by Ms. Abzug and others that "70 percent of Americans live in cities." True, any place with 2,500 people is officially classified as an "urban place." More pertinent facts are: The percentage of Americans living in cities of 250,000 or more is about what it was in 1920. Population density in urban areas has declined from 5,408 per square mile in 1950 to 3,376 in 1970. Thirty percent of Americans live in central cities, down from 35 percent twenty-five years ago. Most parts of most central cities are not in what can properly be called a "crisis."

The national vocabulary contrasts "suburbs" and "cities." But Hempstead, Long Island, like scores of other suburbs, is part of a complex of communities with all the normal functions and attributes of cities, and they are not in "crisis." The "crisis" is, primarily, in some central portions of some older urban areas. It involves, directly, perhaps no more than 5 percent of the population.

Indeed, the United States may become the first developed nation in which central cities—cities as traditionally understood—are important to the nation only as burdens. Writing in a recent issue of *Business Week*, Jack Patterson says:

"Cities, no longer primary manufacturing centers, wholesale-retail marketplaces, or preferred residences for the middle class, are now losing those very activities that . . . have always seemed to belong downtown: the headquarters, 'nerve center' functions. In Connecticut's affluent Fairfield County, for example, the headquarters of at least sixty major corporations, many once located in Manhattan, are now tucked away in office parks or on their own office campuses. This business concentration has achieved a critical mass. It is now pulling in a second wave of subsidiary service companies—advertising agencies, law and accounting firms, consultants, data processors, and other specialists that previously had to be near their clients in the city."

The dispersal of the traditional city has been under way for a long time. In 1895 a wise man, who today would be condemned to be known as an "urbanologist," wrote:

"Three new factors have suddenly developed which promise to exert a powerful influence on the problems of city and country life. These are the trolley, the bicycle, and the telephone. It is impossible to foresee at present just what their influence is to be on the distribution of population; but this much is certain, that it adds from five to fifteen miles to the radius of every large town."

The "future shock" of bicycles and telephones has been followed by even more consequential developments. Professor Edward Banfield notes that automobiles and commuter railroads cut people loose from cities, and trucks cut factories free from central city railheads. Assembly lines and other horizontal manufacturing processes made cheap suburban land attractive, and high-power electricity transmission lines made possible the outward dispersal of manufacturing to such land.

Play follows work. To see the Cleveland Cavaliers or New York Nets play basketball, you drive to suburban arenas. The Detroit Lions play football in Pontiac, Michigan; the Boston Patriots are now the New England Patriots of Foxboro, Massachusetts; and the New York Giants are to be found in New Jersey. If Ms. Abzug wants to find where, increasingly, people live, she should follow the bouncing ball.

[March 24, 1977]

The Hell of Affluence

Considering what the alternatives are, affluence has received rather a lot of disparagement.

"What is called a high standard of living," according to John Kenneth Galbraith, who surely knows, "consists, in considerable measure, in arrangements for avoiding muscular energy, increasing sensual pleasure and for enhancing caloric intake above any conceivable nutritional requirement." Galbraith's most unblinking scrutiny of the hell of affluence is *The Affluent Society:*

"The family which takes its mauve and cerise, air-conditioned, power-steered, and power-braked automobile out for a tour passes through cities that are badly paved, made hideous by litter, blighted buildings, billboards, and posts for wires that should long since have been put underground. They pass on into a countryside that has been rendered largely invisible by commercial art . . . they picnic on exquisitely packaged food from a portable icebox by a polluted stream and go on to spend the night at a park which is a menace to public health and morals. Just before dozing off on an air mattress, beneath a nylon tent, amid the stench of decaying refuse, they may reflect vaguely on the curious unevenness of their blessings."

According to Galbraith, what ails the affluent society is a glut of the wrong goods, private consumption goods as opposed to government services. The market mechanism serves frivolous appetites, not real needs.

The Affluent Society was published in 1958, when a new advertising medium, television, had just become the masses' entertainment. On the basis of no evidence, at least none he cited, Galbraith asserted that advertising subverts consumer sovereignty: manufacturers control consumers' desires rather than make production conform to desires. Madison's Republic depends on Madison Avenue: if advertising did not manufacture "artificial" appetites, this affluent nation would lose its dynamism, all "natural" needs for food and shelter having been satisfied.

Actually, if advertising were as potent as Galbraith thinks, the advent of television, an additional dose of advertising in every living room, should have caused a sharp increase in consumption relative to savings. It did not. People are not so manipulable. And most advertising has modest aims. The aim of a Schlitz ad is not to make America thirsty, or to make people buy beer rather than Buicks, or even beer rather than Dr. Pepper. The aim is to sell Schlitz to people who might otherwise buy Budweiser. The aim is to cancel other beer advertising.

Galbraith's lament about the values of an affluent society may be valid, but it does not explain the disappointment many people feel about affluence. There is a vague feeling that economic growth has not fulfilled its promise. Affluence has not led to a decline in economic cares, to less preoccupation with money. Envy has increased while society has become more wealthy. This paradox is a subject of an intriguing new book, *Social Limits to Growth*, by Fred Hirsch, a British economist.

Hirsch distinguishes between the "material economy" and the "positional economy." The latter is increasingly important and concerns goods, services and jobs that are *inherently* minority enjoyments. As affluence satisfies basic material needs, more income and energy are devoted to "positional competition" for such things as a "choice" suburban home, an "exclusive" vacation spot, an "elite" education, a "superior" job. As affluence increases, competition moves more and more from the material sector to the "positional" sector, where one person's gain is a loss for many other persons. Unlike the demand for radial tires and Right Guard, supplies of which can be expanded indefinitely, each demand for a positional good can be satisfied only by frustrating the similar demands of many other people.

Affluence sharply increases competition for positional goods. So society's economic success increases frustrations and tensions. The feeling that Affluent Man is more harried than ever is sometimes explained in terms of consumption of material goods: the supply of such goods increases while the time to consume stays constant. But an additional explanation is the time and income needed for "positional competition."

Materially, growth has been a leveling force. Cars and air conditioners and other luxuries of one generation have become the "necessities" of the next. But now affluent consumers face increasing frustration because the collective advance of the middle class is *inherently* impossible regarding "positional goods."

Tension, not satisfaction, results when the masses, observing the top strata of society, acquire an appetite for "elite" education, "superior" jobs, beachfront property. Positions of status and leadership, like beachfront, cannot be expanded to become majority enjoyments. The problem of the "positional economy" is social congestion: the desires of the middle class have expanded beyond middle-class opportunities.

The dominant political aspiration of the modern age is equality, but Hirsch's theory of the "positional economy" suggests that, in an important way, the affluent society will become steadily less egalitarian. Pursuit of positional goods will become steadily more important, and those goods are *inherently* restricted to a minority.

The complexity of positional striving is apparent in higher education. Economic growth has made possible the vast expansion of what is called "educational opportunity." But one reason there are so many bored or sullen students is that for many of them college is not an "opportunity." It is what Hirsch calls a "defensive necessity." The sullen student pursues a degree only because it will raise his income above what it would be if others got degrees and he did not.

This scramble for position is an aspect of modern society's mania for "credentials," like the current absurd pursuit of meaningless master's degrees. As Hirsch says, there is a sense in which "more education for all leaves everyone in the same place . . . it is a case of everyone in the crowd standing on tiptoe and no one getting a better view." And the number of persons who are, or (as important) think they are, educationally equipped for "superior" jobs increases faster than the number of such jobs.

Well, maybe not. Today, even more than in 1935, there is much truth in what Bertrand Russell then said facetiously:

"Work is of two kinds: first, altering the position of matter at or near the earth's surface relatively to other such matters; second, telling other people to do so. The first is unpleasant and ill paid; the second is pleasant and highly paid. The second kind is capable of indefinite extension: there are not only those who give orders, but those who give advice as to what order should be given."

[March 21, 1977]

Life at the Waterline

When the *Titanic* steamed into an iceberg, the disaster was not democratic. Fifty-six percent of third-class women passengers died. Only four of the 143 first-class women passengers died. You do not need to ask which class was traveling near or below the waterline.

The social structure of the *Titanic* was like that of society: hard times come first and hardest to persons living close to life's waterline. This fact is relevant to much argument about social policies, including the growing argument about the use of nuclear power plants to generate electricity.

Few things are as subversive of public reasonableness as the misdescription of social issues. Opponents of nuclear power have managed to present this as an "environmental issue." But the dispute concerns the allocation of significant social costs and opportunities.

Hence it is a "social justice" issue. This means that values other than those usually thought of as "environmental" are at stake.

Thirty-four states are considering legislation or referenda to restrict or eliminate commercial nuclear power plants. And California will vote June 8 on a measure that may set a pattern. If passed, Proposition 15 will mean the closing of some existing nuclear plants. Others will be harder, perhaps impossible, to build.

At one level the dispute is esoteric and ideological. Around the nation, and especially in California—home of much high-technology industry—an intense, articulate and growing minority believes that technology, and the economic growth it supports, has gone too far. This idea, which has a pedigree more ancient than steam power, is part of the fuel for the anti-nuclear movement.

But, at another level, the argument is about practical questions of safety. Opponents of nuclear power argue that it involves intolerable risks of catastrophic accidents (release of a radioactive cloud); that the disposal of nuclear wastes is an unsolved problem; that terrorists can build nuclear devices with material pilfered from nuclear plants.

These fears, although not groundless, are not substantial enough to merit an action as severe as Proposition 15. It might indeed (as many of its supporters hope) stop all nuclear power in California and show how to do that elsewhere.

There is nothing inherently insoluble about nuclear waste disposal problems. Terrorists will have many plants to attack in nations where security is more lax than in the United States. And after hundreds of reactor-years of operation, there has not been a single radiation fatality in any of the 162 commercial nuclear plants around the world.

The costs of banishing nuclear power are more certain than the risks of not doing so. Conservation can dampen but not halt the growth of demand for electricity, and the choice is not between demonstrably dangerous and perfectly safe ways of producing it. Against the remote possibility of an unprecedented nuclear catastrophe stands the certainty of numerous deaths attributable to respiratory ailments aggravated by increased use of coal in power plants.

That is just one way of illustrating the point that there are many ways to adversely affect the "quality of life." One sure way is to restrict the life chances of the less well-placed members of society.

When the necessities of life become more dear, the pain is worse for persons whose conditions are most necessitous. Electricity generated by nuclear power is less expensive than that generated in oil or coal plants, and lower income persons are apt to spend significantly more of their disposable income on utility costs than are more affluent people.

In addition, curtailment of nuclear power in California, and elsewhere, would mean slowing economic growth. It is arguable that that would be good for our souls. What is certain is that persons who have farthest to rise in a society have most to lose from the dampening of society's dynamism.

To persons well above society's "waterline," Proposition 15 may look like an "environmental issue." Those at or below the line will be forgiven for thinking that what is at stake is the social environment, that Proposition 15 is a question of social justice. Life at the waterline does concentrate the mind.

[June 3, 1976]

Inoculation by History

One of Cromwell's Parliaments proposed burning all state records so that, having effaced all official memory of the past, life could begin afresh. Disavowal of antecedents is the quintessential revolutionary act, as Napoleon, too, understood.

When shown a genealogy of the Bonaparte family, Napoleon brushed it aside, remarking: "Je suis mon propre ancêtre" (I am my own ancestor). Considering the scale of Napoleon's vanity, and the nature of that genealogy, his rejection of it was understandable. But his rejection also reflected the modern disdain for history as "the dead hand of the past."

Napoleon is the archetypal modern man of action, a bloody nuisance ricocheting around civilization, making history and orphans. He made so much history because he knew so little. He traveled fast because he traveled light, unencumbered by an educated person's sense of limitations, the sense that is the bittersweet fruit of historical understanding.

Such people are tolerable, if you like history as made by Corsican brigands and other modern world-shakers, many of whom have been breathtakingly ignorant of history. But this is of more than academic interest because, according to the Organization of American Historians, the study of history is "in crisis."

The OAH calls the problem "presentism," but that non-word does not obscure the OAH's valid concern. The OAH is alarmed by reports like the one from Illinois, where recent legislation mandating consumer education in public schools is being implemented at the expense of history instruction.

While Illinois is producing students educated in the science of consuming, Hawaii is submerging the study of history in what curriculum designers, speaking in strange tongues, call an "Inquiry-Conceptual Program." The program is supposed to "integrate all the social sciences" so young Hawaiians will be prepared for "taking action on social and civic problems."

The OAH blames the decline of the study of history on the assumption that history "is not a practical subject." But, truth be told, history is *not* "practical" as many contemporary intellectuals understand practicality. It is not a source of tidy lessons for solving social problems. The OAH affirms "the value of historical perspective," but the odds are that some of today's professional historians haven't the foggiest idea what that value is.

Many historians, like many other intellectuals, long to be "relevant" to the specific problems of the day. They flinch from acknowledging that the most useful lesson of history is highly general. It is: things have not always been as they are, and will not always be as they are. This is an especially important—because discomfiting—insight for

Americans, who take for granted freedom and abundance, both of which are, considered in the sweep of history, rare and shortlived phenomena.

History contains more sadness than gladness, more dreams frustrated than fulfilled. But this means that the study of history is, for many historians, unacceptably unfun and unheroic. Like many other intellectuals, many historians want to believe that they are pregnant with the future. They want to dissolve the distinction between thought and action. They are, they think, deliverers: history will yield highly practical "lessons" that will propel mankind to the uplands of happiness.

Many modern intellectuals, like optimists through the ages, recoil from this truth: the best use of history is as an inoculation against radical expectations, and hence against embittering disappointments.

"The trouble with you, Jenny Blair," said Jenny's grandfather (in Ellen Glasgow's novel, *The Sheltered Life*), "is that you do not know the first thing about life. It is only by knowing how little life has in store for us that we are able to look on the bright side and avoid disappointment."

"I can see you're tired, son, and disappointed," said Guidillo to Rocco (in Ignazio Silone's novel, *A Handful of Blackberries*). "You have the sadness of one who set out to go very far and ends up by finding himself where he began. Didn't they teach you at school that the world is round?"

Optimists as different as Marxists and Americans believe in History—Americans call it Progress—as a linear process leading inexorably to higher stages of life. But if historians and other intellectuals were free from Promethean pretensions, young people at school would learn the unfun, unheroic truth that history is circular, like a maelstrom.

[September 1, 1975]

A Little Platoon of Moral Pioneers

INNISFREE VILLAGE, VIRGINIA—Nestled on what was, 200 years ago, an American frontier, this village is an outpost on a new frontier. The frontier is social and moral, not geographical: Innisfree is pioneering a new way of caring for a minority that has been all too easy to neglect.

Innisfree is a self-contained working community with mentally handicapped adults, built on 400 acres of rolling farmland adjacent to the Shenandoah National Forest in the foothills of the Blue Ridge mountains. Today it has residential and work facilities for about thirty handicapped adults and about half as many nonhandicapped co-workers and their children. It is a haven from the complexity and competitiveness of an urban society. But it offers enriching, dignified labor and life.

Villagers provide a substantial portion of their food from their dairy and beef herds, poultry, gardens, and orchards. In their woodshop they produce safe, sturdy toys. In their weavery they produce purses, shawls, scarves. And in their bakery they produce something that is almost as much of an American rarity as is Innisfree: good bread.

These activities are commercially promising and will be more so when the village can afford to expand its facilities. For example, the bakery capacity will expand from 350 loaves per week to 1,500. You can strike a blow for better bread, and other important values, by sending a check to Route 2, Box 506, Crozet, Virginia, 22932.

Innisfree is an example of how small private resources in the service of a private vision can produce a model for public policy.

Approximately 3 percent of American citizens are mentally retarded. When their families are considered, the problem of retardation can be said to directly affect upward of twenty million Americans.

An undetermined but large number of mentally retarded or otherwise handicapped adults are living with elderly parents in an environment that may be over-sheltered today, and may at any time be shattered by parents' deaths. Such parents are haunted by the lack of alternatives to the impoverished living environments of public institutions.

Hundreds of thousands of retarded citizens are in public institutions that offer only what is decorously called custodial care, which often means the warehousing of human beings. Often the warehousing is facilitated by the heavy, regular use of tranquilizing drugs—chemical straitjackets. Many of the retarded or otherwise handicapped could function in community environments like Innisfree, minimizing their handicaps and their cost to society.

The sublime physical setting of Innisfree—a small bear recently wandered in from the forest—is a temptation to romanticism. But that would be disrespectful of the strength of hand and soul required of the saintly (there is no more precise word) nonhandicapped people here who shun distinctions, even salaries, that would put social distance between themselves and the villagers whose personhood they are here to affirm.

The usefulness of Innisfree as a model for public policy is limited only by this:

Government money can purchase professional competence, and can increase society's supply of such competence. But the mysterious dedication that makes Innisfree a community is, like all love, mysterious: it is not a price-elastic commodity, expanding with the size of government appropriations.

Indeed, as the world becomes richer but more secular, handicapped people become more vulnerable. Money can build the large impersonal institutions that are limbos of cool neglect in affluent societies. But the humble cottages of an Innisfree are, like Chartres, manifestations of something like a religious vocation.

A society's ascent from barbarism can be measured, in part, by the care it shows for the defenseless, like mentally handicapped people. It also is true that societies are propelled upward, slowly, by the astonishing energies of moral pioneers operating in little platoons.

Innisfree is one of those little platoons, conquering the nation's inner frontier of caring, unsung except, in a sense, by William Butler Yeats, from whose poem the village took its name:

> I will arise and go now, and go to Innisfree,
> And a small cabin build there . . .
> And I shall have some peace there, for
> peace comes dropping slow . . .
> I will arise and go now, for always night
> and day . . .
> While I stand on the roadway, or on the
> pavements grey
> I hear it in the deep heart's core.

[July 23, 1975]

PART 3

Manners

An Age Beyond Wonder

Few things are as stimulating as other people's calamities observed from a safe distance. So people relish Edward Gibbon's *Decline and Fall of the Roman Empire* and New York City's perils, technological as well as financial.

Much of the nation thinks, not without reason, that the city is sunk in darkness, even at high noon, and that the blackout was a sign of disapproval from above, a foretaste of fire and brimstone and pillars of salt. And let the records show that even the engineers resorted to theology, not physics, when they issued their first explanation in the dark. Consolidated Edison said the trouble, which started with a stroke of lightning, was "an act of God." The theology of the age is that God does not exist and that He manifests Himself in random unpleasantness. "Let us hope," prayed a thoroughly modern cleric (in a Peter De Vries novel involving a flood in suburbia), "that a kind Providence will put a speedy end to the acts of God under which we have been laboring."

Consolidated Edison provides even the humblest of its customers with harnessed power beyond the dreams of Louis XIV. But its customers, like Oscar Wilde, who when shown Niagara Falls said it would be more impressive if it flowed the other way, are hard to please. Consolidated Edison's customers reserve for it that special irritability Americans feel toward utilities and other institutions (such as government) that provide necessary services and then expect Americans to pay for them.

Like an oil embargo, a blackout (a brief but *convincing* energy crisis) demonstrates the fragility of the arrangements, social and technological, on which cities depend. And in this blackout, the nation again saw frenzy clothed as purposeful action. In the 1960s, when some people discovered they could ransack merchants without much risk, they did so. And voices were raised to rationalize the ransacking as a manifesto against the legacy of slavery, etc. Last week the lights went out, and instantly looters came out again, again happily pillaging and burning out small merchants, shattering productive lives. And again the familiar voices were heard, rationalizing barbarism, explaining that the smiling looters were really protesting New York's "inadequate welfare," etc. Thanks to the blackout, sophistries, as well as babies, are gestating.

The fact that extinguishing street lights is enough to crack the thin crust of civilization in whole neighborhoods is just especially vivid evidence that today nothing, not even petroleum, is more essential than electricity, or has done more to transform the world. Before electricity was harnessed a century ago, conditions of life were more like those of Julius Caesar's day than of Jimmy Carter's day.

This transformation began, in a sense, just over the horizon from where, last Wednesday, electricity suddenly seeped away. Ninety-nine years ago in Menlo Park, New Jersey, an inventor formed the Edison Electric Light Co. At Menlo Park, Thomas Edison produced "a minor invention every ten days and a big thing every six months or so." But it took a very different nineteenth-century man to express the sense of wonder inspired by all the "big things." He did it in the greatest American autobiography, *The Education of Henry Adams*.

Born in 1838, great-grandson of the second President, grandson of the sixth, Henry Adams watched with mingled awe and dismay the swift transformation of an agrarian republic into an industrial society, and "found himself lying in the gallery of machines at the Great Exposition of 1900, his historical neck broken by the sudden irruption of forces totally new."

"The year 1900 was not the first to upset schoolmasters," Adams wrote. "Copernicus and Galileo had broken many professorial necks about 1600; Columbus had stood the world on its head towards 1500; but the nearest approach to the revolution of 1900 was that of 310, when Constantine set up the Cross."

". . . to Adams the dynamo became a symbol of infinity . . . he began to feel the forty-foot dynamos as a moral force, much as the early Christians felt the Cross. The planet itself seemed less impressive, in its old-fashioned, deliberate, annual or daily revolution, than this huge wheel, revolving within arm's-length at some vertiginous speed. . . . Before the end, one began to pray to it; inherited instinct taught the natural expression of man before silent and infinite force."

No doubt all this, said about "mere" electricity, seems quaint and overwrought in an age that has discovered the neutron and what it can do. What most distinguishes modern people is that they have so slight a sense of awe about the world around them. But before condescending to Henry Adams, modern people should consider that, in a sense, they take more things on faith than did a thirteenth-century peasant tilling the fields in the shadow of Chartres.

When the peasant wanted light, he built a fire from wood he gathered. Modern people flip switches, trusting that someone, somewhere, has done something that will let there be light. How many switch-flippers can say what really happens, in the flux of electrons, when a generator generates?

The most advanced form of travel for the peasant was a sailing ship or a wagon: the mechanisms were visible and understandable. This year forty-one million passengers will pass through Chicago's O'Hare airport, obedient to disembodied voices, electronically amplified, telling them to get into cylindrical membranes of aluminum that will be hurled by strange engines through the upper atmosphere. The passengers will not understand, and will be content not to understand, how any of it really works. And we think the fourteenth century was an age of faith.

Perhaps ours is the strangest age. It is an age without a sense of the strangeness of things. Of course some people are different. Einstein, for example, had a highly developed sense of the sheer magic of the universe. But New Yorkers are more typical modern people. Their strongest reaction to the blackout was indignation: why was *mere nature* allowed to disrupt technology?

The human race has grown up and lost its capacity for wonder. This is not because people understand their everyday world better than people did in earlier ages. Today people understand less and less of the social and scientific systems on which they depend more and more. Alas, growing up usually means growing immune to astonishment. As G. K. Chesterton wrote, very young children do not need fairy tales because "mere life is interesting enough. A child of seven is excited by being told that Tommy opened the door and saw a dragon. But a child of three is excited by being told that Tommy opened the door." The three-year-old is the realist. No one really knows how Tommy does it.

[July 25, 1977]

The Astrological Impulse

Some intergalactically famous scientists, including eighteen Nobel laureates, have directed some delicate railery against the nonsense called astrology:

"It is simply a mistake to imagine that the forces exerted by stars and planets at the moment of birth can in any way shape our futures.

Neither is it true that the position of distant heavenly bodies make certain days or periods more favorable to particular kinds of action, or that the sign under which one was born determines one's compatibility or incompatibility with other people."

That astrology is preposterous is, of course, obvious to everyone except people with stupendous capacities for the willful suspension of disbelief. But such people are not apt to pay attention to what real scientists say.

"One would imagine," say the scientists from the depth of their innocence, "in this day of widespread enlightenment and education, that it would be unnecessary to debunk beliefs based on magic and superstition." But there is a telling tension between the scientists' belief that enlightenment is widespread, and their correct assessment that "acceptance of astrology pervades modern society" and contributes "to the growth of irrationalism and obscurantism." Actually, the fact that astrology is a huge and high-growth industry says something unflattering about human beings, and about our times.

There is an instinct, implanted deep in the human constitution: when people feel or want to feel intimations of immortality, they look up, beyond Gothic spires, beyond mountain tops, to the stars which seem, unlike people, timeless and autonomous. Astrology is a manifestation of that instinct.

Some people just must believe that the celestial order they think they see has some mysterious significance, and involves some overmastering intention that involves them. Of course there is more than a little vanity in the notion that we wee creatures are somehow connected to those heavenly bodies. This is a way of sneaking in through the back door a belief expelled through the front door in the sixteenth century —the belief that we are the center of a caring universe.

As the science of astronomy matured, we learned that the stars are not all that orderly and timeless. Indeed, there is a lot of wobbling and banging around in the infinitesimal fraction of the universe we observe. We know more about the heavens than the Babylonians did when preparing their star charts, and we know that the heavens are as susceptible to change as is (*pace* Kant) the moral law.

Astrology has a distant, parasitic relationship with perhaps the most beautiful science, astronomy. Astronomy has helped kill the parasite, at least among people open to evidence. But astrology lives on as a specious psychology, enabling people to classify themselves, for whatever pleasure that brings them.

Astrology is the sort of anti-intellectual intellectuality that flourishes in an unsettled age, especially among people foreign to learning but respectful, after a fashion, of what they fancy is learning. Especially since the early 1960s, the world has become a strange place to many Americans who sense a new and distressing unpredictability. Before then it was—or at least they liked to think it was—possible to know the kind of world that was just over the horizon of the future. Astrology bestows on its believers a sense of being not quite completely adrift on turbulent seas. It is odd but true that many people find it soothing to believe that they are under the predictable sway of stars.

Astrology is as old as the hills, but is, in its bizarre fashion, in tune with a major theme of modernist thought. The giants among the makers of the modern mind—Darwin, Marx, Freud—differed on many things, but not on this theme: human beings are subject to common forces. To the extent that people today believe that knowledge is power, it is because they think knowledge enables them to elude or at least swim with those forces, whatever they are. Astrology is an attempt at such powerful knowledge.

The distressed scientists insist, "We must all face the world, and we must realize that our futures lie in ourselves, and not in the stars." But there is no "must" about it.

[September 17, 1975]

Personality Against Character

Having been through the wringers of circumstance for 200 years —revolutionary, civil and world wars; depressions and recessions; television game shows and all the rest—the Republic is pausing to toss watermelon seeds at Aunt Min, toss firecrackers at the dog, and perform other rites of passage. Huzzah!

'Tis the season to look up at the flag, but also down at the good American earth. The theme of our national epic is Hawthorne's theme (in *Seven Gables*): "Life is made up of marble and mud." Our Federal City is rich in sparkling marble, but few Americans today can even imagine the sheer muddiness of the American experience.

Less than the mud of Valley Forge or Guadalcanal I mean the routine mud of common experience, the mud of sod huts on the undulating Nebraska prairie. In the United States even more than in most nations the achievements that stun the imagination were not performed by politicians, or generals with shiny boots, but by plain people with mud on their boots—the sort of people who walked to Oregon behind creaking wagons.

Today we are a nation which in the midst of the severe recent recession still managed to spend millions on "pet rocks." And today we are spending hundreds of millions on citizen band radios for the joy of talking to total strangers in a jargon unsuited for communicating thoughts unrelated to driving.

In such a fatted nation, where football telecasts are interrupted by commercials inciting suburban home owners to buy power saws for trimming the family elm, few can have the faintest idea of what it meant to settle in the middle of a Minnesota forest because the wagon axle had splintered. Or to face the task of clearing a farm, one ax stroke and one stump at a time.

Say what you will about Locke and Montesquieu, our national wagon got rolling because Americans came to believe (as it says in *Huckleberry Finn*) that "All kings is mostly rapscallions." That thought came easily to a man who had just broken his plow on a Connecticut rock, and who would rather buy a new one than to pay a penny to a distant monarch.

We have come a long way from sod huts and muddy boots, to an economy that produces billions of dollars worth of deodorants. And we may be learning what Mark Twain meant: "Soap and education are not as sudden as a massacre, but they are more deadly in the long run." The easing of the physical strain of American existence has proceeded apace with a general loosening of some social restraints.

Gracious, as recently as 1944 the library of the U.S. Naval Academy would not issue the novel *Forever Amber* to anyone of the rank midshipman or lower. American hedonism has come a long way, fast. And there are persons who think America's recent history is summed up by a Thurber cartoon that shows a woman perched on the arm of a sofa, talking animatedly to a circle of enthralled men. Behind her, a disgruntled woman says to another: "She built up her personality, but she's undermined her character."

Many thoughtful persons today think the Republic has more personality and less character than is healthy, that it is afflicted with a weakness—a form of decadence, really—that may be the fatal flaw of developed free nations. The theory is as follows.

The material success of capitalism—to which we owe the marble in our lives—has been made possible by habits of discipline that were reinforced by hardships of life in the Connecticut and Minnesota and Oregon trail mud. But abundance subverts such habits. And the dynamic of our abundance produces—indeed requires—a constant increase in consumption, and in appetites. This dynamic generates a culture of self-indulgence. Such a culture is incompatible with self-government, which is, after all, about governing the self. That is why one stanza of "America the Beautiful" is a kind of prayer: "Confirm thy soul in self-control. . . ."

So there is tension between the economic dynamic that inflames appetites and the need for discipline—political as well as economic—in a free society. Some persons say this is the "cultural contradiction" of capitalism. Others call it the "cultural consequence" of capitalism.

On the eve of its third century, the Republic's most pressing task is to demonstrate that political habits of restraint and moderation are compatible with an economic and cultural ambience that celebrates instant gratification of immoderate appetites. It is a national triumph of sorts that this problem of abundance confronts the descendants of the generations that walked through the mud to Oregon.

[July 1, 1976]

Printed Noise

The flavor list at the local Baskin-Robbins ice cream shop is an anarchy of names like "Peanut Butter 'N Chocolate" and "Strawberry Rhubarb Sherbert." These are not the names of things that reasonable people consider consuming, but the names are admirably businesslike, briskly descriptive.

Unfortunately, my favorite delight (chocolate-coated vanilla flecked with nuts) bears the unutterable name "Hot Fudge Nutty Buddy," an example of the plague of cuteness in commerce. There are some things a gentleman simply will not do, and one is announce in public a desire for a "Nutty Buddy." So I usually settle for a plain vanilla cone.

I am not the only person suffering for immutable standards of propriety. The May issue of *Atlantic* contains an absorbing tale of lonely heroism at a Burger King. A gentleman requested a ham and cheese sandwich that the Burger King calls a Yumbo. The girl taking orders was bewildered.

"Oh," she eventually exclaimed, "you mean a Yumbo."

Gentleman: "The ham and cheese. Yes."

Girl, nettled: "It's called a Yumbo. Now, do you want a Yumbo or not?"

Gentleman, teeth clenched: "Yes, thank you, the ham and cheese."

Girl: "Look, I've got to have an order here. You're holding up the line. You want a Yumbo, don't you? You want a Yumbo!"

Whereupon the gentleman chose the straight and narrow path of virtue. He walked out rather than call a ham and cheese a Yumbo. His principles are anachronisms but his prejudices are impeccable, and he is on my short list of civilization's friends.

That list includes the Cambridge don who would not appear out-doors without a top hat, not even when routed by fire at 3 A.M., and who refused to read another line of Tennyson after he saw the poet put water in fine port. The list includes another don who, although devoutly Tory, voted Liberal during Gladstone's day because the duties of prime minister kept Gladstone too busy to declaim on Holy Scripture. And high on the list is the grammarian whose last words were: "I am about to—or I am going to—die: either expression is correct."

Gentle reader, can you imagine any of these magnificent persons asking a teenage girl for a "Yumbo"? Or uttering "Fishamagig" or "Egg McMuffin" or "Fribble" (that's a milk shake, sort of)?

At one point in the evolution of American taste, restaurants that were relentlessly fun, fun, fun were built to look like lemons or bananas. I am told that in Los Angeles there was the Toed Inn, a strange spelling for a strange place shaped like a giant toad. Customers entered through the mouth, like flies being swallowed.

But the mature nation has put away such childish things in favor of menus that are fun, fun, fun. Seafood is "From Neptune's Pantry" or "Denizens of the Briny Deep." And "Surf 'N Turf," which you might think is fish and horsemeat, actually is lobster and beef.

To be fair, there are practical considerations behind the asphyxiat-ingly cute names given hamburgers. Many hamburgers are made from portions of the cow that the cow had no reason to boast about. So sellers invent distracting names to give hamburgers cachet. Hence "Whoppers" and "Heroburgers."

But there is no excuse for Howard Johnson's menu. In a just society it would be a flogging offense to speak of "steerburgers," clams "fried to order" (which probably means they don't fry clams for you unless you order fried clams), a "natural cut" (what is an "unnatural" cut?) of sirloin, "oven-baked" meat loaf, chicken pot pie with "flaky crust," "golden croquettes," "grilled-in-butter Frankforts [sic]," "liver with smothered onions" (smothered by onions?), and a "hearty" Reuben sandwich.

America is marred by scores of Dew Drop Inns serving "crispy green" salads, "garden fresh" vegetables, "succulent" lamb, "savory" pork, "sizzling" steaks, and "creamy" or "tangy" coleslaw. I've nothing against Homeric adjectives ("wine-dark sea," "wing-footed Achilles") but isn't coleslaw just coleslaw? Americans hear the incessant roar of commerce without listening to it, and read the written roar without really noticing it. Who would notice if a menu proclaimed "creamy" steaks and "sizzling" coleslaw? Such verbal litter is to language as Muzak is to music. As advertising blather becomes the nation's normal idiom, language becomes printed noise.

[May 8, 1977]

The Ploy of Sex

Like flowers groping toward the sun, millions of Americans are groping toward sexual nirvana, and they are guided by a book.

The Joy of Sex: A Gourmet Guide to Love Making has been on the bestseller list for more than a year, 755,000 hardcover copies are in print at $12.95 and the first paperback press run was 600,000.

Modeled on *The Joy of Cooking,* a standard wedding present, this new *Joy* book urges happy couples to satisfy every conceivable appetite.

"Chef-grade cooking doesn't happen naturally. . . . It's hard to make mayonnaise by trial and error, for instance. Cordon Bleu sex, as we define it, is exactly the same situation. . . ." Exactly. No damned nonsense about untutored trysts in the age of textbooks.

This book is edited by a—seriously!—Dr. Comfort and when he says "a little theory makes sex more interesting," you brace yourself for a dose of metaphysics. Metaphysics, in sex as elsewhere, often is, the production of bad reasons for satisfying strong desires.

"The starting point of all love-making is close bodily contact." Right. I'm with you so far, Dr. Comfort. "Love has been defined as the harmony of two souls and the contact of two epidermes." Obviously this is not your run-of-the-muck guide to unspeakable acts and unnatural practices. Indeed, according to Dr. Comfort, there are no such things: "There are no rules, so long as you enjoy, and the choice is practically unlimited."

Well, practically. Even the frisky Dr. Comfort can discover only a finite number of erogenous zones and means of titillating them. But he makes a stab at infinity: "The pad of the male big toe . . . is a magnificent erotic instrument. . . . Make sure the nail isn't sharp." Thanks, Dr. Comfort, we needed that.

Much of the text reads about as joyfully as a Volkswagen owners' manual. It resembles those impenetrable instructions that accompany children's toys: "Snip perforation Q, fold side Y, insert tab T into slot Z." I usually cannot ever *find* tab T. Similarly, this book directs my attention to erogenous zones in places I didn't even know *are* places.

To accommodate slow learners like me, the book thoughtfully incorporates Yogi Berra's epistemology ("You can observe a lot by just watching.") The text is illustrated with what today are called lyrical drawings. That's "lyrical drawings" as in: "Psst, fella, wanna buy some 'lyrical drawings'?"

In every regard, this is a democratic sex book. The man and woman portrayed in the drawings—a real couple, we are told—are quite plain. (Plain. But not normal. They obviously have abnormal stamina and are double-jointed in most joints.) The text insists that all erogenous zones are created equal and all uses of them are of equal dignity.

And now the book is available in paperback. None of that "Let them eat cake" attitude. Now the masses can nibble on each other, with the guidance of an inexpensive book. As Yogi Berra said when he heard a Jew had been elected Mayor of Dublin: "It could only happen in America."

The gourmet analogy is a coy device for stigmatizing any inhibition as a hang-up: "Most wives who don't like Chinese food will eat it occasionally for the pleasure of seeing a Sinophile husband enjoying it and vice versa." So sex and food are equated, and "sex" and "love making" are synonyms.

Such books popularize the image of human beings as mere collections of sensory capacities. Perhaps the pathetic millions who buy this book *are* just collections of sensory capacities. Perhaps sex manuals just add to the public stock of harmless pleasure.

More likely, these books, which sternly warn against making love or mayonnaise by trial-and-error, produce sexual unhappiness. They turn normal lusty people into perfectionists whose fear of failure prevents them from even getting started. How many of us have ever made mayonnaise?

It is no longer enough to be lusty. One must be a sexual gourmet. If you think this is harmless, consider another depressing food analogy. Many people enjoyed plain meat-and-potatoes until some sadist gave them a gourmet cookbook. Now they still eat meat-and-potatoes, but they feel a little guilty. And oddly unsatisfied. A lot more gourmet cookbooks are sold than there are gourmet meals cooked.

[January 29, 1974]

Is That a Red Dog in the Seam?

'Tis nearly autumn, season of mists and mellow fruitfulness, and God as a punishment has sent another football season. The season began eleven weeks ago on June 19 in Lubbock, Texas, with the Coaches All-America game. Game? I mean *classic*. (They're *all* classics, son.)

From sea to shining sea there are little seas of shining sousaphones and trombones. The marching bands are preparing those half-time shows that sound like explosions in ammunition dumps and feature 200 undergraduates spelling out "Ecology!" while playing "Born Free." Large young scholars at Texas A&M and Michigan State are hitting the (play) books, and the overtly professional teams are finishing the "preseason" season that is more than half as long as the college season. It won't be until late January that the last knee cartilage has torn; and ABC's Keith Jackson has regained his composure after the last 2-yard plunge off tackle; and some "color" man (it is hard to tell the colorful rascals apart) has said for the 783rd and final time, "These are two physical teams, so the game will be won in the trenches."

Don't get me wrong. I am a sports fan (if rooting for the Chicago Cubs counts; I won't press the point). I chose to do graduate work at Princeton because it is midway between two National League cities. I am not one of those angry Calvinists who say football—the territorial imperative, Manifest Destiny between the goal lines—caused the Vietnam war. I do not believe big-time football is out of place at universities that offer courses in "packaging." Neither do I believe the masses should watch less football and read more Proust because football does not tax the brain. In fact, my complaint is that football is too cerebral. Like almost everything else, it has become too complicated for laymen.

Some philistines say football is more brawn than brain. Does the "nickel defense" counter the "shotgun offense"? Where are "the trenches"? In 5,000 words or less, distinguish between "blitzing," "red-dogging," "stunting" and "shooting the gap." What does a "rotating zone" do when not rotating? What does it mean to "hit the seams"? (The seams on a rotating zone, silly.) Explain, using diagrams, the difference between "the seams" and "dead space." What do you call teams that capitalize on fumbles? (Stumped? A Louisiana football writer calls them "capitalistic.")

Those are not questions for fans who scored below 700 on their College Board exams. Philosophy may be just a series of footnotes to Plato. And long ago Willie Keeler, baseball's Plato, said all there is to say about his science: "Hit 'em where they ain't." But football is the last frontier of intellect. It is *very* complicated and getting more so, as viewers learned in 1974 when a CBS expert revealed, with a flick of his mind's scalpel, that the Miami Dolphins won the Super Bowl because their defense had "so many different variations." Different variations beat the cleats off unvarying variations every time, and only incorrigible anti-intellectuals will say: variations, schmariations, the Dolphins won the way grade-school kids win, by giving the ball to the big kid (Larry Csonka) whom the other kids can't tackle.

Naturally, technocrats have taken over on the sidelines. Vince Lombardi once was the paragon of coaches. Even in repose he resembled a tiger with a toothache, roaring things like (to a player with splintered ribs) "Shake it off!" and (when he saw room for team improvement) "Jerks! Jerks! Jerks!" Today the George (Blood and Guts) Patton of the playground is Woody (Three Yards and a Cloud of Dust) Hayes of Ohio State who, when angry, which is most of the time, turns colors that are becoming on an Easter egg. But Hayes, who relies on terror rather than science to produce victory, is a relic of football's pre-intellectual era. He is as anachronistic as the prose style of Grantland Rice, whose most famous sports report began:

"Outlined against a blue-gray October sky, the Four Horsemen rode again. In dramatic lore they are known as Famine, Pestilence, Destruction and Death. These are only aliases. Their real names are Stuhldreher, Miller, Crowley and Layden. They formed the crest of the

South Bend cyclone before which another fighting Army football team was swept over the precipice at the Polo Grounds yesterday afternoon as 55,000 spectators peered down on the bewildering panorama spread on the green plain below."

Modern sophisticates may think Rice laid it on a bit thick. But his was the naïve eye of the pre-scientific age, when even learned students of football considered it just a game of excess in which large people shoved one another around, the better shovers winning. Then literary license allowed poets to call football a bewildering panorama of horsemen and cyclones sweeping folks over precipices. Today football, like microbiology and other mature sciences, is the province of icy specialists, like George Allen, coach of the Washington Redskins.

The walls of Allen's inner sanctum are covered from floor to ceiling with charts and lists and other documents of scientific football. Says Allen of his charts and lists: "I certainly believe in visuals." Visuals! What will this Mephistopheles think of next? The Dallas Cowboys have gone beyond even visuals. When selecting players the Cowboys use *computers,* for Pete's sake. Visuals give the Redskins confidence, but the Cowboys have what Coach Tom Landry calls the "confidence factor." Alas, progress has its costs. Rooting for Allen's Redskins is like rooting for Union Carbide and reading about Landry's Cowboys is like reading the *Journal of Econometrics.*

It is, I know, naughty to commit sociology promiscuously, but if you hold up football to the bright light of the social sciences you will see that it mirrors modern life. It is committee meetings, called huddles, separated by outbursts of violence. And in football, as in other branches of higher learning, a special language is the sign of erudition. A university administrator writes to a faculty member: "Having prioritized available funding, your request for staff-support facilities cannot be actuated at present. Student throughput indicators show marked declining motivational values in subsequent enrollment periods in elective liberal-arts choices." Similarly, football's clerisy speaks the arcane language of "visuals" and "confidence factors" and "seams" and zones that rotate.

Modern life is a testing burden for laymen, and there still are lots of them around. If you are one, you may feel inadequate as you sit in front of your television (it is a "visual," by the way) watching the Cowboys, chock-full of confidence factors, split the Redskins' seams with different variations. But science never promised us a rose garden, and if you can't stand the intellectual heat, get out of the seminar.

[September 6, 1976]

Frozen Music in New Orleans

About eight years ago John McKeithen, then Louisiana governor, visited Houston's Astrodome and suffered an attack of a dread American disease, Thinking Big. He exclaimed: "I want one just like this, only bigger."

So today Moon Mullins is standing where third base will be and he is feeling pentacostal. Mullins, a public relations man for the Louisiana Superdome, due to open in April, is suffused with the spirit of Progress, as exemplified by the "sports and exposition facility" that has sprouted, like a gargantuan silver mushroom, from the spongy Louisiana soil.

Gesturing through the chilly gloom at the dome 273 feet above, Mullins says this is the world's largest "unobstructed room," a carefully crafted phrase that obliquely recognizes what the Superdome builders resent: St. Peter's in Rome is larger. But St. Peter's is "obstructed"—it has pillars. And the Pope's "room" doesn't have Astroturf, known here as "mardigrass."

Louisianans amended their constitution to authorize building the Superdome on the assumption that it would cost $35 million and be self-supporting. But the deviously drawn and advertised amendment allowed the Superdome agency to issue state-backed bonds. Today the official "final" cost is $163 million (fifteen times the price of the Louisiana Purchase) and skeptics say it will be closer to a quarter of a billion dollars. The assumption that it will be self-supporting is based on a series of implausible assumptions about its use.

But McKeithen is content. The puny Astrodome would fit inside the Superdome, under the 9.7-acre roof where 97,000 conventioneers, 80,101 football fans, or 64,537 baseball fans can nestle.

Architecture, said Goethe, is frozen music, and New Orleans is a musical city. So the Superdome builders, like God, look upon their work and find it good. Words do not, alas, fail them. A press puff says:

"Over the streets named Bourbon and Basin, St. Charles and Desire . . . over sad jazz erupting into laughter . . . the Superdome rises. . . . It is the depository of Louisiana's belief in itself and a budding, exhilarating, moving certainty that tomorrow can be now. . . . The immensity of the Superdome was founded on the principle and the belief that we inhabit an age when even tomorrow is in danger of obsolescence. . . . Superdome is beyond tomorrow."

Before erupting into laughter about an immensity founded on a principle, consider the budding possibility that the Superdome is, indeed, a glimpse of the future.

A civilization's values are apparent in its most adventurous architecture. Cathedrals expressed in stone and glass Christian civilization's vision. Structural steel and elevators made possible skyscrapers, which are efficient responses to urban land values, and monuments to commercial civilization. Now come Superdomes, the adventurous architecture of—what? A bread and circuses civilization?

Or is this a new cathedral age? McKeithen did once say: "Let's open the Dome with Billy Graham and the Pope. Put one in each end zone. There's enough room." He may have been joking, but I do not know how to tell when a Louisiana governor is joking.

Supporters of the Superdome know their building is no laughing matter. Moon Landrieu, mayor of New Orleans, knows, as Gershwin did, that "they all laughed at Rockefeller Center." Now they're fighting to get in. Mayor Landrieu says: "They called King Ludwig mad, you know, for building all those elaborate castles. And maybe he was, because he killed himself. But now thousands of tourists come to see the castles. So Bavaria's rich, and old Ludwig's a hero again."

And so in the eighth decade of the twentieth century, Mad King Ludwig has become the patron saint of provident government. And supporters of the Superdome are stating that it is "a monument to man's daring imagination, ingenuity and intelligence." Consider, if you have the nerve to do so, the possibility that their statement is accurate.

[November 30, 1974]

On a Hill Above Weimar

Elie Wiesel, the novelist, was a young boy when he, his parents, baby sister and other relatives were deported from Hungary to Auschwitz. His mother, sister and other relatives died there, and he and his father were moved to Buchenwald, where his father died shortly before the Americans arrived. In Wiesel's autobiographical novel, *Night,* the protagonist, a boy, remembers Auschwitz:

"Not far from us flames were leaping up from a ditch, gigantic flames. They were burning something. A lorry drew up at the pit and delivered its load—little children. Babies! . . . Never shall I forget the little faces of the children, whose bodies I saw turned into wreaths of smoke beneath a silent blue sky."

It is a time, thirty years later, for remembering the Holocaust, and for forswearing the innocence that should have died with the millions of innocents. It is time to remember Buchenwald, the concentration camp located on a hill above Weimar, home of Schiller, Liszt and Goethe.

Around noon, April 11, 1945, the SS men left. That afternoon U.S. tanks rolled into the camp. As camps went, Buchenwald was not ambitious. Fewer than 60,000 people died there. The principal killing camps were in the east, outside of Germany. But Buchenwald provided the West with the first shattering sight of what can be done when a modern state is put on the service of radical evil. It is the joining of ancient sins and new forms of tyranny that have made this century a charnel house —the worst century in terms of the quantity of inflicted death, and in terms of gratuitous, ideological beastliness.

The counter-intuitive is always fascinating, and the Holocaust refutes those modern intuitions that flatter men. "What a piece of work is a man!" exclaimed Hamlet, who knew better, "How noble in reason; how infinite in faculties! . . . in action, how like an angel in apprehension, how like a god—the beauty of the world; the paragon of animals!" In 1936 a piece of work called Hermann Göring arrived late at a reception at the British Embassy in Berlin, explaining that he had been shooting. British Ambassador Eric Phipps, who was leaving Berlin and thus could be incautious, replied, "Animals, I hope, Your Excellency." One could not be sure about such things in the middle of the twentieth century in the middle of Europe.

The Holocaust was not just the central event of the twentieth century, it was the hinge of modern history. It is the definitive (albeit redundant) refutation of the grand Renaissance illusion that man becomes better as he becomes more clever. The most educated nation in Europe built modern transportation systems and machines, and transported Jews to machines of mass murder.

The Holocaust, like most modern atrocities, was an act of idealism. It did not make economic sense, and it hindered the German war effort, but it was a categorical imperative for Hitler, and hence worth all the trouble. Genocide requires bureaucratic organization to bring together men and material, railroad rolling stock and barbed wire, Zyklon-B gas and ovens. As the Israeli court in the Adolf Eichmann trial noted, acidly: "The extermination of the Jews was . . . a complicated operation. . . . Not everywhere was convenient for killing. Not everywhere would the local population submit to the slaughter of their neighbors."

The size of the gas chambers defined the issue. Their purpose was not the punishment of individuals for violations of known laws. Rather, their purpose was the liquidation of a people whose crime was existing. A task of that scale required paper work, record-keeping, tidiness: a loudspeaker in one camp announced the request that anyone planning suicide should, please, put a note in his mouth with his number on it. Eventually the bureaucracy tattooed victims of what it called the "negative population policy." It is still with me, the chill I felt on a warm summer night in 1964 in a cafe in Brussels, when I saw the blue numbers on the forearm of the matron at the next table.

There was nothing new about cruelty to Jews and other vulnerable people. For centuries Jews and gypsies (also Nazi victims) were considered "uncanny" and "not belonging," and were hounded through history. There was a time when some Rhineland nobility hunted not foxes but gypsy women. To force the women to run from the baying hounds even when desperately weary, the huntsmen lashed the women to their babies. This occurred in the eighteenth century, the age of reason, and good horsemanship.

[April 8, 1975]

Losing Well Is the Best Revenge

The recession is a little late arriving here.

Gambling profits are running ahead of last year's record pace, when Nevada casino profits increased 14.2 percent, topping $1 billion for the first time. What Nevadans quaintly call "the gaming industry" turns a consistent 19–20 percent profit, which means that about $5 billion was gambled by the thirty million visitors to the state. The visitors gambled, on average, $166, lost $33.

Twenty thousand tourists visit Caesar's Palace on a busy day. The new MGM Grand, which cost $124 million to build, grossed $88 million in its first nine months, $34.4 million in one quarter that produced a pre-tax profit of $9.7 million. The hotel occupancy rate on the Strip is well over 90 percent.

Las Vegas has come a long way since December 26, 1946, when the late Benjamin "Bugsy" Siegel, a California Capone, donned white tie and tails and opened the Flamingo Hotel on what was to become the Strip. But the Founding Father of Las Vegas did not live to enjoy it. On June 20, 1947, some business rivals shot Siegel in the head four times, which was the kind of entrepreneurship that made Las Vegas wince. Respectability is a relative matter, and Las Vegas took a long step toward it in 1967 when Howard Hughes bought six hotels. Now there is $234 million from the Teamsters Union pension fund invested in Nevada casinos.

Local boosters call this "the city of schools and churches." Boosters are the same everywhere. When you get off the train in Oxford, England, you see—or, at any rate, you used to see—a sign proclaiming: "Oxford—Home of Pressed Steel, Ltd." Although Las Vegas schools and churches are numerous and exemplary, the city prospers for three other reasons. The city's industry is well suited to a period of inflation and general economic uncertainty; it is as American as the pursuit of happiness; it satisfies some ache in the human soul.

The federal government is tongue-lashing consumers to keep the money moving, to go out and spend for their country. And in an age of high inflation you don't have to be Ben Franklin to know that a penny saved is ridiculous, and is less fun than a penny gambled.

When "Big Julie" Weintraub was seven, living in Brooklyn, he played poker using Necco wafers as chips. Today he is the Socrates of Las Vegas. He says: "The man who invented gambling was bright; the man who invented the chip was a genius."Most chips are plastic, something an American can believe in, unlike the currency. Softly, softly falls the plastic on the green baize, with never a metallic clank to recall reality. Chips are play money for adults who grew up (and who didn't?) playing Monopoly, every tyke a real-estate speculator, rolling the dice for the joy of grinding kid sister into bankruptcy by foreclosing on Baltic Avenue.

This city's singleminded devotion to money—making it from people who play with it—breeds a democratic belief in tolerance and equality. In his treatise on downward social mobility in the desert *(Fear and Loathing in Las Vegas)*, Doctor Hunter S. Thompson, amateur anthropologist, notes the tradition of tolerance: "If Charlie Manson checked into the Sahara tomorrow morning, nobody would hassle him as long as he tipped big." Las Vegas believes in equality as the eighteenth century did:

> All men are created equal,
> They differ only in the sequel.

Most people are losers in life, most of the time, or suspect that they are. But in Las Vegas, losing is not your fault. It is the fault of Lady Luck (née, the house odds), in whose bright eyes all men are equal.

See that man in the powder blue doubleknit jump suit and white patent leather shoes, the man shoving quarters into a slot machine that he knows is programed to keep 15 percent of his money? Back in Steubenville he wears gray and lives gray, selling aluminum siding and keeping a sharp eye on the petty cash. All year 'round money rules his life. Here he throws some of it away, as a Declaration of Independence. For him, losing well is the best revenge.

[May 20, 1975]

Zarathustra at Forty: Elvis

By 9:30 P.M. the line for the midnight show has snaked around the perimeter of the casino in the Hilton International. Five twenty-foot-high letters in lights out front explain why: "Elvis."

It's oldies but goodies time in Vegas. The Elvis Presley line runs by the Hilton's Vestal Virgin Lounge where another 1950s bad boy, Mort Sahl, social commentator, is telling jokes about Jack Ruby to conventioneers. Down the Strip, the MGM Grand has Fabian (surely you remember "Turn Me Loose," "I'm a Tiger"), living proof of the 1950s inventiveness with durable plastics. But the Hilton has the senior lion of the pop music jungle, the man who rivaled Eisenhower as a 1950s mass phenomenon—Elvis, cultural disturber of the general's decade.

At 12:38 A.M. the warm-up acts are over and the audience is, in its master's words, itching like a man in a fuzzy tree. The orchestra surges into Richard Strauss's *Also Sprach Zarathustra* and—Shazam!—Elvis shambles on to the stage.

Zarathustra is forty now, and forty pounds overweight, with a shape not unlike Henry Kissinger's. There are strains on the seams of his spangled suit, but no matter: the legs are shimmying—Elvis has his act together.

The boy from Tupelo, Mississippi, has come a long way from Heartbreak Hotel to the Hilton, where twice a year he sells out the ballroom, twice a night, for two weeks, the biggest draw in the biggest entertainment factory, a bigger Vegas draw than that durable fragment of the 1940s, Sinatra. But even though Elvis is knocking down well over $100,-000 a week, he's still a good ol' boy.

Wrap him in yellow and orange sequins, bathe him in red and blue lights, back him with flutes and violins, his animal energies still overwhelm even Vegas versions of refinement. He is still Tupelo and Memphis, chartreuse Styrofoam dice dangling from the rearview mirror, a pack of Lucky Strikes rolled in his T-shirt sleeve, running a comb through his DA, asking, "Woncha wear my ring around your neck?"

Between songs a sideman drapes pastel scarves around the beefy Presley neck. With his I'm-cool-about-being-late-for-study-hall saunter, the rolling gait of manchild in the high school corridor, he ambles to the front of the stage, dabs his sweaty brow with a scarf, and awards it to one of the groupies planted front and center. Another groupie hands up a photo of Elvis circa 1957, when he was svelte. He scrutinizes it, bemused, and sighs, almost to himself: "Godalmighty!"

Self-parody is his refuge from the boredom of songs sung too many times. He feigns difficulty getting the left side of his upper lip into its famous curl. He pokes it up with a finger, muttering: "When I was nineteen it worked just fine."

The heavy-lidded stare, the thick-lipped semi-pout, the curled upper lip, more overtly sensual than the androgynous Mick Jagger, more titillatingly antisocial than James Dean, Elvis still stirs those passions that cause people to lose their kinship with the divine.

Twenty years ago Elvis brought the Republic face-to-face with a polarizing preoccupation—hair, recognized through the ages as the dread sign of barbarism. His sideburns crept a provocative inch or so down his cheeks, and his hair was as long as Eric Sevareid's hair is now.

An appearance on the Ed Sullivan show terrified network officials into ordering that all cameras be focused above the subversive pelvis. Today the nation munches newfangled potato chips while staring across the living room into Cher's navel.

At 1:33 A.M., with time running out and his subjects hungering, he throws the red meat: "Don't be cruel to a heart that's true."

"YOUAIN'TNUTHIN'BUTAHOUN'DOG!"

Dy-no-*mite!*

The older members of the audience, those over forty, are content. They stood in line to see a legend, and did. But the younger members of the audience had more fun. Elvis is for the war babies what the madeleine was for Proust. One nibble lets loose a flood of remembrances of adolescence past—not the most pleasant remembrances, perhaps, but the most vivid ones they have.

[March 25, 1975]

Sensory Blitzkrieg

As the American plowman homeward plods his weary way, he is weighed down by the certainty that his offspring, who in a less blessed age would be curled up like little Lincolns on a warm hearth reading *Pilgrim's Progress*, will be huddled around a warm stereo (curse you, rural electrification!) listening to Led Zeppelin.

When Led Zeppelin descended on Washington recently, 18,700 concert tickets were snapped up in three hours. Some people who could not get tickets vented their disappointment just as, perhaps, disappointed Viennese did when they could not get into a Mozart recital. They threw bottles at the police.

The tempestuous behavior by disappointed ticket seekers called to mind the sporadic violence in gas station lines last winter during the oil embargo, the other recent shortage of a life-sustaining commodity. Rock music is to the youth culture what gasoline is to the more adult culture: it is that without which life lacks tang.

Led Zepplin, one of society's vigorously vibrating ganglions, has been a roaring success for six years. That is a Methuselahan age among rock groups, most of whose names (remember The Peanut Butter Conspiracy? The Strawberry Alarm Clock? Stark Naked and the Car Thieves?) are writ on water.

Tom Zito—who, as God is my witness, is paid by the Washington *Post* to know such things—says Led Zeppelin is "world's most popular rock 'n' roll band." That may be an extravagant encomium, but the four Led Zeppeliners are weathering the recession nicely. On a recent two-month tour in their private 707, they streaked like jagged lightning around the Cities of the Plain, siphoning $5 million from American youths (56,800 in one sitting in Tampa).

Before I bought the Led Zeppelin album to which I am listening as I write this, the group had sold ten million albums for reasons that, as I listen, no longer elude me. To understand the group's charm for the young, begin with Mr. Zito's description of the group in action:

"Onstage, Jimmy Page hangs his gold . . . guitar at crotch level, jumps across the raised platform, raises his arm skyward and sends a heavy hundred-decibel wave of metallic sound surging through a bank of amplifiers. Robert Plant's screaming vocals filter through an echo unit and blend with the high-pitched barrage."

According to Mr. Zito's exegesis, Led Zeppelin has a "style built upon the increasing ability of amplified guitars to sound purely electric," the amplification "of the sound of a guitar's metal strings into screaming, twangy, metallic rhythms." The lads specialize in "screaming solo attacks" and "soaring guitar blitzes."

Gracious. Barrages, attacks, and blitzes. It sounds like music to invade Poland by, not to soothe a savage breast. But Led Zeppelin's sound, called "heavy metal" music, is soothing to adolescents, who are notoriously insecure and, for that reason, passionately egalitarian. They derive security from a shared private culture that is unintelligible to adults.

Hundreds of thousands of adolescents happily shivered in the mud at the 1969 Woodstock, New York, rock festival, which became known, revealingly, as "Woodstock Nation," a brief secession from the adult nation. An auditorium full of adolescents, pounded like soft wax by the screaming, twanging barrages of "heavy metal" decibels, experiences transports of bliss, and is transported back to Woodstock Nation, cozy together again.

The "heavy metal" sound blows listeners down like grass before the wind. That might not sound like fun to you but, as is said, different strokes for different folks.

Adolescents stand on the brink of the adult meritocracy. They are constantly examined and graded and ranked. But Led Zeppelin, and adolescents love of it, is immune to critical scrutiny. Led Zeppelin performing, like Mt. Vesuvius erupting, overloads the ear, overwhelms the ability to hear nuances. Indeed, "heavy metal," a sensory blitzkrieg, is the definitive revolt against nuances.

It is impossible for adults to apply their hated standards to this music that is a refuge from the adult world. As regards "heavy metal," all ears are created equal.

[February 15, 1975]

Billy Jack and Other Melancholy Political Fables

I am a grown man, with a happy home, so what am I doing at eleven o'clock on a week night in a nearly empty theater watching a two-year-old film?

Research.

Research about the political fantasies of some young American moviegoers. I am learning that in a world of change, the fantasies are a constant. They are pessimistic and may indicate why so much of youthful dissent and politics consists of theatrical gestures, irrelevant to changing the world. It is quite striking. From one decade to another, the most popular films about the "youthful scene" are melancholy fables about youth's hardships at the hands of older people.

Ten million people, most of them young, have paid $30 million to see *Billy Jack,* which evidently portrays the world as many young Americans gloomily believe it to be. In fact, *Billy Jack* is a third-generation gloom-mongering film devoted to a theme that is perennially popular with large numbers of young Americans. The theme is that young people are very noble, older people are boorish or worse, and older people like nothing better than using their control of the world to force young people to come to terms with an ignoble society.

These are films about the desperate maneuvers by which sensitive youths elude the philistinism and intolerance of older people whose only goal in life is to break every free spirit to the saddle of materialistic, achievement-obsessed America. Young people today, as in the 1950s and 1960s, think the world is picking on them.

Billy Jack is a Zarathustran anti-war ex-Green Beret half-breed Indian who kicks lots of people in the face (barefooted, which evidently is the Green Beret way) and sprays 30–30 bullets at other folks to protect a pacifist "freedom school" on his reservation from the predations of the narrow-minded townspeople.

The school's rule is that everyone "must get turned on by creating something." No one in the lagoon of individualism is turned on by trigonometry or Latin grammar, but everyone plays either the guitar or the recorder and is heavy into "rug weaving, psycho-drama and role-playing, yoga meditation—things the townspeople could never understand."

The townspeople are cousins of those peckerwoods who shotgunned Peter Fonda in *Easy Rider*. Their mood goes rancid when the deputy sheriff's daughter, Barbara, emigrates to the reservation to escape Dad, who was not understanding (he slugged her) when she reported home from a stay at a Zen commune bearing hepatitis, an abscessed tooth and a pregnancy. The spoiled, cowardly son of the local Babbitt has a lot of run-ins with Billy Jacks, who makes him drive his Corvette into the lake. To get even, he rapes the "freedom school's" middle-aged headmistress, so Billy kills him right before killing Barbara's father who has become a nuisance.

Soon things look grim for the school, and for pacifism in general, because Billy is holed up in a shed, surrounded by an army of state troopers. Being part Indian, he wants to die, but being part white man, he agrees to negotiate, and surrenders in return for a government grant for the school. Then he is marched off to jail while the pacifists, their haven secured by his carbine, give him a clenched fist salute.

Billy, evidently, is meant to be a symbol of the price innocent youth must pay for being idealists in a world insanely resentful of their goodness. Freedom is swapped for survival in this town-gown morality play.

Billy Jack is the third in what promises to be an on-going series of movies about embattled youth. In the 1950s there was *Rebel Without a Cause*. In the 1960s there was *The Graduate*. These films appealed to the same age group and have similar themes. Their differences are those one would expect, given their chronology.

James Dean, the hero of *Rebel*, played an unhappy teenager, exquisitely alienated from his affluent parents, a pair of materialistic fossils who wanted him to get good grades in his new high school, where he had trouble communicating with his less sensitive peers. In fact, Dean could not communicate with anyone but Sal Mineo, a studious type nicknamed Plato, whose father, a tycoon, has abandoned him, and Natalie Wood, whose parents were nothing to write home about either. Because his parents were nags, Dean worked out his identity crisis on the edge of juvenile delinquency. In the end Mineo was holed up in a planetarium surrounded by dumb police. They tried to negotiate but after a bit of palaver, Mineo got shot, Natalie got Dean, Dad got a glimpse of how crumby the world is to young people, and Dean got a promise from his Dad that he would try to communicate.

Rebel appeared in the days of Juvenile Delinquency Chic, which was a warm-up for Radical Chic. Norman Mailer discovered true community among Brooklyn's fighting street gangs, and Leonard Bernstein set Romeo and Juliet among Manhattan gangs. Many advanced thinkers are anxious to see delinquents cast as romantic figures, heroic misfits shaping their own destiny on turf carved from a world they did not make. To some, delinquents were as romantic as disaffected college students were to become, as romantic as Indians on the warpath carving "nations" out of South Dakota.

In Hollywood, hard-eyed men, not a tad romantic, knew that romanticism comes in strange guises, and can always be packaged and marketed. By the time Dean died he had demonstrated that you can become a culture hero, and rich in the bargain, by portraying young men nagged by old blisters who think young men should be less moody or idealistic (the two are never clearly distinguished in films) in order to advance within the "system." And along came Dustin Hoffman, Benjamin in *The Graduate.*

Benjamin had graduated and was facing a trauma. His materialistic father wanted him to quit floating around in the swimming pool and get a job, or maybe go to graduate school (which any fool could see was his true vocation). Floating Benjamin knew he did not need and could not perform a job. He was already pampered beyond the dreams of avarice, and was a reverse Renaissance man. His incompetence was catholic.

Like Dean before him and Billy Jack to come, Benjamin was as inarticulate as a minnow. In fact, this jaded twenty-one-year-old seemed comatose until confronted by Mrs. Robinson prancing naked. Benjamin was incompetent at avoiding seduction and at checking into a hotel room. We and Mrs. Robinson soon learned that he had spent four years among the fleshpots of Eastern academia but remained incompetent at what interested Mrs. Robinson.

The movie opened with Benjamin jetting home from a gift (his university education) to receive a gift (a red sports car) which his father was giving him because he endured the first gift. Mrs. Robinson gave herself to him, but she was no dummy. She knew Benjamin couldn't cope with something (like herself) more complicated than a red sports car, so she unwrapped herself and gave Benjamin painstaking operating instructions. But no one would give Benjamin what he wanted, a "meaningful" role in life.

For a long time it seemed that he would only find a "meaningful" role if someone (Dad, probably) floated it out to where he dozed, becalmed in the pool, when not trysting with Mrs. Robinson. But quicker than you can say "coo, coo, cachoo Mrs. Robinson," Benjamin discovered that Miss Robinson could make life meaningful, so he yanked her from the altar, where she was about to marry a dullard who worked within the system, and eloped with her on a city bus. (Incompetent to the end, Benjamin ran out of gas on the way to the church.)

The common theme of these films is that older people are everywhere, so the best thing to do is to hop a bus or get Billy Jack to take off his boots and kick folks in the head.

The world is like school, a useless, unpleasant, achievement-oriented place run by nags. The only imaginably decent world is . . . a school, but a school smack in the middle of an Indian reservation, where some noble savages will be savage enough to keep the nags at bay while the young people get heavy into psycho-drama.

Films like these strike powerfully responsive chords in precisely those young people who fuel protest movements and talk incessantly about the need to change America. But the popular, frivolous message of these films is one of timidity, retreat, escape. They are full of the whiny rhetoric of self-righteous but scared people who never expect to be anything but victims.

[May, 1973]

D Is for Dodo

In Randall Jarrell's *Pictures from an Institution,* the wittiest novel about academia, a foreign visitor exclaims, "You Americans do not rear children, you *incite* them; you give them food and shelter and applause." One recent form of applause for America's college youth has been "grade inflation," the bestowing of high honors on almost everyone. But now there are small portents of significant change. Yale has reinstituted the F grade, and Stanford has reinstituted the D. These and similar developments elsewhere, quiet as snow, may be the first stirrings of a counterrevolution against the strange egalitarianism that had helped produce "grade inflation."

Since the mid-1960s there has been an astonishing rise in under-graduates' grades. At Harvard 82 percent of the class of '74 graduated cum laude or better. Recently the average grade at Stanford was A-minus; at Vassar 81 percent of all grades were A's and B's; at Amherst 85 percent; more than half the University of Virginia student body made dean's list. In a decade the percentage of A students at the University of North Carolina doubled and the average grade at the University of Wisconsin rose from C-plus to B-plus.

Some people (mostly proud parents and students suffering delusions of excellence, or adequacy) argue that grades are not inflated. They say students have suddenly become much smarter. But that would be implausible even if the steep rise in grades had not coincided with a steep decline in the average scores of high-school students on the Scholastic Aptitude Test, a decline now in its twelfth consecutive year.

Some of the real reasons for grade inflation are mundane and practical. In a period of tight budgets, some academic departments use easy A's to lure large enrollments and justify large budgets. Some teachers consider gentle grading a matter of simple prudence now that student evaluations of faculty are weighed when faculty promotions are granted. Some of the minority students pulled into universities by "affirmative action" programs would be swept right out if teachers did not relax standards, and this relaxation tends to raise the "floor" under all grades.

But the primary cause of grade inflation is ideological. It is the egalitarian passion against "elitism."

The word elitism is of recent vintage: it does not appear in the *American Heritage Dictionary* published in 1969. It is a pejorative label for social philosophies opposed to the notion that rigorous egalitarianism is a democratic imperative. And elitist is a label for people (like me) who believe that, frequently, egalitarianism is envy masquerading as philosophy.

Let us clear our minds of cant. Surely a just society is one in which people deserve their positions, and in which inequalities are reasonably related to reasonable social goals. Justice requires a hierarchy of achievement—unless all achievements are of equal social value, in which case all inequalities are arbitrary and illegitimate "privileges." Something like that extreme egalitarianism enjoys a vogue in academic circles, and helps produce grade inflation.

The lowering of academic standards—higher grades, lower admission standards, fewer examinations and required courses, more emphasis on "relevant" materials at the expense of high culture—reflects rejection of the elitist idea that some achievements, and hence some achievers, are more praiseworthy than others. A few years ago two Cornell professors posed this question to that institution's president: "If we prove to you that an arts-and-sciences student can now receive a B.A. degree at Cornell, and thus be presumed to have acquired a liberal education, without having to read a line of Plato, the Bible, Shakespeare, Marx or Einstein, would you consider this to be evidence that there is a crisis in education at Cornell?" They received no reply. Some academic leaders are shocked speechless by the elitist idea that not all pursuits are created equal.

Such anti-elitism is a natural consequence of the decay of academic liberalism into romanticism tinged with a familiar American folly, youth worship. Romanticism holds that each young person contains a creative "self" and that the purpose of education is "self-realization." Hence education is less a matter of putting something in than of letting something out. The romantic assumption is that education should "liberate" young personalities to flower, free from personal inhibitions (once called repressions, now called hangups) and society's standards that are by definition repressive. If Johnny isn't "turned on" by Plato, if Shakespeare isn't "his thing," well, different strokes for different folks.

Anyway, by what standards does anyone judge anyone else's self-realization? Such romantic egalitarianism (like the chemical leveler, drugs) is balm for people who fear feeling either disappointed or ordinary.

My favorite novelist, Peter De Vries, has created a fictional young lady, Geneva, who "sometimes spelled the same word two different ways in the same paragraph, thereby showing spontaneity." Well, why not? According to the new sensibility, spontaneous self-expression is an end in itself. A drama professor learned that in 1969, when he participated in a debate about the "living theater," in which actors, "liberated" from scripts, improvised, and the audience was invited to join the "play." The professor injudiciously said the issue of the debate was "free improvisation versus disciplined skill."

Heckler: What the hell is disciplined skill?
Professor: I believe the theater to be served best when it is served by

supremely gifted individuals possessed of superior vision and the capacity to express this in enduring form.

Heckler: We're all supremely gifted individuals.

Professor: I doubt that very much.

Heckler: Up against the wall.

Everyone is "supremely gifted"? Well, then, grade inflation makes sense: it merely says that almost everyone is above average. And never mind that academic life may come to resemble the "Caucus-race" as explained by the Dodo to Alice in Wonderland:

"First it marked out a race-course, in a sort of circle ('the exact shape doesn't matter,' it said), and then all the party were placed along the course, here and there. There was 'One, two, three, and away!' but they began running when they liked, and left off when they liked, so that it was not easy to know when the race was over. However, when they had been running half-an-hour or so . . . the Dodo suddenly called out, 'The race is over!' and they all crowded round it, panting and asking, 'But who has won?' . . . At last the Dodo said, '*Everybody* has won, and *all* must have prizes'."

[February 9, 1976]

"Branches," I Said, "Do Not Grow on Trees"

Peter De Vries, America's wittiest author since Mark Twain, has published another comic novel, and that is no laughing matter. De Vries could sue American life for plagiarism: for twenty-five years it has imitated his satiric art, which makes his *I Hear America Swinging* ominous as well as hilarious.

Bill Bumpers, the protagonist, is a failed scholar whose doctoral dissertation ("Causes of Divorce in Southeastern Rural Iowa") establishes him as "the Ionesco of sociology" because it reaches the academically unacceptable conclusion that there are no useful generalizations about divorces. So he becomes a marriage counselor in an Iowa town infected with Advanced Thinking.

There a college girl's pregnancy is an academic event: she "takes" pregnancy as a credit course during her "non-resident term." Another student becomes a call girl ("She's majoring in sociology, after all"). As De Vries has said, "A hundred years ago Hester Prynne of *The Scarlet Letter* was given an A for adultery. Today she would rate no better than a C-plus."

Avant-garde art flourishes in this Iowa town. There is a sculpture exhibit that consists entirely of empty pedestals "from one to another of which art lovers strolled reading catalogue copy which ran: 'In his mature period, Kublensky has aimed at progressively more dramatic refinement of the principle of minimal form. Thus an area of virgin space unoccupied by anything save what the viewer himself might imagine it to contain, rather than what the artist has arbitrarily imposed, came to represent to him the ultimate distillation of linear values."

Et tu Iowa?

Many previous De Vries novels are set in suburban subdivisions named after "what the contractors had to eradicate to build them," like "Birch Hills" or "Vineyard Acres." The residents say things like: "I think I can say my childhood was as unhappy as the next braggart's. I was read to sleep with the classics and spanked with obscure quarterlies."

They have thoroughly modern children who consider a mother "a wad of contributing factors," and who "carry their chewing gum in their navels as a protest against bourgeois values." They are partial to art that expresses social concern, such as ballet that depicts "the installation of high tension wires through valleys in which people have hitherto lived in peace."

Their churches are modern enough to consider making divorce a sacrament because "you only get married the first time once." Their pastors coin theologically inventive aphorisms: "It is the final proof of God's omnipotence that he does not exist in order to save us." The Reverend Mackerel of the People's Liberal Church, speaking from his free-form pulpit ("a slab of marble set on four legs of delicately different fruitwoods, to symbolize the four Gospels, and their failure to harmonize") offers this prayer during a flood: "Let us hope that a kind Providence will put a speedy end to the acts of God under which we have been laboring."

For twenty-five years De Vries has been making gentle sport of America's anxiously avant-garde middle class. He is an anthropologist of intellectual faddishness, the instability of a society tolerant of all ideas except old ones. And he has a nice sense of the origin of such tolerance:

"When man was thought to be a little lower than the angels he was quickly censored for the slightest offense. Now everything about him is regarded as a cesspool but nothing is deplored."

But De Vries's message is rarely that obtrusive, and he is always fun. He has perfect pitch for the comedy that explodes from common speech: " 'I wish you wouldn't eat in your undershirt,' she said. 'It's so common. Nobody does that.' " And a man disgusted by masochists says: "They need a good kick in the pants."

Be warned: persons steeped in De Vries's works occasionally lapse into De Vries's drollery.

Recently my two-year-old was on a creek bank exercising his single skill, throwing things in. I tossed a fallen branch. The splash elated him, and he demanded that I find another, quickly. Alarmed by this incontinent pleasure-seeking, I admonished him that instant gratification is not always possible. "Branches," I said, "do not grow on trees." Such bons mots produce parricides.

[May 20, 1976]

Marvelous Marin

Few of the really *now* people—spacy, seeking human beings with their acts together—live here, so I study them in guidebooks, like *The Serial: A Year in the Life of Marin County*. It is by Cyra McFadden, who lives in that hip San Francisco suburb, where most of her novel was first published in a weekly newspaper.

Moderns in Marin try to live down their mothers back in Spokane ("I mean, she makes *casseroles*"), make up bumper-stickers for their Volvos ("Another Glass-blower for Udall"), attach tiny silver coke spoons to their high school charm bracelets, drink at "The Silenced Minority," buy Earth shoes at "The Electric Poppy" and get hair cuts at "Rape of the Locks," where a black militant shampooer harasses the ladies by constantly changing the soul handshake.

Marin's affliction is "French bread thumb," a wound suffered by hostesses who drink too much with hors d'oeuvres and then slice themselves instead of the bread. Marin exercises include Zen jogging, and dressing for tennis.

Marin parents are not on authority trips. They do not get off on value judgments, which they consider a form of child abuse. They nurture the whole child, who acts out hostilities, sometimes with Sabatier cutlery. (It is CBS's fault little Gregor is an arsonist: he saw the Waltons' house burn.) Summer camp is an encounter group where resource persons have M.A.s in behavioral psychology, and campers eat soybeans to save protein for the Third World.

Kate and Harvey Holroyd (Sierra Club, Zero Population Growth) try to stay mellow in a tract house with no Jacuzzi. Their cat, Kat Vonnegut, is "into Meow Mix" (until Harvey totals him with his ten-speed bike). Daughter Joan attends a high school where she learns body language but not the location of Europe, then runs away to become a Moonie and live on broccoli.

The Holroyds have been married longer than any couple they know, which gives them negative vibes. They get into each other's heads and decide that marriage has become mutual dependency, which is a bummer. To redefine the parameters of their interface, Kate tries outside interaction with a psychologist specializing in dysfunctional socialization of rich children. Harvey tries with his secretary, who considers taking shorthand part of the master-slave syndrome and discusses sheep symbolism in Bergman films.

Harvey tries LTRs (living together relationships) with a Safeway checker who says white bread is killing Harvey's enzymes, and with Carol, a member of Kate's consciousness raising and assertiveness training group. Carol is heavy into Carlos Castaneda, considers linear thinking a total shuck, becomes aware of herself as an ecosystem, gets a crew cut, becomes a lesbian.

Kate is in the pits when she leaves home with her Cuisinart and has an unsatisfactory liaison with Phil, who is into sadomasochism. Kate thinks of herself as material for a really laid-back photo-essay on "woman in transition," and decides to become a fully actualized person by getting a job, "but what she had to offer wasn't what she could do, exactly, but what she *was* and there was so much discrimination against women."

Instead, she takes a Spanish course to become less of a WASP imperialist, and enters a commune to write an autobiographical novel about the universal female experience. But in spite of a soulful communard who is into primal screaming, the commune is a downer because of an eighty-five-year-old kleptomaniac who is the commune's statement against ageism in Amerika; a woman named Woman who tells Kate that shaving her legs is a statement against The Movement; and a landlord who is "Germanic" about money because he won't take food stamps.

Fortunately, Kate is blown away by the sight of Harvey wearing a sterling silver biological equality symbol. If that is what is coming down in Harvey's space, she can relate to it. If that is where Harvey is coming from, perhaps marriage isn't an institutionalized gestalt inimical to honest relating.

They renew their marriage vows at a ceremony featuring a rock group (Too Loose Lautrec) and a Unitarian minister who wears "Let's Get It On" T-shirts. He pronounces them cojoined persons. There is sacramental eating of whole-wheat lasagne.

The Serial is a comedy about moderns struggling to keep their chins above the rising sea of their status anxieties. It is a Baedeker guide to a desolate region, the monochromic inner landscape of persons whose life is consumption, of goods and salvations, and whose moral makeup is the curious modern combination of hedonism and earnestness.

[July 31, 1977]

Privacy in the Republic of Appetites

Prudence plucks at my sleeve, reminding me that discussing today's sexual mores is like eating corn-on-the-cob: it can't be done decorously. But as a chronicler of manners I must report that Madison Avenue, nerve center of Madison's Republic, has produced two television commercials for condoms.

Years ago, when the Republic was sunk in the sin of fastidiousness, broadcasters would not advertise "intimately personal products." That category even included what today, in our maturity, we watch Mr. Whipple squeeze and sell as "bathroom tissue." Broadcasters steadily amended their sense of delicacy until, after due deliberations, they approved commercials for hemorrhoid preparations and feminine hygiene products. And soon you may see, as people in Ohio and California have seen, a commercial for Trojan brand condoms.

Broadcasters reflect national values, especially singlemindedness about commerce. Television is a merchandising medium and there is something quaint, and, strictly speaking, un-American, about the idea that there are standards—other than of effectiveness—by which merchandising should be judged. The people who flood our living rooms with a smorgasbord of commercial messages about fetid breath, moist underarms and troubled intestines know this: an appropriate time, place, and manner to sell a product is any that sells the product.

Commercial society regards people as bundles of appetites, a conception that turns human beings inside out, leaving nothing to be regarded as inherently private. Commercial society finds unintelligible the idea that anything—an emotion, activity, or product—is too "intimately personal" for uninhibited commercial treatment.

That idea is a throwback to those dead days when mankind was riddled with what then were called inhibitions and today are dismissed as hangups. Modern man knows a hangup when he sees one, and sees one in the idea that something is too "intimately personal" for a television commercial.

Modern man-turned-inside-out devalues privacy by arguing that people who flinch from exposure to something have a hangup about that thing, and hangups are, by definition, deplorable. This argument is endlessly useful to television, as follows: there is nothing shameful about diarrhea; therefore it is neurotic to wince when television hawks relevant potions in your living room. Behind that word "therefore" lurks a modernist axiom: reticence, like privacy generally, is the refuge of people with hangups.

Actually, some people who wince are worried about tastefulness. But tastes are subjective; only appetites are objectively—which is to say, commercially—serious.

It is not just in commercials that you see the consequences of the doctrine that whatever is not shameful need not—indeed, should not—be treated reticently. Consider the ambience at your neighborhood newsstand amid the array of "men's magazines." There you see the fruit of this principle: sex is natural and therefore (that word again) should be rescued from society's irrational reticence.

The June 1974 *Playboy* magazine was a landmark, of sorts, in the history of American liberty because the lady on the cover had one breast fully exposed. Today that cover would be considered demure, especially compared with what's inside. And what appears inside such magazines appears, in the fullness of time, on their covers.

Now that a stroll by a newsstand is an adventure in the skin trade, it is unrealistic to expect inhibited broadcasting. And condom commercials won't shock whatever remains of the public's sensibilities, as long as the commercials are as oblique as the Trojan commercials are.

The commercials delicately refer to "responsible parenthood" and quote scripture (Ecclesiastes): "To every thing there is a season, and a time to every purpose under the heaven." Perhaps the commercials will be less prim when, as the manufacturer of Trojans promises, they progress to a "comparatively hard sell." According to a Madison Avenue expert, that means commercials "playing up product features."

Progressives who welcome the rapid transformation of sexual attitudes argue that a society with a reticent and cloistered attitude about sex must think sex is shameful. Progressives will welcome condom commercials as another sign that sex has been rescued from Victorian suffocation, and elevated to the respectability of full citizenship in the Republic of Appetites, where nothing is "intimately personal" because nothing is private.

[August 6, 1975]

Depressing Candor

Betty Ford's "Operation Candor" already has caused the Republic to leap like a startled fawn. And she seems to have touched only the fringe of her chosen theme, which is: All You Ever Wanted to Know About the Fords But Were Too Demure to Ask. Her relentless responsiveness to interviewers is a tribute to the prevailing zest for frankness.

Many Americans are still wobbly in the knees after watching the First Lady chat with a CBS interviewer about the First Daughter's commitment to chastity. When the interviewer raised the subject, Mrs. Ford could have glared at him through her lorgnette, if she had one, and if glaring at journalists were permitted by the post-Watergate morality. Instead, she answered with what I gather is widely regarded as "refreshing" candor.

And now comes the September issue of *McCall's* magazine, in which the First Lady says she sleeps with the First Gentleman "as often as possible." The subject came up because Mrs. Ford dragged it up by the scruff of its neck. She volunteered to her interviewer that reporters had asked her everything except how often she sleeps with the President "and if they'd asked me that I would have told them." The interviewer rose to that bait like a healthy trout.

Barrels of ink have been devoted to the debate about whether such talk is appropriate, given that a First Lady is "symbol of American women," and given that American officialdom cherishes its dignified reserve. But since both "givens" are false, the ink has been wasted.

The simple truth is that frankness has been declared a Good Thing, and Mrs. Ford evidently subscribes to the popular maxim that moderation in pursuit of a Good Thing is no virtue. To those who say Mrs. Ford may give candor a bad name, the appropriate response is: why should she be expected to be the second person in Washington with an anachronistic sense of privacy.

A reporter once asked the only such person, Defense Secretary James Schlesinger, why some years ago he had converted from Judaism to Lutheranism. Schlesinger replied, anachronistically: "That's none of your business." Obviously Schlesinger's somewhat straitened construction of the "public's right to know" threatens the journalist's sovereign right to ask, say, Treasury Secretary William Simon if he sleeps in the nude.

You may feel that journalists who ask intimate questions justify Mencken's statement that the average journalist has "the taste of a designer of celluloid Valentines." But journalists, like other retailers, must know their markets, and there is a bull market for gossip.

The Washington *Star* newspaper has included in its refurbishment an anonymous gossip column titled "The Ear." That column has not been informative about the slumbering treasury secretary, but it has reported that "Barry Goldwater wears a red nightshirt" and that author Erica Jong "rarely wears much in the way of underpinnings."

"The Ear" is devoted to the public's "right to know" things like these:

"The Iwo Jima memorial has become the Washington area's favorite necking place for gay military men."

"A handsome, charismatic, married new senator was dating two girls with the same name from the same far-off state. The two girls met at a Capitol Hill party and graciously giggled . . ."

In a forthcoming book John Dean's wife "tells of being chased into her hotel bedroom by this long-armed high-ranker, who's now on Rocky's staff."

"Which husband of which hotshot '60s newswoman was overheard at the Metropolitan Club saying he's making more money than ever these days—in movie porn?"

Perhaps a gifted casuist can distinguish between that sort of "news" and the "news" the journalist from the *National Enquirer* was scrounging for in Henry Kissinger's garbage. But that scrounger moved the *Star*'s intelligent editorialists to some well-chosen words about "predatory journalism" and "journalistic prurience and boorishness."

The *Star* editorialists said that it takes no great leap of the imagination "to see the journalistic hawks as the super-policemen of tomorrow, lording it over a society in which all privacies and intimacies are public." The precursors of such a society are all the people who reject the idea that some things are none of a newspaper's business.

[August 25, 1975]

Thunderblender!

Plymouth, the automobile company, has responded to the energy crisis. Some of its thinkers, unsung heroes of American enterprise, have labored mightily and brought forth a new . . . name.

The name is: Plymouth Valiant *Brougham.*

"Brougham." Roll it around on your tongue. Makes you feel prosperous? Conjures up visions of silver trumpets? It should. A brougham is an elegant kind of carriage named after a nineteenth-century Scottish baron.

For years Valiants have been unpretentious little "economy" cars. But Detroit cannot make much money off economy cars. The big profit margin comes from the "extras," the gadgets and other luxuries that usually go on big cars.

Unfortunately, now more consumers are determined to buy little cars, like Valiants. So Plymouth has dolled up one Valiant model with crushed velour and deep pile rugs and other sybaritic delights, and has hung the name of a baron on the poor thing.

But a Valiant is a Valiant is a Valiant—a humble thing—and a Valiant Brougham is a contradiction in terms. It is as absurd as (say) a Volkswagen "killer whale."

In fact Volkswagen flinches from hyperbole, preferring gentle self-deprecation, like "Super Beetle." It has come to this: we are learning about modest understatement from the Germans.

But we should remember, as Plymouth must remember, the last time Detroit erred on the side of the prosaic in naming a car. On

October 19, 1955, a day that will live in hilarity, a Ford executive wrote to Marianne Moore seeking to enlist the poetess in one of the great business sagas of our time:

"Our dilemma is a name for a rather important new series of cars.

"We should like this name to be more than a label. Specifically, we should like it to have a compelling quality in itself and by itself. To convey, through association or other conjuration, some visceral feeling of elegance, fleetness, advanced features and design. A name, in short, that flashes a dramatically desirable picture in people's minds."

The executive confessed that Ford had a list of 300 names, all characterized by "an embarrassing pedestrianism." Miss Moore promptly put pedestrianism to rout.

Her first suggestion was: "Ford Silver Sword." She helpfully explained that this is the name of a flower that grows 9,500 feet up Mt. Haleakala on the Hawaiian island of Maui. She noted that the leaves "have a pebbled texture that feels like Italian-twist backstitch allover embroidery." Imagine how helpful this was to Ford executives who, we may assume, rarely receive such communications.

Soon Miss Moore warmed to her task and began firing on all eight cylinders. "I feel," she chirruped, "that etymological hits are partially accidental. . . . Let me do some thinking in the direction of impeccable, symmechromatic, thunderblender. . . ."

Yes, yes, *yes*. "Thunderblender." Now *that* was the work of a poetess to whom the muses were singing, nonstop. Breathes there a man with soul so dead, who would never to himself have said, "I want a Thunderblender." And Miss Moore was just getting into gear.

A torrent of suggestions flowed: "Resilient Bullet," "Ford Fabergé" (the reference not to the perfume but to the Russian court jeweler), "Mongoose Civique," "Tonnerre Alifère" (winged thunder), "Magigravure," "Pastelogram," "Tir a l'arc" (bull's eye), "Turcotinga" (an indigo South American sparrow), "Varsity Stroke."

Herewith the full text of one of Miss Moore's letters to Ford: "May I submit UTOPIAN TURTLETOP? Do not trouble to answer unless you like it."

Alas, the poetic spark often falls on damp kindling. Miss Moore was casting her pearls before unreceptive executives. On November 8, 1956, an executive wrote to Miss Moore to break the news that there would be no Thunderblender.

"We have chosen a name out of the more than six thousand-odd candidates that we gathered. It fails somewhat of the resonance, gaiety, and zest we were seeking. But it has a personal dignity and meaning to many of us here. Our name, dear Miss Moore, is—Edsel.

"I hope you will understand."

You can bet she understood.

[April 30, 1974]

Run Over by a Musical Comedy

Seventy years ago Mr. Dooley warned us. The humorist said that when Americans were finished with the English language it would look as if it had been run over by a musical comedy.

Hardly a day passes without some new sapping of the strength of the language as an instrument of straightforward communication between trusting people. For example, I have in hand a full page advertisement from a recent *New York Times Magazine.*

It enjoins: "Live! Really Live!" Where? "On Your Own Ranchette." What, pray tell is a "ranchette?" It is a *half-acre*—or, as the advertisement assures, "a full half-acre"—of Creation 13 miles from the town of Deming in southern New Mexico.

New Mexico is a splendid place and no doubt Deming is a haven of gracious living. But if both a deer and an antelope try to play at the same time on one of those half-acre ranchettes there is going to be a skinned deer shin or a sprained antelope ankle.

The ersatz word ranchette is an illegitimate child of the honest word ranch. But ranchette is used as a label for something the principal characteristic of which is the opposite of ranch-like spaciousness. Thus the very idea of a half-acre ranchette is almost funny. Almost, but not quite.

As an instance of word corruption, ranchette is remarkably crude —more so, for example, than a "personal size" Buick car, whatever *that* is. And because the commercial motive for it is so obvious, no one is apt to be deceived by it. But that does not mean it is innocuous. It is another drop in the endless stream of corrosive tastelessness that is dissolving the language.

Each year we see thousands of abuses like it, and we just pass on, perhaps with a lip curled in contempt for those who stoop to such language, and wishing wistfully that magazines had the gumption to reject advertising on aesthetic grounds when such crudities are involved.

But soon we get numb and lose our capacity for indignation about vulgar uses of the language. We even lose sight of the fact that we should be indignant when people use such language for deceitful purposes.

Sure, we have no trouble discerning the reality that the word "ranchette" denotes, just as we know what our leaders mean when they talk about "free" public education, or about "protective reaction strikes" made in support of "incursions." But we become indifferent to the insult when people reveal their estimate of our intelligence by talking that way.

If we are increasingly confronted and decreasingly affronted by such vulgarity, it is because we have been exposed to it from the cradle up. My first encounter with it came when I was a mere sprig of a lad who had just graduated from Babar books to electric train catalogues. The Lionel Company's catalogue described two devices for uncoupling the little model railroad cars.

The luxury device was electric and automatic. You would place it anywhere in your model railroad layout, park two cars over it, press a button, and the cars were uncoupled.

Lionel's other uncoupling device was just a simple lever mechanism that you had to clip onto the piece of track in front of where you sat to operate the trains. You would park two cars over this mechanism, press down on the lever, and a rod would rise and pry apart the cars. You might as well have used your child-sized fingers for the uncoupling, and forgotten about this second uncoupler. And Lionel had the audacity to label this pathetic device a "manumatic" decoupler.

We play with "manumatic" toys and retire to half-acre "ranchettes" and in-between we wonder where government gets the idea that it can talk to us as weirdly as it does. Cynical and tasteless uses of the language continue to multiply and, as they do, the language becomes weaker, and so does our sense of social cohesion. We develop a general skepticism about all public communication, commercial and political.

When, at last, our language has been reduced to a shapeless, tasteless pulp, it still will have a function. It will be marvelously useful for deviousness, which will proliferate in government and in society as the tortured language becomes an ever more supple servant of it.

[April 13, 1974]

Perils of Parenting

A list of things I frown upon would, I know, be an inventory of the modern age. But I am especially grumpy about a modern literary genre: handbooks on "parenting."

The word "parenting" is evidence of the regrettable modern tendency to turn nouns ("loan," "critique") into verbs. And as I write this, my desk is groaning beneath the weight of books with titles like *Effective Parenting*. That book has a chapter on "Key Words for Parents," such as (I'm not making this up) "there, there, there."

Between Parent and Child is a best seller that "tells you how to talk childrenese, the new way to get through to your child."

Did Tom and Nancy Lincoln talk childrenese to little Abe? Did they read *How to Raise a Brighter Child?*

The Challenge of Child Training lists the twenty "most common mistakes" and includes "draconic severity." I prefer the book inspirationally titled *Dare to Discipline*.

What's a Father for? explains what to do "when your boy wants a doll" and has a chapter titled, "Are you sorry you had a girl? Electra did her mother in." *Liberated Parents, Liberated Children* begins with a controversial premise ("Children are people"). *P.E.T.—Parent Effectiveness Training* concludes with imprudent advice ("How to Avoid Being Fired As a Parent").

It is interesting that *How It Feels to Be a Child* (answer: often crumby) used to be titled *The Myth of the Happy Child*. It is fascinating that the author of *The Primal Scream* has written *The Feeling Child*, which has a chapter on "Intra-uterine Life." It is a scream.

Children generally should be faced with philosophy and managed with common sense, and most parenting books are frequently sensible. Like most parents, most such books do little lasting harm to children. But the profusion of these books, some of which are mischievous, is the result of an American affliction.

Americans are predisposed to believe too much in environmental determinism. They are inclined to regard infants as malleable bundles of "potential"—clay on which determined parents, handbooks in hand, can work wonders.

In the "nature versus nurture" argument, Americans are can-do optimists who believe that skillful nurturing is all-important in shaping individuals. The United States is a manufacturing nation that sometimes seems to regard children as raw material from which ever-better products can be manufactured as know-how increases.

The first edition of Benjamin Spock's *Baby and Child Care* appeared in 1946, when Americans were in the full flush of success. They were anxious to turn from the collective task of history-making to the private task of family-making. Like Studebakers and toothpaste, the next batch of children would be "new and improved."

What Spock and, even more, subsequent authors have done is (as sociologist Philip Slater says) "to encourage Pygmalionesque fantasies in mothers and to stress the complexity of the task of creating a person out of an infant." Yes, "creating": parents, and especially mothers, have been saddled with god-like responsibility for making little Mike into a work of art, like little Mozart.

American women are driven bonkers by the exaggeration of their child-raising, "person-creating" responsibilities. The naggingly insistent voice of much parenting literature says: children become what, and only what, parents consciously make of them. And every creature born of woman can "maximize" his or her potential, given sufficiently attentive parenting, which usually means a sufficiently dutiful mother.

Small wonder that conscientious mothers often feel they are gasping for air. They are being suffocated by the demands of a duty they can never fully dispatch. It is the ever-expanding duty to "maximize" something—their children's "potential."

But, then, parenting is no piece of cake for fathers, either, as I learned one morning recently, when Geoffrey M. Will was still two, and thus still had an excuse for being insufferable.

I ventured a commiserating comment on his wan appearance, which I hoped was the result of his having been up late foraging for stimulation in his father's library. "Geoffrey," I observed, "you look pale."

His face suddenly glowed with pure intelligence, and he shot back: "I'm not a pail; you're a shovel and I'm a sandbox."

You see why I'm bitter about parenting? You try to nurture Mozart, and you wind up with Groucho Marx.

[February 13, 1977]

"Angels and Other Things That Aren't True"

On the corduroy contours of southeastern Pennsylvania, the corn is as high as an elephant's eye. Between the stalks scamper visiting children, most of them blacks and Puerto Ricans from New York City.

This is the 100th summer of a program that began in July 1877, when the first New York children arrived in Sherman, Pennsylvania, invited to a community that had been stirred by a sermon on the perils of growing up urban and poor. Lancaster County began receiving "fresh air children," in 1891, when rural life was no bowl of cherries. The local papers that summer were full of reports like this about country children:

"Clayton Royer . . . fell from a horse about a month since. The horse trod on his right leg. About two weeks ago the lad began to have tetanus. From that time on until his death Saturday the little fellow suffered terribly." To read the yellowed newspapers is to be reminded of how harsh rural life could be.

In 1891 the papers also reported the social terrors of Lancaster County: "Groff's Cash Store Entered—A Kettle of Milk Taken." But the arrival of the "fresh air children" brought forth from the newspapers the robust American prejudice against cities:

"Again we feel constrained to tell the hospitable people of our Borough not to expect too much in the way of angelic disposition, agreeableness of manner, exemplary conduct and the like. . . . These virtues are in many instances wanting in our own children, and surely if they do not possess them we have no right to look for them in those that come from homes of poverty and vice, degradation and misery."

Actually, now as then, most children come from decent families. The children are easy to manage, especially if you are sensible. The Amish disregard the instructions from New York social workers that "positive rewards for good behavior" must be used always, and spanking never, to control children.

The Amish are the ultimate counterculture. They emulate the simplicity of primitive Christianity. They have their doubts about most modern things—cars, television, etc.—and have no use at all for modern notions of childrearing.

Here at the farm of Jonathan and Lydia Lantz, their own boys (six of the eight children), dressed in black britches with suspenders, are keeping tabs on two black visitors from New York. The visitors are a bit younger than most "fresh airs."

Paul, four, who recently saw his mother murdered, doesn't talk much. He crosses a barnyard on the balls of his feet, like Walt Frazier bringing the ball up court. Ann, three, has eyes the size of apricots. A borrowed Amish prayer cap is pinned to her ample hair. She has a regally standoffish attitude toward all animals except one tame pig.

For years the Lantzes took in children from a Puerto Rican family. But the children's mother was murdered recently, and the children are in the custody of an aunt who disapproves of the Lantzes because, Lydia Lantz says dryly, "we teach them about angels and other things that aren't true." So those children are spending the summer in the city where, the aunt may rest assured, they will be safe from talk about or evidence of angels.

For most city children, culture shock concerns matters natural, not theological. One fastidious lad who thought milk came from cartons quit drinking the stuff when he saw it squirt from the real source, which he considered unsanitary.

Even those children whose idea of good eating is potato chips sluiced down by soda pop are good sports. They gamely choke down fresh vegetables and local delicacies such as shoofly pie. At the Lantzes', he took to corn-on-the-cob without coaxing or proper coaching. He liked it even better the second time, when it was cooked.

Time to go. "Give me five, man," say I to Paul, extending my hand, palm up. Instantly: smack-smack, Paul's palm slaps mine and turns up to be slapped in turn.

The Lantz children watch this with the bemused detachment of veteran anthropologists observing the inexplicable ritual of an unfamiliar tribe. They probably are thinking, and not without reason, that the sheer strangeness of urban living produces a lot of odd habits.

[August 8, 1976]

On Her Own in the City

When police, responding to her call, arrived at her East Harlem tenement, she was hysterical: "The dog ate my baby." The baby girl had been four days old, twelve hours "home" from the hospital. Home was two rooms and a kitchen on the sixth floor, furnished with a rug, a folding chair, and nothing else, no bed, no crib.

"Is the baby dead?" asked an officer. "Yes," the mother said, "I saw the baby's insides." Her dog, a German shepherd, had not been fed for five days. She explained: "I left the baby on the floor with the dog to protect it." She had bought the dog in July for protection from human menaces.

She is twenty-four. She went to New York three years ago from a small Ohio community. She wanted to be on her own. She got that wish.

She was employed intermittently, until the fifth month of her pregnancy, which she says was the result of a rape she did not report to the police. She wanted the baby. She bought child-care books, and had seven prenatal checkups at Bellevue Hospital. Although she rarely called home or asked for money, she called when the baby was born. Her mother mailed twenty-five dollars for a crib. It arrived too late.

When labor began she fed the dog with the last food in the apartment and went alone to the hospital. The baby was born on Wednesday. When she left Bellevue Sunday evening, the hospital office holding her welfare payment was closed. With six dollars in her pocket and a baby in her arms, she took a cab home. The meter said four dollars and the driver demanded a dollar tip. When she asked his assistance in getting upstairs, he drove off.

The hospital had given her enough formula for three feedings for the baby. Rather than spend her remaining dollar that night on food for herself and the dog, she saved it for the bus ride back to Bellevue to get her welfare money. Having slept with the baby on a doubled-up rug, she left the baby and dog at 7 A.M. It was 53 degrees, too cold she thought to take the baby. She had no warm baby clothes and she thought the hospital had said the baby was ailing. She got back at 8:30 A.M. Then she called the police.

Today the forces of law and order and succor are struggling to assign "blame" in order to escape it. Her attorney and Bellevue are arguing about how she was released, or expelled, on Sunday evening. Welfare officials are contending with charges that they are somehow culpable for her failure to receive a crib before giving birth, and for her living conditions. (She was receiving payment of $270 a month; her rent was $120.) She has been arraigned on a charge of negligent homicide, but no one seems anxious to prosecute.

Late in New York's U.S. Senate primary, Daniel P. Moynihan, talking like a senator prematurely, said that this case dramatizes weaknesses of the welfare system, and indicated that it also dramatizes the need for him in Washington. Perhaps.

But because cities are collections of strangers, they are, inevitably, bad places to be poor. Not that there are good places, but cities, being kingdoms of the strong, are especially hellish for the poor.

Cities have their indispensable purposes, and their charms, not the least of which is that you can be alone in a crowd. But that kind of living alone is an acquired taste, and not for the weak or unfortunate. They are apt to learn that no city's institutions can provide protective supports like those of an extended family or real community. No metropolis can provide a floor of support solid enough to prevent the bewildered —like the woman from Ohio—from falling through the cracks.

Through those cracks you get an occasional glimpse of what George Eliot meant: "If we had a keen vision and feeling of all ordinary human life, it would be like hearing the grass grow and the squirrel's heartbeat, and we should die of that roar which lies on the other side of silence."

[September 19, 1976]

PART 4

Campaigning

Contemporary American Fiction

Ramsey Clark is a triumph of contemporary American fiction. A Texan who transplanted himself in Manhattan, he seems to have sprung full-blown from the brow of Truman Capote, who created the archetype, the hillbilly who left Tulip, Texas, as Lulamae Barnes and arrived in Manhattan as Holly Golightly, the piquant playgirl of *Breakfast at Tiffany's*.

Like Holly Golightly, Clark fled a past that confined and embarrassed him. (He was Lyndon Johnson's faithful servant, the attorney general who prosecuted baby doctor Spock for antiwar conspiring.) Holly Golightly lodged in an Upper East Side brownstone; Clark goes her one better, lodging in Greenwich Village. He, like she, has redecorated himself, inside and out.

Someone—Capote perhaps—said Holly Golightly was a phony, but a real phony, meaning that the self she created for herself was her real self; it just could not be born in Tulip, Texas. Similarly, Clark, forty-six, the former establishment lawyer who now is shuffling to a different flutist, is an artifact, but not artificial.

A self-creation, he is a phony, but a real phony—a work of art. And having risen on the stepping stone of his dead self to what he considers higher things, he now wants to rise to the U.S. Senate, where a real phony would be an improvement.

Clark's intellect is an invitation to circular reasoning. It is true, that if an idea is liberal, he will accept it. But no idea can safely be called liberal until he accepts it. He is the new Pope of that persuasion.

He thinks President Ford's amnesty program is harsh. He likes forced busing to achieve integration. He thinks maybe the New York City subway should be "free" (that is, people in Tulsa and Omaha and elsewhere should pay for it through federal subsidies, even more than they already are paying). He seems to think that a business making a profit is vulgar. But he wants to tax business profits to finance new social programs.

Today he is wearing Hush Puppies (but not "earth shoes"—he's out of uniform), argyle socks, gray denim wash pants, a suit jacket of a discordant shade of gray, a white button-down shirt, and a dollar tie no more than two inches wide. It isn't clothing; it is a costume.

It is post-Watergate haberdashery, part of the antipose pose, a sincerity gambit from Robert Hall. It is the carefully calibrated "uncalculated" look for politicians cunningly convinced that dishevelment serves the symbolism of candor. This may be Nixon's unwitting revenge on American sophisticates—politicians wearing argyle socks to create the image of people who disdain image.

Clark's opponent, Senator Jacob Javits, seventy, is seeking a fourth term. But Javits is bewildered by the experience of being flanked on the left, and flanked by someone at least as sanctimonious as himself.

Clark, radiating righteousness from every fiber of his humble self, has focused attention on his refusal to accept contributions of more than $100. Javits, who receives large contributions, has become defensive about them, arguing, in effect, "I'm not a crook." (Javits' campaign slogan is: "He's One in a Hundred." That is true.)

Javits, having flown to Castro's police state to prove that his liberalism is impeccable, now is in Manhattan denouncing Clark's 1972 trip to Hanoi, where Clark announced that U.S. prisoners of war in North Vietnam were healthier than he was. (Clark was more accurate than he meant to be.)

Clark is in a telephone booth along Queens Boulevard, shouting above the roar of traffic, so a telephone-loudspeaker hookup at an upstate college can carry his message to a rock concert audience. And a Clark staffer is confiding to me, as every politician's staffers confide to every writer, that the candidate is "really" an "intensely" private, even shy, man. Now private, shy Clark is back in his Dodge bus, careening through Queens, in search of the public.

Clark and Javits, like most politicians who don't disagree fundamentally, say that if the other is elected there will be drought, famine, pestilence, and the death of first born followed by litigation upon inheritances—yea, verily, cry "Havoc!" and let slip the dogs of higher subway fares. Clark and Javits are equally correct.

[October 26, 1974]

Democracy As Vaudeville

In 1968 George Wallace said, "Hell, we got too much dignity in government now; what we need is some meanness." Wallace lost, but the winner satisfied that need. Now Wallace, with bulldog pluck, is preparing his quadrennial crusade against excessive dignity, and other things that offend him.

Only in the South can a man go so far by being an "against" man. In 1912 the socialist presidential candidate, Eugene Debs, finished third in Louisiana, ahead of President Taft, not because many Louisianans favored socialism, but because Debs articulated class resentments and offended comfortable Northerners.

Class resentments thrive wherever status has been linked to land ownership. Resentment of the rich by the poor is an old theme of southern politics, as is the indifference of southern politicians to the material needs of the classes that are nourished by such resentments and not by much else.

Segregationists launched Wallace, but sociologists have legitimized him. This quintessential anti-intellectual is the graduate students' best friend. He has spawned a thousand doctoral dissertations on "Socio-economic Parameters of Aggrieved Ethnicity" and other overinterpretations of the politics of spleen.

In a nation blessed with more scholars than scholarly subjects, no rabble-rouser can escape being studied as a "spokesman." A properly devout behavioral scientist believes that the phrase "legitimate grievance" is a redundancy, and that widespread human cussedness is not an acceptable explanation of a political phenomenon.

If Wallace "speaks for" some aggrieved class, that class has a real grievance about America's supply of spokesmen. Wallace says nothing of substance, and says it dreadfully. His career demonstrates the decline of American rhetoric. He is our most durable demagogue and he has never given a speech or uttered a paragraph or coined a phrase that is eloquent or even memorable.

But his rhetoric is the perfect vehicle for the nonsubstance of his politics. Because he has no truck with ideas, his rhetoric is a lumpy porridge of bald assertions condemning the ever-growing group of

Americans he finds it profitable to resent.

Huey Long, a substantial figure, tried to scare people so he could negotiate with them. Wallace just wants to be scary. Long's famous pledge—"Every man a King"—was blather, but it was a token of his intention to use government to raise the living standards of his supporters. Wallace practices a more austere form of the politics of envy, promising only to pull down people (for example, pointyheaded professors) who have risen too high.

Long was a kind of Nasser, a thoroughly unpleasant bully who nevertheless had some practical ideas and programs that he really thought would improve the lot of the people to whom he addressed his rhetoric. Wallace has Long's vulgarity and energy without Long's purposefulness. Louisiana was a laboratory for Long's experiments with power. Alabama is an off-Broadway stage where Wallace perfects his national act.

Like Spiro Agnew, Wallace runneth over with opinions. It would be absurd to ask his opinion of (say) monetary policy, but he is hot for law-and-order, and has been since he quit defying courts.

I cannot think of Wallace without remembering James Hamilton Lewis, U.S. senator from Illinois (died 1939), who disguised his toupee by sprinkling fake dandruff on his shoulders. Lewis, like Wallace, had a feel for the theater of politics. But Wallace has built a political career on nothing else.

Of course there is nothing new about reducing democracy to vaudeville. Our ninth President, "the roarer of Tippecanoe," William Henry Harrison, was recommended to the electorate on the ground that he drank hard cider. And Champ Clark of Missouri, speaker of the House of Representatives, nearly won the Democratic presidential nomination in 1912 (he had a convention majority but needed two-thirds) with little more program than his song "Ol' Hound Dawg":

> Everytime I come inna town
> The boys start kickin' my dawg aroun'
> T'ain' no matter if he is a houn'
> They gotta stop kickin' my dawg aroun'.

As an analysis of America's problems in 1912 Clark's song lacked a certain rigor. But it compares favorably with Wallace's words, which are not good-humored and don't rhyme.

[April 20, 1975]

An Anointing

If God wanted humans to fly, He would not have created the Eastern Airlines shuttle to express His disapproval. It is an instrument of torture that moves people between Washington and New York, back and forth . . . and back and forth . . . and back and forth, world without end, amen. As businesslike as a sardine tin, it offers neither food nor drink for the businesslike people who link the capitals of government and commerce, often to the detriment of both.

The routine bad news has just been announced by a disembodied voice from the cockpit. (Do airlines select pilots for their voices? Do pilots go to Gravelly Voice School?) We will circle Wilmington, Delaware, until the winds subside in New York. Such announcements are part of the warp and woof of campaign life for the party upfront, Senator Henry Jackson and aides.

At about this time Jimmy Carter also is headed for Jackson's destination, the St. Patrick's Day parade. There, the two will exchange the glaciating smiles of contestants who are, from now on, in a zero sum game: one man's happiness means the other's sorrow.

Jackson's next stop is a press conference in the kind of hotel room that usually is called something like the "Rousseau Salon." Every hotel has a room like this, heavy with chandeliers and busy wall coverings, and furniture rendered in a style that is, approximately, Louis XIV via Las Vegas. Today it is the scene of an anointing. The vanguard of the proletariat, in the form of seven labor leaders, will endorse Jackson.

Deployed in a heavy, stolid row, the labor leaders resemble raw material for another Stonehenge. Like the candidate they favor, they are not given to snappy one-liners. They are as basic as the room is gaudy, and they bestow their political blessing with the delicacy of a drop forge.

When questions are invited from the press, up pops a sweet young thing with long amber hair and an elaborate grievance against bourgeois civilization. She has a multi-part question-cum-statement that promises to run slightly longer than *Moby Dick,* and she delivers it in the sing-song diction that is the distinctive style of the well-bred radical.

The gravamen of her charge is that Jackson is an instrument of some conspiracy (run by the Elders of Zion? the Bavarian Illuminati?, she tantalizingly doesn't say) to make New York City a pawn of jack-booted capital. As she does her number, the mighty men of labor are getting restless, and several of them seem disposed to place their cal-loused hands around the young lady's indefatigable larynx.

But nothing perturbs a veteran of the Eastern Shuttle. Besides, Jackson has been hearing congressional testimony for thirty-five years, so he has learned to endure nonsense spun by experts. Eventually he says, as politely as possible and as firmly as necessary, "You don't know what you're talking about." She sits down.

Great God of Battles! Is it legal to talk to children like that? What a lovely moment. The young lady probably was raised by thoroughly modern parents who applauded her every solecism. She probably at-tended an impeccably modern college, where the faculty would have endured slow torture unto death rather than let logic impede the free flow of her soul.

And now, at last, and not a moment too soon, she has encountered a proper adult. Never let it be said that presidential campaigns lack educational value.

By evening the candidate is speeding through the illimitable tracts of Queens to a restaurant overflowing with men in dinner jackets adorned with green carnations. They have gathered to drink Scotch and eat ravioli and honor an Irish saint.

Around a small table in the middle of a large adjoining room, a group of matrons is standing, indifferent to the political transactions next door. Their stances are midway between attention and prayer, listening as a recording echoes through the room. First, the disembod-ied voice of a balladeer sings nice thoughts about Ireland. Then another recording, the disembodied voice of a monsignor, pronounces grace. It is a strangely affecting tableau; the saint would approve.

The candidate has come and gone before the ravioli is gone. He is churning toward the Polish Community Center in darkest Yonkers, world without end, amen.

[March 25, 1976]

Ford's Weakness

Political campaigns are not seminars; the truth gets its hair mussed. It is getting thoroughly tousled by Ronald Reagan, who opposes surrendering U.S. sovereignty over the Panama Canal. He argues that the Canal Zone is "ours" because we bought it.

Actually, it is ours, sort of, because we stole it, not from Panama, which we created to facilitate our larceny, but from Colombia. But the fact that we stole it is not sufficient reason for giving it back, unless we are prepared to give back our backyards. We are going to hear from Indians, Mexicans and others if we start yielding all real estate (beginning with the ground under Plymouth Rock) that we acquired with more exuberance than due process. But no one advocated yielding *all* our ill-gotten gains, so what is the canal ruckus about?

Obviously the canal is a symbol, and Reagan's sharply defined stand is a scalpel for surgery on the nation's soul. It is comparable to a 1960s antiwar refrain.

Then we heard: "The way to get out of the war is to get out." That analysis lacked rigor, but it fit nicely on bumper stickers and galvanized many people (some of whom thought the United States should get out of Utah).

Today many Americans are vaguely afraid that the nation is in retreat around the world. They say: "The way to stop the retreat is to stop." They see a Panamanian dictator making hay by making anti-American noises, and they say: the soft tropical soil along the canal is suited to the digging in of our national heels.

Reagan's argument, like much political argument, involves simplification, even caricature. But however flawed, it at least is a *political* argument, an attempt at public pedagogy. Reagan hopes the canal issue will cause a vague anxiety to crystallize into an idea. That is the first step in manufacturing a national resolve. He is using the canal issue as Andrew Jackson used the national-bank issue, as a piece of sand around which he hopes nacre will form, producing in time a pearl beyond price, a broad new consensus about national purpose.

I believe he is mistaken about the canal, but at least he knows what a political campaign is for. That is more than can be said for his opponent.

To refute Reagan (and to refute Ford, who recently said the United States "never" will yield canal defense and operational rights; perhaps Kissinger has not told him that we are negotiating the terms for yielding them), Ford unleashed his dove, Barry Goldwater, to charge that Reagan does not understand international subtleties. Goldwater (who as a presidential candidate said, "I would turn to my Joint Chiefs of Staff and say, 'Fellows, we made the decision to win, now it's your problem'," and who said of Cuba: "Do anything that needs to be done to get rid of that cancer. If it means war, let it mean war") now wonders whether Reagan has "a dangerous state of mind." Ford has found an unlikely kettle to call the pot black: Goldwater is one of thirty-seven Senate cosponsors of a ferocious resolution demanding that the United States "in no way cede, dilute, forfeit, negotiate, or transfer" any "sovereign rights" over the canal.

Ford's personal contribution to the nation's understanding of large issues has included a demand that Reagan release his 1975 tax returns. There being no decent reason for this petty demand, it was left to stand as an innuendo. It would be nice to be able to say that the small men in the White House understood Treasury Secretary William Simon's tart reminder that a citizen's tax returns are private. In this episode, as in his video-taped appearance with NBC's vulgarians on the "Saturday Night" show, Ford made the least of his major asset, his office.

Ford's Wednesday-morning quarterbacks say that everything would be hunky-dory if Reagan had not attracted Democratic crossover votes. As an alibi that ranks with the British official's response to criticism of the breakdown in military medicine during the Crimean War: "The medical services would have been perfectly adequate if it had not been for the casualties." True, Reagan is guilty of expanding participation in Republican primaries. It also is true that many New Hampshire Democrats cast Reagan write-in votes that were not counted. If New Hampshire Democrats could have voted for a Republican in February as they can in November, Reagan would have won there. That would probably have propelled him to victory in Florida, and Ford would be out of the race today.

The nation is suffering neither war nor recession, but Ford has been more roundly humiliated in intraparty competition than any President since Taft in 1912, and he may become the first President since Arthur in 1884 to be dumped by his party. Ford is the most conservative

President since Coolidge, and he is being nibbled to death by a rival only marginally more conservative, which suggests that Ford's problem concerns something besides philosophy.

That something involves the dual nature of the presidency. Ford is a passable head of an administration, but an unsatisfactory chief of state.

A President is the articulator of collective aspirations, or he is not much. He is articulate, or he is inadequate. He must have some of (in Bagehot's words about Pitt) "the successful power to give in a more than ordinary manner the true feelings and sentiments of ordinary men." But more, he should be able to elicit and, in doing so, shape public sentiments that otherwise would be inchoate.

There never has been a great inarticulate President. Ford is the most inarticulate President since the invention of broadcasting. (Coolidge was taciturn, which is different.) And it will not do to argue that Ford's weakness is "merely rhetorical." Rhetorical skills are not peripheral to the political enterprise; and they are among the most important skills a person can bring to the presidency.

The presidency is inherently, meaning constitutionally, a weak office. Without the cooperation of other people and institutions, Presidents cannot do much. But the presidency can be, as Teddy Roosevelt found and said, "a bully pulpit." The only expansive presidential power is the power to persuade. That is why, in our wired nation, with its cumbersome central government, an inarticulate President is like a motorcycle motor installed in a Mack truck.

Only a President can persuade this impatient nation to accept short-term pains for long-term gains, which is the essence of government. Without a persuasive, articulate President many important government decisions do not get made, and those that do are in a subjunctive mood. That is the serious significance of Ford's faltering, unserious campaign, which lacks a theme, dignity and persuasiveness. The national mind is a vaporous thing from which comes, occasionally, precipitates of thought. A President, *especially* in his campaign, should be a precipitator. Ford is not.

[May 17, 1976]

Exorcising the "God Issue"

Because it is the wittiest book in English, and because 1976 is its bicentennial (and a year when wit is especially welcome), many people are reading Gibbon's *Decline and Fall of the Roman Empire.* Readers are brooding about what, if any, evidence suggests that our Republic is on Rome's road to ruin.

Well, there is good news, of sorts. We are not threatened by what Gibbon blamed for the wreck of Rome. Gibbon was nothing if not blunt about the moral of his story: "I have described the triumph of barbarism and religion."

It was as he "sat musing amid the ruins of the Capitol, while the barefoot friars were singing vespers in the temple of Jupiter, that the idea of writing the decline and fall of the city first started to my mind." Through the years of scribble, scribble, scribble, Gibbon developed his indictment of the religion of those friars. Christianity was the culprit: "A pure and humble religion gently insinuated itself into the minds of men. . . ."

So, gentle reader, rest easy. Our Republic seems safe from the terminal sickness of Rome. There is no immediate danger of a volcanic eruption of Christianity.

But some farsighted sentries on the watchtowers of civilization, scanning the horizon, have espied a cloud that is now slightly larger than a man's hand, and is bearing down on us, fast. The problem, put plainly, is that Jimmy Carter believes in God.

No, on second thought, that statement of the problem is inadequate. After all, every politician believes in God, or at least in the currency, which is close enough. On every dime, right below FDR's chin, it says: "In God We Trust." The special problem with Carter, according to those who think it is a problem, is that he has the disconcerting habit of letting his religion affect his behavior. For example, he prays—a lot.

There is a Washington doctrine about the appropriate way to pray. Prayer is fine, if done in moderation, and in the East Room of the White House, or over Rice Krispies at a congressional prayer breakfast where, obviously, prayer belongs. But Carter prays in church, and even at home and while campaigning, for Pete's sake.

What is even more disconcerting to the disconcerted is that he prays, he says, "about twenty-five times a day, maybe more"—whenever, you might say, the spirit moves him. This is probably why Stuart Spencer, a Ford campaign aide not previously known for his theological interests, announced that Carter is a "fundamentalist." The burden of campaign duties prevented Spencer from elaborating on this insight, but evidently he thinks "fundamentalism" is not a virtue.

Carter's praying, church-going, and religious feeling antedate his presidential ambitions. That is why some people suspect that he is guilty of sincerity. This apparently is disconcerting to some, including many of the people ("progressives," they are called this spring) who claim loudly and often that Carter "doesn't believe anything."

To give the disconcerted perhaps more than their due, let us assume that they fear what Gibbon lamented: a religious enthusiasm that is, they feel, inconsistent with intellectual seriousness and emotional balance. But it is not clear why, in this most bloody century, drenched with secular fanaticisms, Christianity seems like a clear and present danger.

Perhaps Carter's religious fervor disturbs some people because they are comfortable only with politicians who have no spiritual processes more complex than calculation, politicians who can be trusted to obey the First Commandment (revised): Thou shalt worship naught but the Gallup Poll.

Sixteen years ago candidate John Kennedy's Catholicism was an issue. The issue, supposedly, was his potentially "divided loyalty." The fear was that Kennedy's loyalty would be divided between the national interest and the temporal and spiritual claims of the Catholic church.

Today *Newsweek* magazine contains a report accurately headlined "Carter and the God Issue." The "God issue" is real, and it is, I suspect, this: there is something vaguely distressing about a politician who might occasionally doubt the axiom, *vox populi vox Dei* ("the voice of the people is the voice of God").

[April 1, 1976]

Sauerkraut Ice Cream

You are a better than B-film actor if you can recite the following lines with a straight face:

Richard Schweiker, who is liberal Republicanism incarnate, has "the same basic values" (Ronald Reagan's words) as Ronald Reagan, who is the conservative word made flesh. Together they will rescue the GOP from Gerald Ford, whom Reagan wants to debate in order to dramatize the grave defects of Ford's conservatism.

Reagan could have risen from slapstick to drollery if he had justified his embrace of Schweiker on the ground that if the Reagan-Schweiker ticket wins in November, President Reagan will not have to contend with that predictable nuisance, Schweiker. As senator, Schweiker has made a career of opposing proposals Reagan supports. Instead, Reagan prudently left to Schweiker the gymnastic event of explaining things.

Schweiker says he succumbed to Reagan's seduction for a number of weighty reasons. One is that Reagan shares one of Schweiker's most cherished convictions: candidates should select running mates before the convention. Schweiker, until recently a Ford delegate, has not until recently seemed obsessed by that principle. But it is now clear that his attachment to that principle is so passionate that it outweighs all his demonstrated disapproval of Reagan's views about how the nation should be governed.

For his selection of Schweiker, Reagan has what is, in his profession, an impeccable excuse: desperation. And if he wasn't desperate before elevating Schweiker to glory, he now should be desperate to find condiments to make palatable the words he must eat. There are choice words about Walter Mondale, whose liberalism is a sliver more or less advanced than Schweiker's (more or less, depending on whose micrometer does the measuring). And there are some delectable Reagan words about how crumby it would be for Reagan to pick a running mate from the opposite wing of the party just to garner votes.

With a single stroke Reagan freed Ford from the burden of selecting a running mate acceptable to Reagan. Who could Reagan now claim to find unacceptable? Elliot Richardson, whose views are measurably more moderate, and immeasurably more mature, than Schweiker's?

Imagine the interesting thoughts dancing in the head of Nelson Rocke-feller, whose views on national security are much closer to Barry Gold-water's than to Schweiker's, and who was thrown overboard to appease Reaganites.

But Reagan and Schweiker have not exactly contributed to the public stock of harmless pleasure. Their caper is another subtraction from the dignity of the political vocation.

The vice presidency is an office for advocacy. It is for rhetorical soldiering on behalf of a President's policies. Schweiker has been as consistent in criticizing as Reagan has been in defending expanded defense budgets; and Schweiker has been as determined in defending as Reagan has been in criticizing domestic spending. Schweiker cannot become Reagan's advocate without stimulating lively speculation about whether he was most sunk in opportunism before or after Reagan anointed him.

Neither Reagan nor Schweiker understands how to broaden a polit-ical base without cracking political foundations. No one expects politi-cians to go through life bent double beneath the weight of iron princi-ples. But surely they should have for their own most fervently expressed positions at least a small portion of the respect they incite their audi-ences to have for those positions.

In spite of Reagan's and Schweiker's effort to make it seem so, politics is not a Jackson Pollock painting, a random scattering of ele-ments on the public canvas. And a "coalition" is not just any joining of individuals by the sticky secretion of their ambition. Surely a coalition worthy of the name is the coming together of groups on a common ground of public purpose.

If the Reagan-Schweicker ticket is a political coalition, then sauer-kraut ice cream is a culinary coalition. It is the joining of ingredients that are both respectable but do not belong together. There are a few things you do not get from persons who are serious about eating, or politics.

Schweiker, having worked strenuously and successfully in the Sen-ate to prevent the dark night of conservatism from descending upon the land, now says he will apply himself selflessly to the task of making Reagan leader of the land. But this will be a short collaboration between Reagan and Schweiker.

This marriage of convenience will be even briefer than most mod-ern marriages. It will last only about twenty-four days, but will be long remembered as evidence that many politicians need to be taken seriously, but not because they all are serious about things other than themselves.

[July 29, 1976]

A Runner That Renown Outran

For some persons, politics is nothing but pursuit of the swiftly darting spotlight. It is a debilitating narcotic for addicts who, at the end of the day, would rather be objects of sport in the spotlight than anything in the shadows. Harold Stassen's problem is that he carries the shadows with him.

The other day he held a press conference three blocks from the White House. Then he dashed to Chicago to massage the uncommitted delegates. Then he was off to Kansas City to instruct the Republican platform committee, and set up his headquarters. He has written letters to all delegates, with an enclosure:

"Informal samples of opinion taken with the enclosed ballot indicate that 14 percent of the voters, who are Independents and Democrats, would swing from Carter to me, and this makes the crucial difference for victory."

When Henry James described "the demoralizing influence of lavish opportunity," he had in mind the late Venetian painters. The description also fits the busy, busy candidate, Harold Stassen, sixty-nine.

He was thirty-one in 1938 when he was elected to the first of three terms as governor of Minnesota. He was the "boy wonder," a rising rocket, but his second stage never ignited. He was forty-one in 1948 when he became a casualty of presidential campaigning in the broadcasting era.

Before the Oregon primary the former district attorney from Minnesota sought a debate with the former district attorney from New York. Stassen's flat Midwestern delivery was suited to Minnesota audiences of sons and daughters of Scandinavia. Thomas Dewey's euphonious voice had helped him earn his way through Columbia University by serving as a cantor in synagogues. The debate, heard by the nation, was decisive.

Stassen lost, and fell through a crack in time. When you ask him today about his foreign policy, he talks about that postwar chimera, the

United Nations: "When Arthur Vandenberg and I drew up the charter . . ."

How often has he run? Only once, officially, in 1948, he insists. Not true. In December 1951, he announced for President. Robert Taft of Ohio wrote to a friend: "The reporters seem to think he is off his beam." He beat Eisenhower in the 1952 Minnesota primary, but Eisenhower nearly won with write-in votes, and this strong showing moved him closer to an open candidacy. Stassen got twenty votes at the convention.

Did he run in 1960? "Nooooo, I don't think so. . . ." He's no longer sure. But he remembers 1968 at Miami Beach, when his nephew, a Minnesota state senator, put his name in nomination, an act of family piety shouted over the roar of an inattentive convention.

Stassen is (in A. E. Housman's words) one of the

> Runners that renown outran,
> And the name died before the man.

It is said that Stassen is a joke. But what he is doing, although harmless, is a bit dreadful. And he is not the only person in humiliating pursuit of the spotlight. Is there really a much larger residue of dignity in Richard Schweiker than in Stassen?

Schweiker, too, seems to be suffering disorientation. He sheds old convictions and dons new ones with the frantic energy of a tourist who, having hit at blackjack for $500, dashes to the Las Vegas K-Mart to refurbish his wardrobe of doubleknit leisure suits. How little the loss of self-respect hurts when one receives, in exchange, Secret Service protection, and other symbols of consequentiality.

"It appears," says Stassen in his letter to the delegates, "that notwithstanding the lack of a campaign, and notwithstanding the humor and ridicule from some of the media, the people do recognize . . ."

He is right in insisting that he is no joke. It is inexpressibly sad to see anyone sacrifice so much for the trophies of politics. Every time an American enters a race for a public office, however lowly, he should be issued a copy of John Webster's *The Devil's Law Case* with this passage underscored:

> Vain the ambition of kings,
> Who seek by trophies and dead things
> To leave a living name behind,
> And weave but nets to catch the wind.

Stassen, never a king, ever a weaver, is whistling down the wind, heading for Kansas City.

[August 12, 1976]

Unanimity Without Enthusiasm

Eight years ago Hubert Humphrey brought down upon himself an avalanche of ridicule for advocating the "politics of joy." Today, in what but for an unruly 1968 convention would be the final year of Humphrey's presidency, Democrats have placidly nominated the Cheshire Cat of American politics. His luminous smile lingers in the air, suggesting substance that sometimes isn't there.

Governor Roger D. Branigin, rest his soul, should be smiling among the angels. He also should be patron saint of this year's convention.

In the 1968 Indiana primary, Robert Kennedy and Eugene McCarthy crisscrossed Branigin's turf, poaching his delegates. To counter this impertinence, Branigin ran as a favorite-son presidential candidate. This pilgrimage took him to the banks of the muddy Wabash, where he suffered this muddy thought about the Vietnam war that was fracturing his beloved Democratic party:

"This is a sad war. The only war that was a happy one was World War One. People used to dance in the streets to see the boys go away. I was there, and I remember the songs . . . 'K-K-K-Katy,' and 'How You Gonna Keep 'Em Down on the Farm, Now That They've Seen Paree?' But we haven't had a happy war since then."

And Democrats have not had a happy convention since the 1964 convention sent forth Lyndon Johnson to rescue the country from Barry Goldwater and the threat of war abroad and discord at home. The 1968 and 1972 conventions were sufficiently disheveled, physically and intellectually, to discredit the party's claim to competence.

Conventions are a party's only visible corporate existence. They are the most vivid moments in the quadrennial process whereby parties reconstitute themselves as transitory collections of coalitions that are, in some cases, short-term recruits. As the late Alexander Bickel accurately noted, "No American political institution is more visible than the convention, or more often visibly shoddy, and none is less visibly constituted and managed."

But this year the proliferation of primaries did to the Democratic convention, and may yet do to the Republican convention, what the development of the party system long ago did to the electoral college. Primaries have reduced the convention to a pro forma, lifeless mechanism for registering decisive choices made elsewhere.

Thus, in the Babylonian ambience of midtown Manhattan, menaced by atmospheric impurities and beset by ladies of loose morals and other naughty citizens, Democrats assembled to blacken the name of the Republican party, to feign indignation about the plight of the Republic, and to try to agree on reasons for saying that their candidate is all that stands between the Republic and perdition.

Deep inside those Democrats who have a deep inside, there are seedlings of doubt about that last point. Many Democrats have little passion to invest in deploring the state of the nation or celebrating the virtues of their candidate. Indeed, the most telling facet of the convention is the striking contrast between the amount of unanimity and the amount of enthusiasm for the man who is the beneficiary of the unanimity.

Ambrose Bierce once defined a President as one of two men about whom we know this: many millions of Americans do not want him to be President. The same principle applies to the nomination process.

Of the 150 million Americans eligible to vote, about 25 million, or one-sixth, voted in this year's primaries. Slightly less than 16 million voted in Democratic primaries, only 6.2 million, or 39 percent, for Carter. So a substantial majority—61 percent—of the small majority of Democrats who cared enough to vote in primaries preferred someone other than Carter.

A convention where journalists outnumber participants does remind us that there invariably are more Boswells than Johnsons. Nevertheless, delegates usually manage to experience extravagant certainty about the marvelous qualities of the man they gather to anoint. But this year there is markedly less than the usual torrent of amiable hyperbole. The party seems to have embraced Talleyrand's maxim for sound politics: "Surtout pas trop de zèle"—above all, not too much zeal.

But what it means is this: like an Olympic-class broad jumper, the Democratic party has leapt, in just four years, from ideological extremism and procedural obscurantism to the cheerful determination that this time, by cracky, there will be no nonsense about divisive things, such as ideas.

[July 15, 1976]

Odd Man in

News vendors in the Americana Hotel displayed toward their guest, Jimmy Carter, the solicitousness due a presumptive President. They put paper strips over photographs of the otherwise undraped ladies on the covers of those magazines, the sight of which could turn a Georgia Baptist into a pillar of salt. But this sacrifice of commerce on the altar of deference may have been just a bit premature as an anticipation of the Carter presidency.

Democrats today radiate an emotion that is impossible to misread, confidence that theirs is an idea whose time has come. And at least while operating within his party, Carter fits Bagehot's definition of the successful statesman, "the one who most felicitously expresses the creed of the moment." But the party has had its creed of the moment for decades. And it is primarily Carter's reluctance to dwell on it that many Democrats find especially felicitous, and prudent.

When confronted with their platform, some sensitive Democrats heatedly denounce as unfair not only anyone who has read it, but also the notion that the party might try to implement it. Yet the platform is significant, if only because it was tailored to serve the candidate's convenience by expressing the party's consensus. It distills to a four-letter word: "More." More public jobs, more revenue sharing, more subsidies for older cities, more spending on health (compulsory national health insurance), more direct subsidies and subsidized loans for housing (and for farmers, and other rural folks, too), more spending on environmental research, transportation, veterans, artists, the elderly, etc.

The closest the platform comes to the dismal subject of paying for these bounties is a pledge to "save"—the Democrats' word—taxes: "A responsible Democratic tax-reform program could *save* over $5 billion in the first year with larger *savings* in the future."

The emphasis is mine. The words are from politicians who know how to use language to conceal thoughts. The pledge to "save" billions in taxes is a pledge to increase the government tax bite by billions.

While Republicans have been explaining that invasions are "incursions" and bombing raids are "protective-reaction strikes," Democrats have been honing their own euphemisms. Once they favored "affirmative action" as the name for reverse discrimination on behalf of minorities and women. Now they call it "compensatory opportunity." And they call forced busing "mandatory transportation." Their new notion that raising taxes is saving taxes is an especially Orwellian twist. The taxes the Democrats propose to "save" for us will, the platform hints, flow painlessly from the closing of "loopholes," those exemptions that benefit thy neighbor but not thyself.

Obviously Democrats have no serious plans for raising the huge sums that would be needed to pay for what they promise to do. And there already is a large and widening gap between the government's resources and commitments. Thus if Carter wins, the 1976 campaign may be remembered as a reversed image of the 1932 campaign.

Then FDR was usually vague, and when he wasn't he was inaccurate in suggesting what his administration would do. Campaigning, he praised conservatism; governing, he invented modern liberalism. Carter's candidacy is in the liberal tradition; a Carter presidency will face a conservatizing reality.

In 1932 FDR, who was to pioneer new dimensions of public spending, denounced Hoover's spending and on one occasion even suggested a 25 percent reduction in spending. The future architect of the New Deal's "alphabet agencies"—NRA, WPA, CCC, etc.—castigated Hoover for piling "bureau on bureau, commission on commission."

Today Carter is running as, and with, a liberal in the New Deal mold. He is a budding architect, anxious to add new wings to FDR's mansion. But President Carter, like all who govern, would be governed by numbers, large ones, written in red ink.

Oh, well. Recalcitrant fiscal reality is, as always, for later. For now Democrats are happy warriors who do not mind marching under old banners. Their party is proud of having been, until recently, the party of ideas that challenged conventional wisdom. But today it is, in a sense, the complacent party. If you seek tangy political argument about fundamentals, you must seek it in the Republican party.

Democrats, unlike Republicans, are wholly at ease with the premise of modern American government. This is not surprising: the Democratic party formulated the premise four decades ago.

FDR's New Deal broke with nineteenth-century liberalism (which is what passes for conservatism today) by abandoning the premise that *society*, as distinguished sharply from *government*, produces the elements of happiness in life, and that government's role is merely to maintain a framework of order in which people *pursue* happiness. What was new about the New Deal was the notion that government has a duty to *provide* people with some, and more and more, of the tangible elements of happiness.

FDR declared that in simpler days "government had merely been called upon to produce *conditions* within which people could live happily." (Emphasis added.) But now it is expected to produce and deliver happiness. Government, FDR said, "was not instituted to serve as a cold public instrument to be called into use after irreparable damage had been done." It "has a final responsibility for the well-being of its citizenship." That those words now seem unremarkable is evidence not that they did not herald a revolution in attitudes, but that the revolution was swift and thorough. Today Republicans no less than Democrats, but less serenely, practice Rooseveltian government, in this sense: they must, as a practical matter, accept that the average voter holds any administration accountable for the nation's economic performance, and for his or her enjoyment of the product.

Long and forever gone are the days when it was thought that well-being, economic and otherwise, should be solely the result of the individual's ability to cope with *society*, with social forces that government could not or should not regulate. The New Deal changed, irreversibly, Americans' expectations, and the legal and psychic relationship of Americans to their government.

This year some GOP conservatives seem to be trying, again, to turn an election into a referendum on the propriety of those expectations and that relationship. It is unclear how the GOP can benefit from so straight-on a challenge to the settled habits of mind of the American majority, which accepts the Rooseveltian premise that government should supply crucial elements of happiness.

This GOP challenge is a risky tactic against Carter, who calls to mind Disraeli's recommendation: Tory men and Whig measures. Carter is an unmistakably conservative *person*. The values he obviously cherishes and repeatedly invokes—piety, family, community, continuity, industriousness, discipline—are the soul of conservatism. The appeal of Carter to conservatives is in his aspiration to use government vigorously in the service of conservative values.

Some conservatives assert that this is a contradictory aspiration, that a large government such as ours is an *inherently* liberal device, *inherently* hostile to conservative values. But this is true only if freedom is the only conservative value, and if freedom means only the

absence of restraints applied by government. And if that is what conservatism asserts, it is a stone-cold anachronism, as dead as the nineteenth-century liberalism it resembles. It is a political philosophy irrelevant to the government that has evolved in response to Americans'
desires.

Because Carter's political persona is a blend of liberal measures and
conservative values, he is a baffling foe for Republicans. Thus they may
make his personality their target.

He burns with an unfamiliar religiosity; he is a workaholic, ascetic
and abstemious; he seems to have few friends his own age; he reportedly does not suffer criticism, dissent or defeat gracefully; he has a
mania for punctuality, an obsession with efficiency, and a short fuse that
burns down quickly when reality is not as frictionless as he thinks it
should be. These traits, taken singly or even together, might not give
Americans pause. What is troubling is these traits combined with his
evident humorlessness.

I refer not just or even primarily to the fact that eyewitnesses
report that Carter launching a joke is no laughing matter. Rather, I
refer to what many people take to be Carter's severe case of the politician's occupational disease, an unsleeping and, in a sense, unsmiling
concentration on self.

Self-promotion is part of the politician's job. Few politicians are
shrinking violets; none has the agreeable insouciance of the lilies of the
field. But because politicians are especially susceptible to the sin of
self-centeredness, it is vital that they evince character traits, beginning
with a sense of humor, that make politicians safe for democracy.

C. S. Lewis, from whose witty Christian apologetics Carter, like all
of us, can profit, believed that a sense of humor is a saving grace. It saves
people from the sin of pride because "humor involves a sense of proportion and a power of seeing yourself from the outside." The absence of
this saving grace is bad for the soul. And the absence of it in a politician
can be a public nuisance, or worse.

Americans know that some of their recent misfortunes were
related in complicated ways to the complicated inner lives of Presidents
Johnson and Nixon. Americans are not unperceptive; nor are they glut-
tons for punishment. If on November 2 voters believe that they are
being asked to choose between a candidate who is even a little bit
strange and one who is not, other considerations may pale to insignifi-
cance. It is possible that Carter is just the person to transform Ford's
uninspiring but unquestionable normality from a liability to an asset.

Carter's admirers say he is like the planet Earth, temperate at the surface, molten underneath. Certainly his surface is as hard as Georgia granite. But hardness can mean brittleness. Republicans hope that they or other irritants will puncture his surface, causing scalding lava to pour forth. They base their hope on the fact that Carter really seems to believe that Morris Udall was beastly to him with those barbs in the late primaries. Certainly Carter may find the autumn unpleasantly stimulating.

Carter is the varsity. He is intelligent, tough, lean and hungry. But there still is a campaign course to be run, and there *are* unresolved issues of policy and personality. Even if it is true, as is frequently and casually said, that the country hungers for "change," the country may not find enough change in Carter's policies, or a pleasing kind of change in his personality. That is why he was premature (although entirely in character) when he recently sought federal financing for a study of how most efficiently to manage the post-election transition to the White House. He still has miles to go before he sleeps in Mr. Lincoln's bed.

[July 26, 1976]

"Disposable Income!"

Republicans, it is said, should feel like veal scallopine, pounded thin and about to be chewed up. Certainly Carter, who claims to enjoy celestial intimacies, has created the impression that all the universe is striving toward a Carter administration. But Carter should remember the song of that southern skeptic, Sportin' Life: "It ain't necessarily so."

Conventions, like groves of aspen, tremble in the slightest breeze, and Republicans arrived here braced for heavy weather indoors. But in the sauna called Kemper Arena, and in various watering spots, rivals mingled, managing polite forbearance from signs of disgust, and calling to mind De Gaulle's description of his meeting with FDR at Casablanca: "We assaulted each other with good manners."

The Ford-Reagan battle produced the jest that the GOP is a minority trying to divide itself into a majority. Actually, the fight was a splendid tonic. Far from leaving the party prostrate, it quickened its pulse, stimulated participation and kept the party on page one. And if Ford and his organization have been tempered, like steel, that is because

Reagan stoked the furnace. Furthermore, Ford's principal handicap has been a faint aura of illegitimacy. He came to power too easily. But campaigning across the continent has been a legitimizing ordeal.

With ten weeks to go Carter is still a political projectile of uncertain velocity. He lost eight of his last fifteen contested primaries. He was shellacked in New York and California. If Hubert Humphrey had Carter's twenty-three-point lead in polls, the lead would be solid. Voters have had time to decide what they think of Humphrey. But a year ago 90 percent of the public had not heard of Carter. And today, because Carter seems to believe that the way to keep knowledge pure is to keep it scarce, almost nobody knows what Carter plans to do. Thus his lead probably is soft.

In Robert Dole, the dark, slender gunslinger from Matt Dillon country, Ford has found his tongue. Pointing out other people's errors is not a duty from which Dole shrinks. He is a speaker whose gifts make him a swan among ducks in the brown Senate pond. He is well equipped to perform the duty frequently assigned to a vice presidential candidate, that of behaving like a pyromaniac in a world of straw men. He will be point man in the GOP attack which will portray Carter as a politician of exceptionally elastic convictions, a Jell-O man who fills many molds but doesn't keep any shape.

Carter sought the nomination as a centrist. Since his acceptance speech, which embraced advanced liberalism, he has not said much, aside from his remarkable promise to make "the family" a responsibility of the federal government. But since the selection of Walter Mondale, Carter has been choosing a staff ideologically suited to advocate and implement the Democratic platform's Mondalean proposals to expand scores of federal programs and agencies. Some call this staff "liberal," others "progressive," according to the resources of their vocabularies. There is no doubt what to call the Ford-Dole ticket: conservative.

The banner of Ford-Dole conservatism should be a loaf of bread and a stick of butter rampant on a field of green emblazoned with the words DISPOSABLE INCOME! The first (and sometimes it seems the only) goal of contemporary conservatism is to increase the individual's disposable income.

It is not clear what yeast is quietly at work in the nation's soul. But Republicans are wagering that the bread-and-butter issue is the price of bread and butter—inflation. That and other costs of government are natural Republican issues. Politics involves dressing up narrow self-interest in the fine cloth of broader motives, so theoretically inclined

Republicans argue, correctly, that what costs money also costs freedom. Money is congealed labor. When government conscripts money through direct taxation or regulation or the surreptitious taxation of inflation, it conscripts the *time* of our lives, and limits our ability to make free claims on the world's resources.

Republicans hope that as the election nears, voters will focus on the fact that the transition from candidacy to presidency is a transition from lofty assertion to laborious doing, and they will prefer the prose of Ford's message to the poetry of Carter's presence.

But there is a danger. Thoughtful people will not warm to Ford's message if it seems to reduce citizenship to consumership, as Nixon's did in 1960 when he declared, "It's the millions of people that are buying new cars that have faith in America."

Be that as it may, democratic conservatism like Ford's usually bears some resemblance to that of Britain's former Tory Prime Minister Harold Macmillan. In 1963, at the end of his career, he gave this summation of his political life:

"I usually drive down to Sussex on Saturday mornings and I find my car in a line of family cars, filled with fathers, mothers, children, uncles, aunts, all making their way to the seaside. Ten years ago most of them would not have had cars, would have spent their weekends in their back streets, and would have seen the seaside, if at all, once a year. Now— now—I look forward to the time, not far away, when those cars will be a little larger, a little more comfortable, and all of them will be carrying on their roofs boats that they may enjoy at the seaside."

That may strike you as a dispiriting summation of a great man's career. But there is idealism implicit in such democratic materialism. The assumption is that affluence—more disposable income, if you will —means more freedom for the individual to achieve a better self. To be sure, the freedom may not be used for noble purposes. But that is not generally the politicians' business.

Ford's conservatism, like Macmillan's, is criticized as banal, which it is. Of course, banal politics is not the worst kind. Ford, like Macmillan, is criticized because he does not give people a "sense of purpose." Ford can reply, as Macmillan did: "If people want a sense of purpose they can get it from their archbishops." That puts politics in its place, which should not be at the center of the human drama.

Still, there is a sense of something missing in the conservatism that marches beneath the DISPOSABLE INCOME! banner. In my mind's ear, I hear Peggy Lee singing, "Is That All There Is?" What is missing is political leadership that summons individuals to citizenship, to the pursuit of something in addition to the expanded personal freedom that disposable income conveys. What is missing is a politics that appeals to what Lincoln called "the better angels of our nature." Lincoln, the fountain of Republicanism, is a reminder that even a muddy stream can have snow at its source.

[August 30, 1976]

Fishers of Men

At a lectern adorned with a portrait of FDR, and using rhetoric borrowed by the bushel from JFK, Jimmy Carter launched his campaign by emphasizing that he is a man of the future. A fisherman and a fisher of men, Carter knows you do not catch trout with clever argument.

On that first official day of campaigning, he went to the stock car races and invited the drivers to the White House for dinner. It was like February 27, 1960, in Bloomer, Wisconsin, when Kennedy told seventy-five farmers that "the American cow is the 'foster mother' of the human race and a great asset to the nation." Carter promised a "new generation of leadership" (he only omitted saying that a torch is being passed to it) and he said that it is time to get the country "on the move again." Ted Sorensen, Kennedy's speechwriter, should be getting royalties from the Carter campaign.

Back at the White House President Ford countered Carter by putting on a vest and golf shoes (not at the same time, of course). Political cognoscenti translated this as "acting presidential." Ford is running as Eisenhower against Carter running as Kennedy.

And so as we (in the words of Chicago's Mayor Richard Daley) "look with nostalgia at the future," it is reasonable to wonder if American politics ever changes. But of course some things have changed.

Having determined that the nation should get "moving again," John Kennedy's first marching order was that it should strengthen its strategic arms arsenal which had, he said, suffered from the dangerous laxity of the Eisenhower administration. When Kennedy spoke the United States enjoyed unquestioned strategic superiority. Today it enjoys "rough equivalence" with the U.S.S.R., and that phrase will not long be applicable. Carter's imitation of Kennedy stops short of Kennedy's call for a national effort at preparedness, although such a call would be more timely today than it was then.

The world "leadership" comes easily and often to Carter's lips. But it never seems to refer, as it occasionally did when Kennedy used it, to the act of telling Americans that they must do some things they would rather not do. Having prospered in the primaries by telling the American people that they were in no way to blame for Vietnam and Watergate and inflation and other recent unpleasantness, Carter has learned that no politician fails to win favor when appealing to the self-satisfaction of his audience.

Ford's campaign, if he decides to have one, almost certainly will resemble Carter's campaign in one particular. It will reflect the diminished preoccupation of American politics with production. Roger Starr, a gifted student of politics, explains the transformation this way:

"Production politics have been primarily concerned over the years with whether the nation should favor agricultural over industrial production; slave-labor over free-labor production; unionization or management prerogatives; production by cartel over freedom to compete. . . . But a large and growing fraction of the American population has passed beyond the consummate interest in production that once preoccupied it. . . . The 'sides' in politics may be shifting from an identification with conflicting producers' interests, to an identification with consumers against producers generally. . . ."

Carter's central campaign promise is to make the government as good, decent, honest, compassionate, etc. as you and I are. His is a free-floating politics of virtue, cut loose from (among other things) the traditional moorings of a politics preoccupied with "production" rivalries. His idiosyncratic idiom may be appropriate for addressing an electorate in transition toward new preoccupations.

More Americans work in service industries than in manufacturing and agriculture. One in six working Americans works for government. Most of these people do not have the sort of concerns that are traditional to our "production politics." Many students and elderly Americans (persons over fifty-five comprise 30 percent of the total voting population) are not involved in production at all.

Carter and Ford are both groping for new ways of gripping this electorate. Carter's latest suggestion, that FBI Director Clarence Kelley should be fired, is part of his strategy of turning the election into a referendum on righteousness. (He's for it.) That has almost nothing to do with the problems of governing, but then neither does Ford's vest.

[September 12, 1976]

Missionary Work

Germany, too, is suffering an election campaign, and Helmut Kohl, the more conservative candidate, is promising to restore "cleanliness, punctuality, dependability, savings, and hard work." *Cleanliness?* Something strange is happening when a national politician makes an issue of his nation scrubbing behind its ears. Something similar is happening here, with President Ford advertising his thoughts about Susan's sex life and Jimmy Carter advertising his thoughts about God's thoughts about adultery.

Politicians are not a race remarkable for daring conversation. But Kohl, Ford and Carter are responding to similar vibrations from their similar publics. They sense an unfocused but strong public hunger for reassurance. At a time when more and more people have less and less confidence in the institutions of government, there is a tendency to evaluate politicians less in terms of the actions they promise than the values they embody.

Whatever their politics, most people are conservative in the sense that they feel a kind of spiritual vertigo when the traditional moorings of life—cultural and religious values—seem to be turning into ropes of sand. Thanks to national instruments of public-affairs communication (broadcasting networks, *Time, Newsweek, Playboy*) national leaders are conspicuous, even intrusive in our daily lives. They are insufferably so when perceived to be unsympathetic toward traditional values.

Increasingly, there is a widespread feeling that if government can't be as useful as once was hoped, at least it need not be offensive. And the feeling is that if government is incomprehensible, at least we should have leaders who are utterly familiar. Thus, politicians tell us rather more than is sometimes seemly about themselves.

The risk, of course, is that in telling much they will say the wrong thing, as Carter did in *Playboy* magazine. Pornography, when as slickly packaged as *Playboy*, is an acceptable adornment on many American coffee tables. But Carter cannot appeal to its readers without offending many nonreaders. *Playboy* got Carter to discuss *Playboy*'s raison d'être: lust. He tried to say that he is not invincible against the temptations of Satan. But he clothed that thought in some salty words that do not usually issue from presidential aspirants.

Carter is said to have the literary conscience of a Flaubert: he picks his words with notable precision. So analysts fell to analyzing his *Playboy* words. Why, they wondered, did Carter use words the *New York Times* found unfit to print? Is he after the vulgarian vote? Will the genius of American pluralism produce an organization of "Dionysians for Carter"?

But before we drown in analysis let us note that Carter has invited the deluge. He understands the primacy of the personal in contemporary candidacies. In the primaries he was disposed to regard campaigning as an exercise in autobiography (indeed, his most detailed campaign document *is* his autobiography). And he is indisposed to believe that the nation might become surfeited with knowledge about Carters, large and small.

Carter's staff has bravely asserted that his *Playboy* words will offend only people who are so conservative they won't change their socks or eat new potatoes. But they are ignoring the fact that the problem is less with what he said than with where he said it. He has worked skillfully to make his character—his inexpressible *goodness*—the central issue of the campaign. Hence, the one thing he cannot afford to do is act out of character.

Ah, well. To those who say that *Playboy* is a lousy place to do Baptist missionary work, Carter can respond that piety frequently has been expressed in strange ways. When Henry VIII, Defender of the Faith, went to war against Francis, Most Christian King of France, Henry ordered twelve huge guns, and named each for one of the apostles. And when Cecil B. De Mille's *Ten Commandments* was showing in New York twenty years ago, the theater's souvenir stand sold children's coloring pencils inscribed with the words "Thou Shalt Not Commit Adultery."

It is a shame when a serious man, a serious candidacy, and the nation are distracted by nonsense like the remarks in *Playboy* on adultery. Carter should hand out pencils, and not so many interviews.

[September 30, 1976]

The Sneeze

In the top half of the first inning of the debate, Jimmy Carter called for more "202 programs." Most of the vast viewing audience had not the foggiest idea what he meant, *and that is why he said it.* It is a curious fact that incomprehensible jargon is considered reassuring evidence that the speaker, unlike most people, understands the government. Responding, President Ford faulted Carter for not being sufficiently "specific."

In the debate Carter, the anti-Washington candidate, reassured Democrats that he is FDR's legatee, a practitioner of laundry-list liberalism, with a program for every constituency. A President is the closest thing the nation has to a chief of state, and Ford reassured Republicans that he thinks the state is too much with us, but that he has no unsettling plans for it. In ninety minutes there was not a flash of wit or bit of eloquence. It was a debate between two ordinary politicians doing what Democrats and Republicans ordinarily do. They probably gratified their firm supporters, who made up about two-thirds of the audience.

To the audience, the debate was as satisfying as a completed sneeze. Those of us steeped in the literature of small children can say poetically what it meant to the debaters:

> The cow kicked Nelly in the belly
> in the barn,
> Didn't do her any good, didn't do
> her any harm.

It was less a debate than a joint press conference featuring two veterans of thousands of press conferences. So, not surprisingly, it contained no surprises. It could hardly have been the decisive event of the campaign unless one of the debaters had come unglued under the pressure. But both demonstrated their professionalism. The public admires professionalism, but takes it for granted.

Johnny Unitas was quarterbacking the Baltimore Colts in the final minute of a game, down by 4 points. He threw a long pass toward a receiver in the end zone, then turned away and began casually walking off the field before the ball reached the receiver. In the winner's locker room an awed reporter asked Unitas why he didn't bother to watch to see if the receiver caught it. Unitas snapped: "He's paid to catch it." That is a quintessentially American attitude, and the debaters were judged accordingly. Both are, in a sense, paid to perform well under pressure and neither will get substantial new support for having done so.

The debate did not fully indemnify the nation for the dismal campaigning that preceded it, but it was useful in showing that each candidate is up against the painfully sharp edge of a fact. For Ford the uncomfortable fact is that people are out of patience with the government. For Carter the uncomfortable fact is that the government is out of resources.

Patience? It has been well said that today the three least credible sentences in the English language are:

1. "The check is in the mail."
2. *"Of course* I'll respect you as much in the morning."
3. "I'm from the government and I'm here to help you."

More and more Americans feel they are having more and more contacts with government, and that most contacts are involuntary and unpleasant. Ford cannot elude the fact that he works at the tippy-top of the bland leviathan that irritates many people. And he will not benefit from what is wrongly called anti-government "conservatism" in the electorate.

What is called "conservatism" might better be called infantilism. Those of us blessed with small children recognize childishness when we see it. Increasingly the nation, like a child, wills the end without willing the means to the end. The end is a full platter of government services. The means to that end is the energetic government that does the inevitable regulating and taxing. Today's "conservatism"? The average voter has looked into his heart of hearts, prayed long and hard, and come to the conclusion that it is high time the government cut his *neighbor's* benefits.

The tendency to will the end without willing the means is one reason why the government is out of resources, and why Carter has problems with his arithmetic. The United States is no exception to the rule that in democracies the appetite for government benefits is increasing, and the willingness to pay for them is diminishing. Carter's arithmetic reflects this. No matter how imaginatively he fudges his assumptions, he cannot plausibly argue that he can deliver his laundry list of blessings while balancing the budget and avoiding substantial new taxes.

This accounts for the least attractive aspect of his campaign and the most unbecoming aspect of his debate performance. He is driven by simple arithmetic to the specious suggestion that substantial new revenues can be painlessly wrung from "special interests" such as the "rich corporations" and "the rich," especially those "big shots" who consume "fifty-dollar Martini lunches."

It is naughty to suggest that the Democratic platform's proposals can be paid by revenues derived from measures like making business lunches non-deductible. And before Carter imposes new taxes on corporations he should consider Irving Kristol's cautionary thought:

"There is a powerful argument to the effect that corporations, by their very nature, do not ordinarily pay taxes so much as *collect* them. After all, where does a corporation get the money to pay its taxes? There are only three possible sources. (1) It can get it from its stockholders by holding down or cutting their dividends—in which case, we are talking about a concealed tax on dividend income. (2) It can get it from its customers in the form of higher prices—in which case, we are talking about a concealed sales tax. In the normal course of events, these are the ways corporations do raise their tax money—or, to be precise, these are the means of collecting it.

"But if, for any reason, a corporation cannot lower or omit its dividend or raise its prices without crippling the business, it does indeed pay taxes instead of merely collecting them. It then (3) gets the money from retained earnings which would otherwise be re-invested in new plant, new processes, etc."

Given Carter's proper concern about unemployment, he cannot be serene about increased reliance on option 3, because capital formation is job creation.

Such issues may not be decisive. The public mind is not a micrometer; it is a mill that will slowly grind to fine flour the corn the debaters poured into it. Ford went to Philadelphia plagued by the electorate's vague feeling that he (as Archbishop Laud said of Charles I) has "not it in him to be or to be made great." He left Philadelphia secure only in the knowledge that this year the electorate cannot insist on greatness. Carter must have gone to Philadelphia full of sympathy for the local baseball team. Like the Phillies he stumbled through a miserable September, dissipating most of a huge lead. Most, that is, but not all of it.

[October 4, 1976]

Hitting Bottom

On the equitable principle that you should praise those most in need of praise, let us now praise Robert Dole. Until Dole took wing in his debate with Walter Mondale, it was unclear when this campaign would hit bottom. But surely it sank as low as it can sink when, responding to a question about Mondale's use of the Watergate issue, Dole said:

"It is an appropriate topic, I guess, but it's not a very good issue any more than the war in Vietnam would be or World War II or World War I, or the war in Korea, all Democrat wars, all in this century. I figured up the other day, if we added up the killed and wounded in Democrat wars in this century, it would be about 1.6 million Americans, enough to fill the city of Detroit."

Pass over Dole's small-minded use of the word "Democrat." That contraction of the word "Democratic" is supposed to be frightfully clever and wounding. It is the sort of thing that substitutes for thought among politicians unaccustomed to having thoughts. Which brings us to Dole's thoughts about "Democrat wars."

It would be comforting to believe that Dole's thoughts stirred so little comment because the public was shocked into silence. But a more plausible explanation is that, thanks in part to this numbingly base campaign, the public's threshold of disgust is now so high that Dole's thoughts could not shock.

It is, of course, absurd to consider knowledge a prerequisite for elective office. But for Dole's information, here are some facts about Republican support for U.S. wars.

It is true that every Democratic administration since the second Cleveland administration has initiated American participation in a war (counting as one the Kennedy-Johnson administration of 1961–1965). But in no instance did participation begin dishonorably, and in every instance opinion on the war was divided along sectional and ethnic lines, not party lines.

World War I? President Wilson backed into war not because he was cunning, but because he was reluctant. After the *Lusitania* was sunk, in 1915, Republicans pushed for more aggressive protection of U.S. citizens and shipping. In the spring of 1916, as Wilson was preparing to seek reelection on the slogan "He Kept Us out of War," his secretary of war committed an American rarity, a genuine protest resignation. He was protesting what he considered inadequate U.S. preparedness. Theodore Roosevelt, who still represented much Republican thinking, was raging about Wilson's unwillingness to join the war.

World War II? Perhaps Dole thinks Japan attacked "the Democrat Party," not the United States, or that, in any case, it was wrong for FDR to seek a declaration of war after Pearl Harbor. Perhaps Dole doesn't know that Hitler declared war on the United States *before* the United States declared war on him. If Hitler had not done so, FDR might have faced some fierce opposition—much of it from Republicans—to U.S. intervention in the European theater.

After Pearl Harbor the GOP disposition was less isolationist than "Asia First." The GOP base was in the Midwest among Germans, Irish and Scandinavians who were delighted to have turned their backs on Europe, and whose somewhat populist democratic feelings were expressed in anti-English prejudices. But the Midwest was nationalistic. Nationalism seeks an outlet, and after Pearl Harbor the Pacific was it. Thus FDR in 1940 prepared for the inevitable war in Europe by appointing Republicans Henry Stimson and Frank Knox as secretaries of war and the navy, respectively.

Korea? When President Truman responded to North Korea's aggression, Senator Robert Taft, "Mr. Republican," said: "The general principle of the policy is right," and that there was "no alternative to what the President has done." In Korea, as in Vietnam, the most important arguments at the outset concerned the correct strategy for prosecuting the war, not the legitimacy of U.S. participation.

Speaking of Vietnam, which Dole would be well advised not to do, the GOP can hardly be portrayed as a sheet anchor of resistance to escalation.

Dole probably turns vermilion with rage when Democrats run against Herbert Hoover. Yet Dole is running against Woodrow Wilson. By doing so Dole is displaying partisanship and intellectual standards comparable to those that led the Nixon White House to attempt to revise history by concocting fake Vietnam war cables.

People who lie about history deserve to be forgotten by it.

[October 21, 1976]

The Disease of Politics

The Thirty Years War began and ended at Prague, the English civil war at Powick Bridge, World War I at Mons. The presidential campaign is ending where it began, in the Slough of Despond. Many people, including many who will vote for him, feel about Jimmy Carter the way a Roman patrician is said to have felt when he first heard Brutus speak: "I know not what this young man intends, but whatever he intends, he intends vehemently." And many people who will vote for Gerald Ford will do so as a "damage limitation" tactic: at least *he* can't run again in 1980.

But it is probable that by 1980 American politics will have changed only by becoming more so. This campaign has reflected less the limitations of the candidates than the nature of the nation's social and intellectual condition. American society and its governance are becoming more complicated much faster than the public mind is becoming more complicated. The social pyramid is becoming steeper. Candidates operating at the top and striving to communicate with those at the base distill their purposes into simple concepts.

In 1960 the winning concept was heroism, "Now the trumpet sounds . . . a long twilight struggle." In 1964 it was idealism, the Great Society and all that. In 1968 and 1972 it was recuperation (from heroism and idealism). In 1976 the concept is cleanliness. Professor Alasdair MacIntyre correctly diagnoses it as a kind of "moralizing politics, one that neatly divides questions of political morality on the one hand and questions of political substance on the other.

"According to this distorting view, political morality is about the personal character of politicians: do they take or give bribes, what are their sex lives like, will they tap our telephones? So the question of whether a politician is a good or bad man becomes a logically distant question from the question of where he stands on substantive issues. But in politics, where someone stands on certain issues is a central measure of his or her moral substance. The kind of moralizing that disguises this all too easily preempts the detailed discussion of the issues for the professional politicians, for government, and leaves to the voter the empty husks of moral judgment."

What is said and done during these interminable campaigns has little to do with what happens in the subsequent four years. This disjunction between campaigning and governing has occurred because candidates have abandoned the idea that campaigns can be exercises in public pedagogy. But it is idle to blame candidates entirely for the sterility of campaigns. Candidates move heaven and earth to get forty seconds of television film of themselves among Pittsburgh Poles or Iowa cornstalks, and they do not do this because it is fun. They do it to satisfy the public's sad superstition that the conduct of public affairs can be televised and thereby made somehow accessible to all. Of course this gives rise to an absurd notion of "the conduct of public affairs."

Another reason American politics has become thin gruel is that it lacks the nourishing, healthy contention of public philosophies. Liberals and conservatives are in intellectual culs de sac. Professor Roger Starr says of liberalism:

"At all times, those in the liberal position advocate more widespread distribution of whatever seems most valuable—goods, services, status, education. This generous cast of mind can be justified only if one accepts two assumptions about nature and man. First, one must believe that there is plenty in nature to fill all human requirements. Second, one must believe that the shortage from which some suffer is imposed neither by the state of nature nor by man's lack of productive inventiveness. One must believe that those who suffer from deprivation are the victims of the greed of others, or of institutional failure, or of historic but temporary underdevelopment; these accidents apart, men must be conceived to be roughly equal in capacity."

If nature is not as bountiful, or men's capacities as equal, as once was assumed, then equality must be forced on men. That is a paralyzing thought for liberals, whose philosophy derives its name from the word liberty.

Conservatives are comparably disarrayed. True conservatives distrust and try to modulate social forces that work against the conservation of traditional values. But for a century the dominant conservatism has uncritically worshiped the most transforming force, the dynamism of the American economy. No coherent conservatism can be based solely on commercialism, but this conservatism has been consistently ardent only about economic growth, and hence about economies of scale, and social mobility. These take a severe toll against small towns, small enterprises, family farms, local governments, craftsmanship, environmental values, a sense of community, and other aspects of humane living.

Conservatism often has been inarticulate about what to conserve, other than "free enterprise," which is institutionalized restlessness, an engine of perpetual change. But to govern is to choose one social outcome over others; to impose a collective will on processes of change. Conservatism that does not extend beyond reverence for enterprise is unphilosophic, has little to do with government and conserves little.

Lacking more coherent philosophies, candidates will not rise above merely competing to purchase the allegiance of private interests. Ford has been a more flagrant purchaser only because he is in a position to be. For example, he recently announced (in Texas, naturally) tighter quotas on imported meat. It might seem odd that anyone who wants to be President of 215 million hamburger eaters would act on election eve to bolster beef prices. But Americans tolerantly accept a politics that has little substance other than a thousand transactions like the one between Ford and the cattlemen. One effect of such campaigning is that no one has any excuse for becoming disillusioned *after* the election.

Politics should be citizens expressing themselves as *a people*, a community of shared values, rather than as merely a collection of competing private interests inhabiting the same country. Instead, politics has become a facet of the disease for which it could be part of the cure. The disease is an anarchy of self-interestedness, and unwillingness, perhaps by now an inability, to think of the *public* interest, the *common* good. This disease of anti-public-spiritedness is not a candidate's disease. It is a social disease.

There is little evidence that Americans want a more elevating politics. It is as though they have taken too much to heart the moral of the story about Charles Parnell, the adored Irish leader. When Parnell met an old man who was working on a road, the old man exploded with enthusiasm. "Calm down," said Parnell. "Whether I win or lose, you will still be breaking rocks."

[November 1, 1976]

Sowing in the Autumn

Electorates are regularly advised that the next election will decide whether there will be desolation and despair, charnel houses and spectral voices wailing in the air, muffled drums of tragedy, heresies, schisms, tempests, tumults, brawls, hurricanes, floods, plagues and shipwrecks, or whether the air will smile, the earth will exult, and the land will flow with milk and honey and nectar.

But America has come of middle age. It is hard to excite or disappoint. And neither candidate seems genuinely passionate about any issue. As Murray Kempton has said, the absence of honest passion is an attribute of professional wrestlers.

Both candidates have experienced fits of generosity of the sort that seize characters in Charles Dickens' novels. In the final debate they were asked what sacrifices they would ask Americans to make. President Ford replied that the nation must stiffen its spine and bravely endure a tax cut. Jimmy Carter said that his administration would require less sacrifice than Ford was contemplating. This was an appropriate ending to a campaign that began with the candidates competing to see who could most emphatically assure farmers that national policy never will be allowed to interfere with the farmers' profitable business of alleviating the Soviet Union's food shortage.

The most interesting question of this political season is how this came to be a close election. Since 1928 Republicans have won just four presidential elections, twice with a war hero and twice when the Democratic party was drawing and quartering itself. If the 1968 election had been two days later—or if LBJ's Vietnam peace initiative had been made on the Saturday before the election rather than on Thursday; that is, if it had been made too late to allow President Thieu time to reject it before election day—Republicans probably would have held the White House only eight of the last forty-four years, and then only thanks to a five-star general who could as easily have been elected as a Democrat.

The GOP limped into the 1976 campaign crippled by the worst political scandal in U.S. history. (Surely, the remarkable fact is not that Walter Mondale has tried to make an issue of Watergate, but that only he has tried, and that he has had to try so hard.) Then came a bitter and protracted nomination battle.

Ford would have lost the nomination if, in February, 800 New Hampshire Republicans had switched to Reagan; or if Reagan had begun his better television appearances, or his slashing attacks on Ford's foreign policies, a few weeks earlier. Having won the nomination, Ford saddled himself with a running mate whose campaigning has been something of an embarrassment. Yet today Ford and Carter are stumbling toward the finish line so close together that it is reasonable to believe that any number of other possible Republican nominees would have beaten Carter easily.

The Democratic candidate, like his rival, has been brought low by, among other things, a great leveler, the grotesque campaign process.

During the 1952 campaign, Dwight Eisenhower, just seven years after commanding the greatest military force in history, sat in a New York television studio. The face that had felt the sting of sea spray off Normandy was now caked with pink make-up, and a flack was reciting a script for a commercial. Eisenhower sighed: "To think an old soldier should come to this."

The 1976 campaign has been about five times as frenetic and generally degrading as the 1952 campaign was. It has demonstrated what such a process can do to candidates whose public stature is rather less than Eisenhower's was.

Whoever wins will be washed into office by the smallest wave of enthusiasm since the wave that deposited Franklin Pierce years ago. This matters. To the extent that the government has an engine, the presidency is it; to the extent that the nation has a political will, the President expresses it. The presidency is made formidable by the respect and affection of a public that today feels little of either for either candidate. So, unfortunately, the field in which the weed of weak government will flourish has been ploughed and fertilized by this campaign. The harvest will come.

[October 31, 1976]

To the Survivor

Half a league, half a league, half a league onward, they galloped full tilt through a punishing year that drained their stamina and lacerated their spirits. They spent as much of their energy, and other people's money, as the laws of nature and the nation permit. Each was spurred on by the peculiar severity of the American system; unlike in Britain, where the person who does not become prime minister does become leader of Her Majesty's loyal opposition, the loser in a presidential election instantly becomes yesterday's person.

The campaign ended as a blitzkrieg of television commercials, some of which came perilously close to plagiarizing Pepsi-Cola's jingle ("We're the Pepsi people, feeling free, feeling free"). These commercials were attempts to seize the attention of a nation that long ago developed sophisticated mechanisms for filtering out such bombardments. For me the campaign reached a peak of sorts in a surrealistic moment in the final week. In the comfort of my living room I watched a hoodlum smash a doctor's face and next the bones in the doctor's hands. Then turning from routine American entertainment to public affairs, I saw a toe-tapping, finger-snapping Ford commercial featuring a cast of thousands and a chorus singing "I'm feeling good about America, I'm feeling good about me!" I'm OK, you're OK, and now back to breaking bones.

Today a convalescent languor has settled upon the nation. This is not just the result of the campaign, which was more disappointing than maiming, like a glass of indifferent Beaujolais. True, the campaign was worthy of the late Mayor O'Keefe of New Orleans who, when a group of matrons asked him to build a theater in the Greek style, declined because, he said, there were not enough Greeks in New Orleans.

But no presidential campaign could have indemnified the nation for the last thirteen years, during which public affairs have been too much with us. The period of trouble began (I am *not* suggesting a cause) eleven weeks after CBS decided there were enough television sets, and enough interested viewers, to justify the first thirty-minute evening newscast. The trouble began thirteen years ago this month, in Dallas, but it also began simultaneously in all the nation's living rooms. The assassination (and the murder of Oswald, broadcast live and on instant replay) was the nation's initiation into the dismaying intimacy of public affairs in the television age.

During those four November days television was a great unifier, the instrument of an invaluable national catharsis. And television performed a service in the 1960s by bringing into living rooms Bull Connor's police dogs; and the heavy smoke as the underprivileged burned Watts and the overprivileged burned campuses; and body counts from a war served up nightly at mealtime. But there is such a thing as sensory overload from exposure to public affairs as well as to rock music. And before the "living-room war" was over, the nation began fourteen months of the televised collapse of an administration.

The discomforting immediacy of public affairs in the television age has left the public enervated and longing for nothing so much as psychic "breathing space." And the retreat of the public from public affairs is being accelerated by something else. That something is the depressing sense that the weave of the world's fabric is so dense that it defeats even noble motives.

Strenuous and generally large-hearted involvement in the difficulties of the postwar world left Americans feeling uncomfortably like Mrs. Jellyby in Charles Dickens's *Bleak House*. She had "handsome eyes though they had a curious habit of seeming to look a long way off. As if . . . they could see nothing nearer than Africa!" She practiced what Dickens called "telescopic philanthropy," neglecting her ill-fed and ill-clad family while worrying about "the natives of Borrioboola-Gha, on the left bank of the Niger." Prompted by unpleasant experiences abroad, Americans decided that charity should begin at home. But they have found that even at home the charitable impulse has become complicated, as Max Ways explains:

"At the time when Saint Francis impulsively gave his fine clothes to a beggar, nobody seems to have been very interested in what happened to the beggar. Was he rehabilitated? Did he open a small business? Or was he to be found the next day, naked again, in an Assisi gutter, having traded the clothes for a flagon of Orvieto? These were not the sorts of questions that engaged the medieval mind. The twentieth century has developed a more ambitious definition of what it means to help somebody."

But the reason government cannot act with Franciscan impulsiveness is not just that it sometimes defines what "help" is too ambitiously. Some government attempts to provide unambitious help cause injury.

For example, society's charitable impulse, as feebly expressed in the existing welfare system, often is destructive of families. A kind of "ambitiousness" also is a problem.

If you were to make a list of clear government successes in the last, say, forty years your list might include the Tennessee Valley Authority, the Manhattan Project, Project Apollo, and the interstate highway system. These were successes in the limited sense that government accomplished what it intended to accomplish. There would be other things on your list, but most would be like these, programs that delivered clearly defined durable goods. Similarly, the great successes of nineteenth-century government were in causing canals and railroads to be built, and in distributing a natural bounty, land. But government becomes discouraged, and discouraging, when it tries to deal with concepts more complicated than dams and highways, when it tries to deliver "meaningful jobs" for adults, "head starts" for children and "model cities" for all.

Yet increasingly government is asked to deliver such complicated commodities. That is why, if power is the ability to achieve *intended* effects, government is decreasingly powerful.

That thought only seems perverse if you equate size with power, which admittedly would make sense if government were a machine. But as President Taft wearily exclaimed to no one in particular when a zealous aide began lecturing him about the "machinery" of government, "The young man really thinks it's a machine!"

Fortunately, there is time enough to worry about what government is and what it is good for. Why, shucks, there are almost twenty-four months before the next presidential campaign begins in deadly earnest. What matters for now is that the long gallop has produced a, well, survivor. To him I offer the encouraging words of a philosopher (actually Casey Stengel): "They say you can't do it, but sometimes that isn't always true."

[November 15, 1976]

PART 5

Governing

Doctrine of Disappointment

The Carter administration is still a cloud on the horizon no larger than a man's hand. But already persons who profess to know the shape of things to come say the administration will have a shape very like a whale.

Or maybe a weasel. Or perhaps a camel. Which is to say that much of the current speculation—a deluge of words and a drizzle of thought —calls to mind poor, dotty Polonius when Hamlet was making him look ridiculous:

Hamlet: Do you see yonder cloud that's almost in shape of a camel?

Polonius: By th' mass, and 'tis, like a camel indeed.
Hamlet: Methinks it is like a weasel.
Polonius: It is backed like a weasel.
Hamlet: Or like a whale?
Polonius: Very like a whale.

The Carter administration will be a very smart whale. A *New York Times* article about Jack Watson, a Carter aide, says that Watson's undergraduate record indicates that he has a "formidable ability to excel at almost anything." Gosh, "formidable" hardly does justice to an ability that would have been the envy of Leonardo da Vinci.

These are breathless days for many gentlepersons of the press. Some of them recently found themselves noisily displeased with Carter's press secretary because he was, they thought, insufficiently forthcoming with details about the Carters' Thanksgiving menu. This struggle for the "public's right to know" about the cranberry sauce corresponds to the "English muffin" phase of press coverage of the Ford administration. Similarly, the *Times*'s apotheosis of Watson corresponds to the heady discovery, sixteen years ago, that the Kennedy administration was going to be uncommonly bright.

Such monomania about the trivial details of a President's life, and childish exaggeration of the virtues of the persons who orbit around him, are symptomatic of the prevailing conception of politics and history. According to this excessively "voluntarist" conception of history, human volition—will power—determines social realities. Thus, the history of a nation is merely a record of "decisions" made by politicians, or about politicians by electorates, exercising free will.

Unfortunately, the "voluntarist" concept of history exaggerates the extent to which social realities are alterable by "decisions." So it is a doctrine of the perpetually disappointed. But some of the giddy journalism of the current "transition" is undeniably attuned to it. The transition is a period when people choose to ignore the fact that relatively little of a stable nation's life and destiny actually changes when a tiny fraction of its government's personnel changes.

The education of Presidents almost always consists in large measure of altering their excessively "voluntarist" view of the world. That Carter's education has begun is apparent in his much tempered post-election expectations for lowering unemployment in the near future.

Already some of his followers are beginning to think as Robert Strausz-Hupé does: "Whenever a politician-in-office says that the situation now calls for realism, he is about to ditch those who voted for him because he appealed to them in the name of idealism." But the "problem" is only that Carter is now face-to-face with responsibility. That tends to concentrate one's mind on what one can and cannot be fairly blamed for not achieving.

Journalism is, inevitably, part of the process of apportioning blame. It is frequently said that journalism is "history written as it happens." Journalism is, understandably, preoccupied with the vivid present and, thus, with politics.

But the best history is distinguished by an awareness that there is much more to the lives of nations than the decisions of politicians and electorates. Such history, and the best journalism, is sensitive to the fact that free will matters, but within the narrow limits of given physical and intellectual forces, and beneath the weight of accumulated traditions.

Presidents begin by encouraging and profiting from "voluntarist" journalism, which exaggerates their range of freedom. But they come to desire an "historical" journalism, which would convey the fact that history is a rolling river that can be resisted but not controlled. Soon enough they feel threatened by writing that does not portray their administrations as they know them—more shaped than shaping.

[December 2, 1976]

Authoritarian Short-cuts and the Litigation Tactic

Residents of the other forty-two states that levy income taxes may take pleasure from the fact that New Jersey is no longer exempt from that torment. But there should be national dismay about the process that produced this result.

The enactment of New Jersey's income tax demonstrates how "progressives" and their relentlessly result-oriented allies in the judiciary circumvent democratic processes.

A New Jersey court imposed the tax. Technically, the legislature enacted the tax. But it acted under judicial duress, degraded from a representative institution to a passive instrument of the judicial will.

The legislature repeatedly rejected income-tax proposals. But recently the state court seized upon a clause in the state constitution that says the legislature shall provide a "thorough and efficient" educational system. From those three words the inventive court extracted this extraordinary judgment: different per-pupil expenditure rates in different school districts are unconstitutional because they violate the "thorough and efficient" clause.

Obviously the judges were not seriously interested in "thoroughness" or "efficiency." Rather, they were interested in exploiting vague constitutional language to legislate their social goals. As the judges were undeterred by the absence of a real constitutional warrant for their action, so, too, they were undeterred by the absence of pertinent evidence, such as evidence that the quality of education varies directly with the dollars spent on it.

In fact, per-pupil spending differences are generally not large in New Jersey. Some poorer communities match the spending of some richer communities by subjecting themselves to higher tax rates. And it is possible that this is what bothered the judges. They may have used the constitution's language about "thorough and efficient" schools as a pretext for attacking the "problem" of communities making unequal tax efforts.

The court decreed that to comply with the empty constitutional phrase the state must spend on education sums that could be raised only by an income tax. And the court ordered the state's 2,500 schools closed to nearly 100,000 summer students and staff pending enactment of the tax. Thus did the court usurp the legislature's revenue raising and appropriating functions.

It is arguable, and probably true, that New Jersey needed an income tax. But it certainly is true that such arguments about taxation should be settled in institutions of representation.

A crucial premise of popular government is that democratic decision-making procedures confer legitimacy on the results of those procedures. But today "progressives" are reversing this logic. They behave as though a meritorious result, which New Jersey's income tax may be, legitimizes whatever process brings it about.

The New Jersey episode is part of a national pattern. Advocates of prison reform, abortion law reform, housing reform, coerced integration, and numerous other causes, are abandoning the process of political persuasion and adopting the authoritarian short-cut of seeking decrees from ideologically sympathetic judges. As Solicitor General Robert Bork says:

". . . there is now before the States . . . the Equal Rights Amendment, which provides that it shall be primarily the function of the judiciary to define and enforce equality between the sexes. The amendment, we are assured, does not mean that no distinctions whatever may be made between men and women. . . . Yet it is proposed . . . that the Supreme Court rather than Congress or the state legislatures make the necessary detailed and sensitive political choices to write a code for the nation. In that sense, the amendment represents less a revolution in sexual equality than it does a revolution in constitutional methods of government."

The surest sign that a political persuasion has become anemic is a dependence on litigation rather than legislation in pursuing goals. In the 1930s the debilitation of conservatism was apparent in its reliance on the Supreme Court's irresponsible activism. Unable to sustain their opposition to the New Deal in electoral and legislative arenas, conservatives relied on the obduracy of appointed judges.

Recently roles have been reversed as "progressive" judges have become the principal weapon of persons impatient with democratic decision-making. But that is a weapon that will cripple any cause that depends on it.

Having adopted litigation as a substitute for political persuasion, "progressives" will awaken too late to the fact that their powers of persuasion have atrophied. They will discover this when they need their powers of persuasion to assuage the resentfulness of the national majority that the litigation tactic was designed to circumvent.

[July 25, 1976]

A Man of the People

In the acid rain of sexual accusations against congressmen, and the non-denials from the accused, there has been one remark of more than prurient interest. Representative John Young (D-Tex.), discussing a former secretary's charge that he paid her a handsome salary in exchange for sex, delivered a sermonette on the folly of great expectations:

"When a man is holding public office, the greatest thing they can say about him is that he's a man of the people. Then when they find out he is, that's when the trouble starts."

Actually, "the trouble starts" not when the public learns that a politician has led an other-than-blameless life, but when they learn that his personal comportment has been base enough to make his public piety even more hypocritical than they had hitherto suspected. The interesting thing is that so little trouble starts.

Representative Young's lament is an expression of the vertigo he evidently is feeling. Poor him. His vocation is to be a "man of the people." But now some people, in their perversity, have rounded on him and suggested that he should be a gentleman. Representative Young intimates that something about this is unfair. He has half a grip on half a point. The point, put in a way Young undoubtedly would prefer not to have it put, is this:

A nation that feels a democratic imperative to celebrate the lowest common denominator sooner or later will get the lowest common denominator everywhere, including its legislatures. The empty-headed celebration of the common man will produce many "leaders" who are, to be polite, common.

Americans have never insisted that their leaders satisfy Scriptural standards by being wise as the serpent and harmless as the dove. But there was an era, albeit brief, when the premise of American government was that uncommon men should rule.

Today Americans wonder how it came to pass that a tiny collection of thirteen loosely related communities, with a population of about three million free persons, could produce the generation of American founders who, at Philadelphia in 1776 and 1789, accomplished history's most stunning feat of political creation.

The answer is not, or at least not primarily, that an accident of history blessed the colonies with an extraordinary number of sage and decent men. A better explanation is that there was then a habit of deference to excellence in public life. After all, the most remarkable thing is not that the Founding Fathers existed, separately, but that the political process brought them together in Philadelphia.

But there have been changes in the theory and, hence, the practice of American democracy. The changes began with the "Jacksonian revolution" in democratic thinking.

In his first message to Congress, in 1829, President Andrew Jackson said: "The duties of all public offices are, or at least admit of being made, so plain and simple that men of intelligence may readily qualify themselves for their performance."

The duties of public office are "plain and simple" only if government problems are only problems of administrative technique. But such a simpleminded conception of politics is blind to the political virtues of judgment, prudence and courage.

The devaluing of the political vocation has been followed in our century by a related degradation of the state. Today the state exists to be "responsive." Politicians exist to respond like simple mechanisms to impulses recorded from demanding constituencies. This "plain and simple" task requires no uncommon virtues. Indeed, to be vigorously servile to all demands, a politician should be (in Representative Young's words) "a man of the people," prepared to serve democracy by representing its common denominators, including (perhaps especially) the lowest.

The modern servile state possesses, at most, utility—never dignity. Not surprisingly, the public evidently thinks it would be unreasonable to expect dignified politicians, and that, anyway, dignity is irrelevant to the politicians' low function. Perhaps that is why early indications are that constituents of Representative John Young will not cast him into outer darkness but instead will cast him back into Congress.

[June 17, 1976]

Government by "The People"

Jimmy Carter has done well, if not necessarily good, by telling the American people that they are wise and wonderful. But he was not strenuously correct when he said in the debate: "Every time we've made a serious mistake in foreign affairs, it's been because the American people have been excluded from the process."

"The people" were, if anything, more jingoistic and less discerning than the government in the Spanish-American and First World Wars. "The people" rejected the League of Nations before Congress did. President Roosevelt, not "the people," recognized that Hitler was intolerable, and FDR's most noble acts were in weaning "the people" from their morally obtuse isolationism. After the war, "the people" *demanded* the disastrous pell-mell demobilization that incited the Soviet Union. Today the nation's cherished alibi is that Vietnam was a "presidential" war foisted on the nation surreptitiously, in the dead of night. Actually, Lyndon Johnson's policy was about as stealthy as a steam calliope. Congress funded it and "the people" supported it.

Nevertheless, Carter's confidence in "the people" is limitless: "If we can just tap the intelligence and ability, the sound common sense and the good judgment of the American people, we can once again have a foreign policy that will make us proud instead of ashamed."

Such populism is more than a bit murky on the subject of leadership. Carter frequently speaks about leadership, President Ford's incapacity for it and his own desire to exercise it. But for anyone whose populism is as pure as Carter's evidently is, there is not much for a leader to do, aside from drinking deep drafts of "the people's" wisdom. Put plainly, a populist leader keeps his ear to the ground. Never mind that, as Churchill said, it is hard to look up to leaders who are detected in that somewhat ungainly position.

Georges Clemenceau, no populist, said that the voice of the people is the voice of God and the leader's job is to follow that voice *shrewdly*. Clemenceau packed into the word "shrewdly" a lifetime of hearty misanthropy and cynicism that he didn't bother to conceal. But if you take Carter at his word, there is not a skeptical, let alone a cynical, atom in his body, and the nation can "depend upon it" that he means just what he says. He says the first thing he would do to involve "the people" in foreign policy would be to "quit conducting the decision-making process in secret."

Political principles are like spilt needles. They make it risky to step or sit. And it is fun to imagine how Carter would live with the principle that foreign policy "decision-making" should not be secret. Would the press be invited to cover National Security Council meetings? Would our negotiating options in the strategic-arms talks be published? Would intelligence estimates, which are the basis of many decisions, be released regardless of what they would reveal about intelligence sources? Of course the answer is, *"Of course not."*

When pressed to be specific about how he would involve "the people" in the decision-making process, Carter also said: "I would restore the concept of fireside chats, which was an integral part of the administration of Franklin Roosevelt." Actually, FDR averaged only about two "chats" a year. But even if Carter intends to hold many more, it is stretching things a bit to advertise this as "involving the people" in the decision-making process. Will he invite the sovereign public to write in, expressing its opinion on commodity agreements, the capitalization of the World Bank and the correct negotiating position regarding the Soviet SS-20 missile? What *is* the will of "the people" regarding a 600-kilometer-range limitation on Cruise missiles? These subjects are not peripheral technicalities. Foreign policy *is* decisions about things like these.

Carter's professed belief in the paramount importance of involving "the people" in foreign-policy decisions brings to mind the cartoon on page 209. It originally appeared in *The New Yorker* about fifty years ago. Although the artist did not intend it as such, it is, I think, a brilliant *political* cartoon.

When Einstein wrote the words that are quoted as the cartoon's caption, he obviously was not speaking literally. He used the word "people" as it frequently is used, to refer to a group implied by the context in which the word is used, in this case the tiny group of physicists and mathematicians competent to understand what Einstein was talking about. The cartoonist's inspiration was to see the splendid absurdity in taking literally Einstein's reference to "people." And there is, increasingly, something comparably comic about most political references to "the people," especially references to their role in "decision-making," particularly in foreign policy.

Carter is surfing on the rising wave of "anti-elitist" feeling. That

Drawing by Rea Irvin; © 1929, 1957 The New Yorker Magazine

'People slowly accustomed themselves to the idea that the physical states of space itself were the final physical reality'
—Professor Albert Einstein

feeling is becoming stronger, and more irrational, as more people flinch from the obvious fact that modern government inevitably is the vocation of specialist elites. Claiming a "personal" relationship with "the people," and pledging that soon "the people" will rule, Carter is closing in on his prize. But when dawn breaks cold and gray along the Potomac on January 21, Robert Strausz-Hupé will still be right: "All government is rule by the few over the many. That, in an oligarchy, the ruling class is more exclusive than it really is, this is aristocratic pretentiousness. That, in a democracy, the ruling class is more open than it really is, this is egalitarian cant. Both conceits are necessary for the welfare of the respective establishments."

[October 18, 1976]

Senator Pell Makes His Mark

The name of Senator Claiborne Pell (D-R.I.) will not dominate histories of our age, but he has left a mark. Thanks to him, President Carter was able to nominate a new head of the National Endowment for the Humanities, the foundation most important to American scholarship.

Pell's importance to the life of the mind in America flows from his chairmanship of the authorization subcommittee that is sovereign over NEH. Pell's most notable recent exercise of sovereignty was in blocking for more than a year, until after the election, a vote on President Ford's nomination of Ronald Berman for a second term as head of NEH.

Brought to a vote, the nomination, which enjoyed broad and distinguished support in the academic community, would have been approved overwhelmingly. But Berman is a scholar and a Republican. To give Pell his due, he was offended by the former, not the latter.

Pell has one idea and it is philistine. It is that NEH funds too many "esoteric" projects that do not "reach out to the length and breadth of our country." He thinks NEH should be more like the National Endowment for the Arts, which Nancy Hanks runs in a way designed to win the admiration of Congress.

Pursuant to the exacting standards Carter set for appointments ("Why not the best?"), the White House conducted a seven-month search for a nominee. After several candidates declined, Carter reached into his administration for Joseph Duffy, one of the early supporters of Carter's candidacy. Duffy's mandate includes cleansing NEH of what Carter calls its "elitist image."

Duffy, forty-five, describes himself as "a political person." He is a minister from Connecticut, where he received a Ph.D. from Hartford Seminary. He taught there and, briefly, at Yale before running unsuccessfully for the Senate, administering the American Association of University Professors, and serving as an assistant secretary of state. His wife is an assistant secretary of commerce. It is possible Duffy is "the best," or just the best person to deal with Pell.

Pell thinks NEH has been a "pale shadow" compared with the Arts Endowment, which he says has "generated more momentum" at the "grass roots." Pell thinks NEH should support "lumberjacks," "grocers" and "shoemakers." Presumably (Pell is a bit vague on this point) each would do his thing in the humanities field of his choice.

Surely Pell is pleased at least by the fact that NEH spent just $500,000 to bring the exhibition of the treasures of King Tut to six million museum visitors. NEH spent just $250,000 to bring "War and Peace" to 20 million television viewers. That is .0125 cents per viewer, a statistic that should satisfy Pell and others who think such cost analysis is a sufficient analysis for evaluating investment in culture.

Pell falls easily under the spell of statistics, and is powerfully affected by the fact that in a recent fifteen-month period the Arts Endowment issued 5,050 grants from $115 million while NEH issued only 2,045 from $111 million. Perhaps Duffy will be inclined and able to make Pell understand that the aim of NEH under Berman was excellence, whereas under Ms. Hanks the aim of the Arts Endowment seems to be the satisfaction of the largest number of applicants from the largest number of congressional districts.

The Arts Endowment can give $500 to a voter in Leadville, Colorado, who wants to be subsidized when playing the recorder. The Arts Endowment can rationalize this in terms of a populist, democratic doctrine that "art" is almost any instantaneous enjoyment of "self-expression." Needless to say, congressmen understand the charm of this.

But most worthy humanities projects—for example, scholarly research and translations—involve a more demanding standard of excellence achieved over time. Hence grants for humanities projects generally must have a larger "critical mass" than grants for "art" as the Arts Endowment can conveniently define it.

Because of the nature of the disciplines it encompasses, NEH is inherently more comparable to the National Science Foundation than to the Arts Endowment. But one should not dwell on this fact, lest it kindle in Pell and Carter anxieties about the Science Foundation's "esoteric" projects and "elitist image."

[August 11, 1977]

Looting

The morals of piranha are especially unpleasant when adopted by people, as during New York's blackout. But Andrew Young, ambassador to the U.N. and sociologist to the world, explains that, "If you turn the lights out, folks will steal. They'll do that in Switzerland, too, especially if they're hungry."

To say that even in staid Switzerland mob violence would follow a power failure is to imply that such violence is normal, natural. To suggest that New York looters were hungry is to imply that they were justified.

The fate of liquor stores suggests that the looters were thirsty. And even the unemployed probably are not hungry in the city with the nation's most generous welfare payments. But the attempt to legitimize looting is not new.

It began with the 1965 riot in Watts, a community of adequate single-family homes in Los Angeles, a city which the Urban League then rated first on a list of sixty-eight cities in terms of how blacks fared. The trouble started after a routine traffic arrest, and consisted of violence by a minority of the community against the community. But around the nation progressives peered at Watts through ideological binoculars and declared that the riot was a spontaneous protest against injustice, a "black Bunker Hill." For example, civil rights leader Bayard Rustin said the riot was for an "express purpose"—political expression.

A more detached and informed study later concluded that it was a "collective celebration in the manner of a carnival, during which about forty liquor stores were broken into and much liquor consumed." Fifteen percent of those arrested were children. One-third of the adults arrested had been convicted of major crimes, another one-third had been convicted of minor crimes.

But the earnest search for extenuations was under way. So (not coincidentally) were the riot years. By 1967, a presidential study was feeding the fires by announcing that rioters had "let America know" what reforms were called for. Similarly, today a television reporter announces to a network audience that New York looters were motivated by "desperation." The *New York Times* says the looters acted in "rage" and the nation should "heed" them.

New York's looters stimulated a depressed manufacturing industry: the guilt industry. It responds to crimes against society by manufacturing indictments of society. Therefore, it is useful to look back to that industry's last boom, a decade ago.

In Detroit in 1967 the riot began with a raid on an after-hours joint. In his essay, "Rioting Mainly for Fun and Profit," Professor Edward Banfield reports that, at first, "Negroes and whites mingled in the streets and looted amicably side by side. . . . Stores having things that could be consumed directly . . . were looted no matter who owned them. . . . Buildings symbolic of the 'white power structure'—banks, public offices and schools—were untouched. As one of the rioters, a child, explained, 'There was nothing to steal in the school. Who wants a book or a desk?' "

Banfield classifies four kinds of riots: "the rampage," an outburst of animal spirits; "the foray for pillage"; "the outburst of indignation" triggered by an incident of perceived injustice; "the demonstration" to advance a principle. Most riots have been, primarily, compounds of the first two.

At 11 P.M., September 9, 1919, in Boston, the first day of the police strike, a brick was thrown through a cigar store window. The rampage was under way. (In 1711 Boston felt compelled to decree punishment for looters "taking advantage of such confusion and calamities" as fire.) In 1863 participants in New York's misnamed "draft riots" found time to loot Eleventh Avenue.

Obviously what happened in New York's blackout was not peculiar to that city or this age. But neither was it the normal, natural reaction (for Americans or Swiss) to an opportunity to pillage. Ambassador Young's formula—"If you turn the lights out, folks will steal"—equates an opportunity with a cause.

Young flinches from facing this disturbing fact: unlike Switzerland, the United States has within its urban population many people who lack the economic abilities and character traits necessary for life in a free and lawful society.

Unlike Switzerland, the United States has within its urban population many people who lack the economic skills and acquired civility necessary for life in a free and lawful society.

[July 21, 1977]

Intensity and the Filibuster

James Allen of Alabama is an unlikely man to be the Senate's most gifted practitioner of the noble art of using a filibuster to block Senate action. His is not the medieval fervor of an emotional soul, and his forte is not flashing badinage or the deftly aimed *mot justes*. Nothing—not badinage, not *mots, justes* or otherwise—flashes from Allen, whose demeanor at all times is that of a man who has been prematurely aroused from a summer snooze to which he hopes to return soon.

But he actually is about as sleepy as a well-rested fox. He is like a lot of those wily men the South has been sending North ever since it lost the Civil War and began exercising a disproportionate influence in Congress.

In recent weeks, as he did last December, Allen filibustered to block a vote on a bill that would provide public financing of presidential and congressional campaigns. Some editorialists and columnists and other low forms of pond life have called Allen obstinate, meaning that he finds resistible their advocacy of a position inconsistent with his.

But Allen can be gloriously, usefully obstinate without violating the letter or the spirit of the Senate rules.

Unlike many of his colleagues he had the good fortune to have humble political beginnings as a state legislator. The Alabama legislature's rules, like those of the Senate, are based on Jefferson's manual of parliamentary practice. Jefferson wrote the manual during his tenure as Vice President and President of the Senate (1797–1801). Never since has a Vice President used his copious spare time as fruitfully.

Jefferson noted that while "it is always in the power of the majority, by their numbers, to stop any improper measures proposed on the part of their opponents, the only weapons by which the minority can defend themselves" are "the forms and rules of proceeding" which help a minority check the majority's "wantonness of power." The filibuster, permitted by Senate rules, is a shield for minorities.

Pure democratic theory demands the political equality of all individuals. But prudent men, like Jefferson, try to accommodate this fact: not all individuals feel equally *intensely* about all issues. So our system has provisions to diminish the number of occasions when intense minorities will be intensely frustrated by casual majorities.

For example, "special majorities" are required before certain things can be done. It takes two-thirds of both houses of Congress to propose an amendment to the Constitution, and three-quarters of the states must ratify an amendment. And the most important (because the one most central to the regular business of government) "special majority" provision is Senate Rule XXII. This requires a vote of two-thirds of the Senate to shut off debate on a particular matter and bring it to a vote.

Filibusters are the traditional means by which intense Senate minorities make it hard for majorities to get their way. But today's filibusters have a new look. Gone are the days when Senator Strom Thurmond (then a Democrat) held the floor to pour forth twenty-four hours and eighteen minutes of uninterrupted rhetoric—a record. For such filibusters a senator needed the intrepidity of a Spartan at Thermopylae, and kidneys of cast iron. Today filibusters involve little talk, other than a senator saying "no" at the right times.

The Senate runs, when it runs, with the unanimous consent of its members. If a single senator present in the chamber refuses to consent to a request to set a time for voting on a bill, no vote occurs. Technically, the "debate" on the bill continues.

Usually the Senate goes on to other matters, and the filibustering senator must stay near the floor so he can jump up and object if anyone tries to get a "unanimous consent agreement" to vote on the bill he wants to block.

Eventually, if the majority favoring the bill is large enough to muster two-thirds of the Senate, and if it is so *intensely* in favor of the bill that it is willing to violate the Senate's proud tradition of unlimited debate, it votes to shut off the filibuster. This is what happened to Allen. But Allen, with his filibuster, asserted a principle, the Jeffersonian principle that intense minorities should not be ridden over by slender or lukewarm majorities.

[April 11, 1974]

Treason in the Woodwinds

As promptly as could be reasonably expected, but more than three decades late, the House of Representatives has administered euthanasia to the Internal Security Committee, known in its halcyon days as the Un-American Activities Committee. In the Kremlin hard-eyed men must weep for those vanished allies who helped give anti-communism a bad name. No day when Martin Dies (D-Tex.) and J. Parnell Thomas (R-N.J.) jousted in public with international Bolshevism was considered a total loss by people anxious to discredit popular sovereignty.

As late as 1967 Chairman Edwin Willis (D-La.), in a kind of committee *Areopagitica*, explained that he did not oppose "honest and responsible dissent from American policy by patriotic Americans" as long as that dissent would enhance the public's "appreciation of the basic correctness of the policy our government is pursuing." But by 1967 the lads no longer had fire in their bellies.

It had been different twenty years earlier, when an assistant to the secretary of the Navy begged a congressional investigator: "If you see J. Edgar Hoover, tell him I'm a good American." Those were the days when Aware, Inc., "the anti-Communist organization in entertainment, communications, and the fine arts," feted the committee with an event billed as "Cocktails Against Communism."

Having thus restored their tissues, committee members could listen to their favorite witness, Harvey Matusow, who once made fourteen telephone calls to the New York Yankees, using fourteen different voices, and persuaded the Yankees to keep Yogi Berra from appearing on television with a leftwing actor.

Because the committee was in the entertainment business, it was obsessed with its private sector competitors. Its greatest publicity bonanza was an investigation of Hollywood, which John Rankin (D-Miss.) thought was a nest of Jews and, hence, an outpost of the Comintern. Rankin warned that "loathsome paintings" by leftists had "got into the home of Charles Chaplin, the perverted subject of Great Britain who has become notorious for his forcible seduction of white girls."

Gerald L. K. Smith gave the committee ambiguous news about Hollywood: "I am convinced that Frank Sinatra is not a naive dupe."

And witness Gary Cooper disemboweled communism: "From what I hear, I don't like it because it isn't on the level."

The committee owed much of its prominence to its critics' hysteria. But columnist Murray Kempton knew how to burn the committee with the dry ice of his contempt. When the committee descended on New York to expunge subversion from the Metropolitan Music School, Kempton wrote:

"The director and registrar . . . took the Fifth. . . . A clarinetist testified that he had been a Communist and left, naming in the process ten musicians who had been Communists with him. One was a violinist, another a bass player. The rest were all woodwinds; the brass section appears free of treason."

The committee's style of Socratic questioning had a distinctive tang: "Would you say that Stalin is the Genghis Khan of the twentieth century?" "Would you say that (the New Deal) was more Fascist or more Communist?" But the committee was ludicrous without being funny. True, the committee frequently was well matched with the Communists and other totalitarians and zanies it tormented. But tormenting is not a constitutionally enumerated power of Congress. And like most bullies, the committee was happiest when attending to (in Kempton's words) "the degradation of unimportant little people."

[January 21, 1975]

Committing Acts of Government

Few sounds are as gamy as the carnal noises of consenting adults committing acts of government in private. Such sounds were recorded March 21, 1971.

That day the sap was rising in some public servants (Richard Nixon, John Ehrlichman, Treasury Secretary John Connally, and others) who met in the White House. There, under the eye of Eternity and in the range of a Sony tape recorder, they went about the business of government, which consisted of giving succor to the strong. After light persiflage, the conversation turned, as is only natural, to the subject of milk, and Connally took the floor.

In midseason form Connally is a modern Chesterfield, noted for polished manners. That day he was the soul of discretion in drawing the attention of Nixon away from problems of international statecraft, and to crass considerations of practical politics. Connally counseled Nixon that "looking to 1972, it appears very clear to me that you're going to have to move strong in the Midwest. You're going to have to be strong in rural America, and particularly that part of the country." But there was a problem.

His shiny black shoes glistening with drops of dew from the grasslands, Connally reported that the Midwest, the buckle that binds the Union together, was groaning under the weight of an injustice: federal price supports for milk were too low.

Connally knew this because the Associated Milk Producers, Inc. told him. He reported that AMPI had educated Congress and that Representative Wilbur Mills, then considered healthy, and Speaker Carl Albert supported a bill to raise milk price supports.

Nixon concluded, correctly, that Congress, given time, would pass this bill, and concluded, politically, that he didn't want to veto it. A veto, he said, "would be just turning down the whole of Middle America." So Nixon bought Connally's argument. He decided to raise price supports on his own, beating Congress to the punch. This cost American consumers many millions of dollars, many of which came from George F. Will, whose two boys, taken together, are fifty-one pounds of congealed milk.

For Nixon the clinching consideration must have been Connally's discussion of AMPI's patterns of gratitude:

"They're going to make their association and their alliances this year and they're going to spend a lot of money this year in various congressional and senatorial races all over this United States. And you don't want to be in a position . . . [where] people think they forced you into doing something for them. . . . If you do something for them this year they think you're doing it because they've got a good case and because you're their friend. If you wait till next year, I don't care what you do for them, they're going to say, 'Well, we put enough pressure on them this election year and they had to do it.' And you, you get no credit for it."

This tape-recorded civics lesson was played recently at Connally's trial. He is charged with accepting two $5,000 bribes in 1971 in return for helping persuade Nixon to raise the price supports.

If Connally received $10,000 for arguing as he did, he committed a crime: a bribe is an illegal reason for acting.

If Connally was motivated only by the political considerations he expressed, then he is (speaking quaintly) innocent. If he did it for power, not money, then he committed government, not a crime.

Of course that is, increasingly, a distinction without a difference. But it is the moral fiction that sustains politicians in both parties as they go about their business, which is, increasingly, using public power to broker private privileges.

Government in the Subsidy Society is a mixture of Augustan rhetoric and Hogarthian behavior. Politicians lay down a dense fog of rhetoric about the "public interest" and concern for the weak. And behind the fog politicians scramble to see who can be first to raise the price of milk in Harlem. The real scandal is not the way a few politicians break laws, but the way most politicians make laws.

[April 10, 1975]

Agnew Takes a Page from Wilde

Not to be outdone by his former master, who celebrates Disraeli's government, Spiro Agnew, uninterested in government to the end, expended the nub of his burnt out candle in homage to the writer who made his first splash emulating Disraeli's dandyism—Oscar Wilde.

In his essay on "The Decay of Lying" Wilde praised "the true liar, with his frank, fearless statements, his superb irresponsibility, his healthy, natural disdain for proof of any kind." Wilde reserved special contempt for the United States, "that country having adopted for its national hero a man who, according to his own confession, was incapable of telling a lie."

There are two possible explanations for Agnew's "farewell" speech. Either he was forced to make it, as part of the bargain struck concerning his plea, or he is a true artist, using his speech to stoke the dangerous fires burning in America's "constituency of resentment."

This constituency is the free-floating mass of the disgruntled. It is without allegiance to party and, in the spirit of its penchant for patriotic themes, it is fueled by the hostile suspicion about our institutions that suffused Agnew's speech.

It includes the Wallace vote. It also included the Agnew vote, a submerged but significant component of Mr. Nixon's otherwise tepid constituency.

Its members believe they are not getting their just deserts from

American society, and that being victimized is necessary and sufficient proof of virtue. Now in a sudden flash they have seen their most favored imaginings deliciously confirmed. Agnew, the symbol of resentment armed, has been done in by establishment forces symbolized by Elliot Richardson.

Agnew ended speaking about "America's strength and her glories," asserting that "our democracy" is "working better than ever before." But the core of his speech was the insinuation—he did not have the courage to boldly state his dreary convictions—that the American system of justice, from the top of the Justice Department on down, conspired to force him from elective office.

He said he wanted to avoid "a paroxysm of bitterness," but he used his speech to incite bitterness among the resentful by charging that there was no way for him to get a fair trial. He distributed blame for this across the public and private sectors, the government and the press. (The economic interpretation of history is dead: "faceless editors" have replaced "faceless financiers" in populists' nightmares.)

Agnew dismissed as "insinuations" the forty pages of hard allegations—specific as to names, dates, places, amounts of cash—published by the Justice Department.

He complained that the allegations were never "proven" or "independently corroborated or tested by cross examination." But such proving, corroborating and testing is what *trials* are for. And he begged Congress and then bargained away his office in a desperate campaign to prevent a trial.

The hollow heart of his speech was the assertion that it was "not realistic" to say that he was the "initiator" of extortion, and that "any experienced political observer" would laugh at the notion that the witnesses against him were "innocent victims of illegal enticements from me." In the annals of American rhetoric, this offering from a practiced rhetorician will not rank with "Caesar had his Brutus. . . ."

But the speech may be enough to keep the "constituency of resentment" contentedly chewing a new rancor. This constituency is perpetually hankering for a solitary figure of passionate convictions whose solitude will make him conspicuous, and whose convictions will guarantee him martyrdom. This fate always confirms the sour belief that the American deck is stacked against those whose grievances are the wages of their virtuousness.

Perhaps it would be best for the "constituency of resentment" to swallow Agnew's speech. The constituency might be less inflamed by

Agnew's "martyrdom" than by having to face the fact that, even in his final speech, far from being the pillar of faith in an administration of malleable men, Agnew was the disposable bottle into which Nixon poured whatever cheap wine he chose.

[October 19, 1973]

Government by Western Union

A wind has howled through this city. The gale—temporarily abated—has consisted of bellowed demands for President Nixon's resignation. Those who want Nixon to yield to this gale are proposing a lawless solution to Nixon's lawlessness.

The answer to them is in a passage in *A Man for All Seasons,* in which Sir Thomas More, his wife Alice, his daughter Margaret, and her fiancé Roper, have a heated exchange concerning how to deal with the sinister Rich, who reeks of betrayal.

Alice: Arrest him!
Margaret: Father, that man's bad.
More: There is no law against that.
Roper: There is! God's law!
More: Then God can arrest him.

Alice: While you talk, he's gone.
More: And go he should if he was the Devil himself until he broke the law!
More: So now you'd give the Devil benefit of law!
More: Yes. What would you do? Cut a great road through the law to get after the Devil?
Roper: I'd cut down every law in England to do that!
More: Oh? And when the last law was down, and the Devil turned round on you—where would you hide, Roper, the laws all being flat? This country's planted thick with laws from coast to coast—Man's laws, not God's—and if you cut them down—and you're just the man to do it— d'you really think you could stand upright in the winds that would blow then? Yes, I'd give the Devil benefit of law, for my own safety's sake.

The Constitution was designed to temper, deflect and sometimes just frustrate winds of passion. It insists that even the strongest and most persisting passions must flow into legal channels. So if the American people have a passion for "dealing with" Nixon, they should deal with him through the constitutionally provided channel, impeachment.

It may give many citizens vertigo, cognitive dissonance or something equally unpleasant, but the fact is a fact: Nixon—acting no doubt, for purely self-interested motives—has, at long last, actually served the cause of law-and-order. He has done so by not resigning, *yet*.

Had Nixon allowed himself to be blown out of office by a gale of telegrams and a gust of editorials, he would have set a precedent even more injurious to the nation than his dreadful presidency. In the future, every time a divisive issue seized the nation's attention, our new political mores would cause citizens to ring up Western Union to "vote" on the future of the regime. We would have created an unholy hybrid by grafting an informal "no confidence" vote procedure onto our tripartite governmental system.

In fact, if Nixon had resigned after the first gust of telegrams and editorials, no future "informal" referendum-by-telegram would have been truly informal. And while cashiering Nixon for his many affronts to the Constitution, the nation would have moved a giant step toward that to which the Constitution is an ingenious alternative—plebiscitary democracy.

As we do not want government by "plumbers," so, too, we do not want government by Western Union. Nixon and his men have made the United States seem like a banana republic. Those trying to force resignation *now* would make the United States seem like the French Fourth Republic.

The lawful way to topple a regime—impeachment—is not a "happy" solution to the current commotion, but nothing is. Nixon's resignation *at this point,* before any charges have been filed *formally* against *him,* would inflame resentments without answering questions.

But if the House of Representatives votes impeachment, filing formal charges with the Senate, due process will be served, and Nixon will have two options. He will be able to fight for vindication in his Senate trial, *producing all evidence he thinks will help him.* Or he can resign rather than fight. If he does resign, he will spare the nation a divisive period, and he will be able to say he is resigning for that purpose. But his resignation will be construed—reasonably—as a *nolo contendere* plea. There is a fresh and useful precedent for this.

[November 13, 1973]

Richard Nixon: Too Many Evenings

As in all true tragedy, we see in Richard Nixon's ruination the ravages of a failing to which all men are prey.

Nixon's sin, like all sin, was a failure of restraint. It was the immoderate craving for that which, desired moderately, is a noble goal.

It is a terrible curse to want anything as much as Nixon wanted power. He wanted it more than he wanted friends. Indeed, he wanted it with a consuming passion that left no room for friendship. And when, in his final extremity, he looked around for friends to grapple to his soul with hoops of steel, there were no friends there.

A heart weighed down with the weight of woe to the weakest hope will cling, and for two years Nixon clung to the wicked hope that the rule of law could not reach up to him. His final hope was that the task of breaking a President to the saddle of law would tax the American people's composure to the breaking point. The dashing of all such hopes is the happy issue of our Watergate affliction.

Nixon probably is not quite as thoroughly bad as, caught in the tangled web he could not stop weaving, he came to appear. And no one else is as good as they may now be tempted to feel. It would be wrong for people—journalists, politicians, judges—to preen themselves on their performance during this protracted sorrow. No one did more than his duty, as professional and citizen, and many people did less.

In the end Congress was driven to the brink of doing its duty to protect the Constitution. Many journalists did what they are paid to do, reporting things that had been improperly concealed. And the judiciary construed and administered the law. But no one deserves a garland for doing his duty.

Although the Nixon White House ran amok as no other has done, and its abuses were uniquely lurid and sinister, there is a sense in which the kind of work we have been doing is work without end. As our megagovernment grows, its potential for abuse of power grows. Keeping the government lawful is a task comparable to painting the Golden Gate bridge. It is endless: you just get to one end and then you have to go back and start again from the beginning.

But surely, now, we can and must relax a little. Oscar Wilde's aphoristic criticism of socialism—"It would take too many evenings"—meant that it is uncivilized to allow politics to become a dominating preoccupation. Watergate has taken too many evenings. Now there are books to be read, children to be played with, and other humane and civilizing pursuits that have been neglected because the task of getting the government back on the leash demanded a hideously large slice of the Republic's energies.

Life under these conditions has not encouraged the softer emotions, but one would have to be dead to all human feelings not to feel deep regret for the suffering endured by Nixon's brave family. I am thinking especially of Julie Eisenhower. Filial devotion is always moving, as is courage, and plain spunk. Ms. Eisenhower's brave combativeness on behalf of her father provided the nation with something valuable, an example of strong and noble character.

Her ordeal, like the Republic's, is over and this year autumn, the season of mists and mellow fruitfulness, will be spring, the season of rebirth and renewal. Nixon, by resigning, has struck the Watergate fetters from Uncle Sam's wrists. And President Ford, like a healing zephyr, arrives, his decency and goodwill settling like a balm on our lacerated feelings. Now, at last, there is a stillness. The angry drumbeat of contention dies away "and silence, like a poultice, comes to heal the blows of sound."

[August 8, 1974]

Deus es! Deus es!

When Charles V's aides warned him against exposing himself to danger at the battle of Pavia, he sniffed: "Name one emperor who was ever struck by a cannonball." Backstage after his onstage appearance at the Boston burlesque theater, Wilbur Mills said: "This won't ruin me . . . nothing can ruin me." As Byron understood, power, fragile political power, makes men vulnerable to the delusion of invulnerability:

> A king sate on the rocky brow
> Which looks o'er sea-born Salamis;
> And ships, by thousands, lay below,
> And men in nations;—all were his!

> He counted them at break of day—
> And when the sun set where were they?

Today, while wishing Mills peace, it is well to consider the immunities men of power enjoy, and how unhealthy they can be.

John Ehrlichman, testifying before the Senate Watergate Committee, argued with Senator Lowell Weicker (R-Conn.) about the White House use of Anthony Ulasewicz to investigate the habits of some politicians. Ehrlichman said a politician should be criticized "both in terms of his voting record and in terms of his morals. . . ."

I know of . . . incumbents . . . who are not discharging their obligation to their constituents because of their drinking habits . . . and there is a kind of unwritten law in the media that that is not discussed, and so the constituents at home have no way of knowing that you can go over here in the gallery and watch a member totter onto the floor in a condition which . . . would preclude him from making any sort of sober judgment. . . ."

Weicker responded with fustian: "I always thought we settled these matters (elections) on the basis of issues . . . but to sit here at this moment in time and tell me that we are going to settle our elections on the basis of sexual habits and drinking habits and domestic problems and personal and social activities." But Weicker's response could obscure the fact that Ehrlichman had raised a legitimate issue.

A senator who is a sexual athlete after dark can still discharge his formal duties. A senior White House aide who is drunk during office hours cannot. You may believe, as I do, that when such sensitive matters are involved, the public's "right to know" should not be construed to extend beyond its "need to know." But it is certainly arguable that the public needs to know when dissolute, self-indulgent people are wielding power. Many personal habits of public people are not private matters.

When a drunk senator was first exuberant and then nearly comatose on the Senate floor, news reports referred delicately to the "high spirited" senator. Mills' eccentric behavior did not begin October 7, with the Tidal Basin incident. The Washington press corps is said to be carnivorous, but Mills' behavior did not attract comment until it became publicly bizarre. No wonder a politician can come to feel invulnerable—that "nothing can ruin me."

For years Mills' vanity has been noteworthy even on Capitol Hill, which is no garden of shrinking violets. He has exercised power with a willfulness bordering on capriciousness, a disdain for limits, a daytime willfulness not unlike and probably not unrelated to the nighttime behavior that destroyed him.

Some political titans surround themselves with servile staffs and other servants. These and other perquisites of power insulate them from the humanizing limitations—like the need for manners, and conformity to other community standards—that help mere mortals define their selves, and keep in touch with reality.

There once was a Pope—Urban IV, I think—whose retinue would greet him with the chant, "Deus es! Deus es!" ("Thou art God!"), to which he once replied, "It is somewhat strong, but really very pleasant." But the addictively pleasant perquisites of power, like other narcotics, can be deranging.

Elevated by power beyond restraints, people lose their senses of place, time, self. Then they suffer the crippling restraints of derangement. The phrase "power mad" is not always just a metaphor.

[December 10, 1974]

An Appropriations Question

Say what you will about Congressman Wayne Hays, he has never pretended to be what he isn't. Nature did not design him as a gentleman, and he has respected nature's sense of limits. Beneath his forbidding exterior is a forbidding interior. In the House of Representatives, where his strength is the strength of ten, his razor tongue is supplemented by the power to dispense parking places, office space, and other comforts under his control as chairman of the Administration Committee.

Let it not be said that his comportment will shatter the American belief that elected officials are the flower of American manhood. Americans long since concluded that politicians have many vices (and some virtues) on which hypocrisy is an improvement.

Americans have been so relentlessly tutored in cynicism by their politicians that they will not fret about the possibility that Hays does not set an elevating example for the nation's youth. Indeed, some Americans will welcome the Hays contretemps the way columnist Murray Kempton welcomed the elevation of Sonny Liston, a strikingly unrehabilitated ex-convict, to the heavyweight championship:

"He . . . can, if he is as true to his essential character as he shows every sign of continuing, be a heavyweight champion who will be the discredit his profession has needed all these years. . . . He will help us grow up further if he destroys the illusion that a man whose trade it is to beat another man senseless for money represents an image which at all costs must be kept pure for American youth."

Lying is the homage certain politicians pay to what they are afraid may remain of society's capacity for disgust. So Hays devoted a Sabbath to sprinting from one television station to another, devoutly looking network audiences in the eye and lying, vehemently if briefly. Of his mistress he said, "I never had a relationship" with her.

Soon he noticed that he was getting the worst of a monologue. So he tried the normal secondary tactic, just enough truth to give truth-telling a bad name. He said, yes, she was his mistress, and he apologized for "not presenting all"—a nice touch, that—"the facts." Then, as Spiro Agnew did when the "post-Watergate morality" began to chafe, Hays demanded a House investigation. Throwing himself on the mercy of his colleagues was not a high-risk strategy. The quality of their mercy is not strained.

Recently Jerry Landauer of the *Wall Street Journal* discovered in published records of the House a rich lode of evidence that some congressmen have been filing false claims for travel expense reimbursement. Some of the trips may have been fictions. Others were made by air, but the lawmakers filed for the more lucrative twenty-cents-per-mile reimbursement for automobile travel.

When in the shadow of Capitol Hill a black teenager illegally lays his hands on rather less money than the congressmen are grasping for, it is called grand larceny. Grasping by congressmen evidently is considered petty larceny, forgivable because proportional to the men involved.

The response in the House to Landauer's reporting has been a gently reproving letter reminding congressmen that they should not fib, and inviting them to return any money taken under false pretenses. The letter was from the chairman of the Administration Committee, Wayne Hays.

Most Americans are dutiful liberals now, in the sense that they believe that what consenting adults, including public figures, do behind closed doors and drawn shades is not public business, if it is not done at public expense. This is sensible if, but only if, facets of personality can be hermetically sealed from each other. It is not sensible if, as I believe, the arrogance and capriciousness of a public figure in the conduct of public business may be related to a long habit of ignoring society's standards of honorable behavior in private matters.

However that may be, it is apparent that any further investigation will treat the Hays affair as an appropriations question. It will be about whether Hays's mistress was truthful when she said she was paid $14,000 a year for sex, and did not work during office hours. Evidently the public issue raised by the Hays affair is not whether a powerful legislator was a knave. Rather, the question is: does his mistress type, or doesn't she?

[May 30, 1976]

"Technically Sweet"

In 1918 Ernest Rutherford, a physicist, missed a meeting of experts advising the British government on anti-submarine warfare. When criticized, he replied: "I have been engaged in experiments which suggest that the atom can be artificially disintegrated. If it is true, it is of far greater importance than a war."

Twenty-seven years later, with another war raging, an American physicist, J. Robert Oppenheimer, staring at an atomic fireball rising over the New Mexico desert, was haunted by a passage from the Hindu epic, *Bhagavad Gita:* "I am become Death, the shatterer of worlds."

In an incredibly few years, while the mass of men were preoccupied with unstable politicians, currencies, and societies generally, a few dozen scientists demonstrated the annihilating instability of matter itself. Einstein had postulated the equivalence of matter and energy, and other scientists proved it by transforming matter into energy.

So on July 25, 1945, the cruiser *Indianapolis* (which was torpedoed and sunk less than four days later) anchored in the Mariana islands to unload a twenty-four-inch-high cylinder. It was the core of "Little Boy" —the first atomic bomb. Early on August 6, thirty years ago, the B-29 flying "Little Boy" toward Japan received a radio bulletin from an advance scout plane: weather would permit hitting the preferred of three target cities. The pilot told his navigator: "It's Hiroshima."

When this new instrument of war exploded, the secretary of war was Henry L. Stimson who, in 1929, shut down the State Department's code-breaking operation because "gentlemen don't read each other's mail." This is not a gentlemen's century. This century's principal enterprise is war. The century's most novel modification of this ancient enterprise is the obliteration of the idea of noncombatants. The century's most novel cultural achievements are gentlemen's codes containing no unbreakable rules about incinerating cities.

The firebombing of Tokyo in March (shortly after a similar raid on Dresden) had killed 140,000, twice the number killed in Hiroshima. It was too late for qualms, other than those that caused Stimson to oppose dropping the bomb on Kyoto, which has famous shrines. Surely Hitler and Stalin ("A single death is a tragedy, a million deaths is a statistic") are in one terrible sense the great men of the century. These two pathbreakers, who pioneered new dimensions in the theory and practice of mass death, have done more than all the century's artists and philosophers to alter, probably forever, the way human beings feel.

The world conflagration that began with Hitler and Stalin as allies dramatized early-on the rapid, uneven evolution of the technologies of war. The episode was tragicomedy—a charge of Polish cavalry, with sabers, against Nazi tanks. The war ended with atomic thunderclaps in the Orient. In six years the victors, joined tardily by the century's most accomplished killer, Stalin, had matched the losers' willingness to destroy noncombatants, and had taught the losers a stern lesson in the hazards of lagging behind in the science of mass death.

The scientists who built the bomb were understandably fond of Einstein's aphorism that the world has more to fear from bad politics than from bad physics. But science usually is the subservient partner, and weapons often are the issue, of a marriage between science and the modern state. This relationship was tidily summarized in an incident in the New Mexico desert the morning the atomic age dawned. A scientist lamented that the unexpected violence of the explosion had destroyed his measuring instruments. A general soothed him: "If the instruments couldn't stand it, the bang must have been a pretty big one. And that, after all, is what we wanted to know."

The unresisting subservience of most science to the purposes of others—usually the purposes of the subsidizing state—is a predictable consequence of what began 300 years ago. Then, scientific academies first banned, as distracting, philosophical debate about the consequences of scientific endeavor.

Oppenheimer illuminated this facet of the modern temperament when he said scientists were fascinated with building the bomb because the concept was so "technically sweet." That is what physicist Enrico Fermi thought when he admonished his morally troubled partners in building the bomb: "Don't bother me with your conscientious scruples. After all, the thing's superb physics."

[August 1, 1975]

Governing Appetite

President Carter, who knows the hymn, has been sowing in the morning, sowing seeds of kindness, sowing in the noontide and the dewy eves. Since January 20 he has spent more than seven hours talking gently to the nation on radio or television, helping the nation feel more comfortable with him than it felt when rendering that equivocal election result. But now come the unkindest of cuts. Carter has promised, with characteristic panache, that on April 20 he will give the nation an energy policy as serious as its energy problem. He says the policy may cost him the sheaves of popularity he has harvested in the last three months.

There have been three presidents in the three and a half years since the Arabs did this nation the favor of making the energy crisis visible. President Nixon, who had other worries, invoked the chimera of "energy independence." President Ford opposed a stiff tax on gasoline to cut demand: "I was interested in a poll that was published today which indicated that 81 percent of the people don't agree with the various people who are advocating this. I think I'm on solid ground." But all ground seems solid when your ear is to it, and as Churchill said, it is hard to look up to "leaders" seen in that position. Carter seems prepared to *govern*. He knows popularity is capital that cannot be banked forever. After hearing Carter on energy, the nation may feel that someone big has hit it hard with something heavy.

The energy crisis is, basically, an oil crisis. It is a pure Malthusian problem, the exponential growth of demand against an exhaustible supply of an indispensable commodity. To state the problem is to comprehend the essence of Carter's policy: conservation. In the short run, it is cheaper to save oil than to produce it. In the medium run, no practicable incentives will cause production to increase as fast as demand will increase unless governments cut demand. And in the not-so-long run, all production incentives will collide with the exhaustion of recoverable reserves.

Among major industrial nations only the Soviet Union is currently self-sufficient in energy. The United States cannot meet essential energy needs from sources secure from military or political interdiction. So the stabilizing power of the non-Communist world is vulnerable to production decisions made by a few Saudi leaders. They have a $20 billion trade surplus. They have no economic reason to increase the pace at which they exchange oil of increasing value for paper of decreasing value.

Weak governments, afraid to compel conservation, have printed too much money in order to prevent standards of living from declining as a result of the export of purchasing power to oil-exporting countries. Fragile countries have borrowed enormous sums rather than allow energy costs to curtail development. The result is inflation. And producing countries will not cheerfully expand production of a wasting asset in exchange for currencies of wasting values.

In the 1950s the world consumed more oil than had been consumed in all history. In the 1960s it did again. Today the non-Communist world uses fifty million barrels a day, the Communist world ten million. Saudi Arabia has the only substantial capacity for increased production: in a few years it *could* go from ten million to twenty million barrels a day. But world demand growing at 5 percent annually would slurp up such a Saudi increase in four years. Mankind has used 360 billion barrels. There may be 1.5 trillion left. But even if use grows only 2 percent a year, oil will be gone in less than fifty years, and severe shortages will begin much sooner.

Carter's energy ace is James Schlesinger, a man of somewhat cumbersome architecture, slow in movement but quick in comprehension. He has a smile like pale February sunshine, and precious little to smile about, considering that soon businessmen, environmentalists and people who own houses or automobiles may want to tear him into thin shreds and scatter him across the Potomac to teach him a lesson. His recommendations (more dirty strip-mined coal, more nuclear plants, higher prices) will be *democratically* disagreeable.

The policy must influence billions of decisions by millions of Americans, and exhortations to voluntarism will not suffice. This is, in part, because there is no current shortage and there will be a conspicuous surplus when Alaskan oil begins flowing this summer. But the main

reason why conservation demands determined government action—
the coercion of regulations, taxes and rising prices—is that conservation
must conflict with two cherished American values, comfort and convenience.

Less than half of U.S. energy is used in the production of goods and
services. Most of it is used in individuals' consumption, especially in
furnaces, air conditioners and, of course, automobiles.

Through the ages there have been essential commodities the demand for which is relatively "price inelastic." That is, demand dropped
only slightly even when price rose steeply. Salt was one such commodity. Today gasoline is another because automobiles are essential for so
many cherished things like Sunday drives, cross-country vacations, low-
density suburban living. It is politically risky to start using government
to revise the American Way of Life. But it is irresponsible not to.

Americans tend to misplace reality; often they lose it at the movies.
I suspect that movies like *Giant* shaped American attitudes about the
limitlessness of oil. You may remember Jett Rink (James Dean) poking
around in dusty Texas earth and striking oil in his own backyard. Soon
he had something like Spindletop drenching him with oil.

But in 1972 a Texas official said of the Texas oil fields, "This old
warrior can't rise any more." The familiar word is *depletion* and it may
mean the end of a lot that is familiar. The consumer civilization of the
postwar period was made possible by $2-a-barrel oil. For a century—
about half the life of liberal democracy—we have been passing through
what scholars call an "energy bubble" of cheap oil. Viewed against the
vast backdrop of history, this experience has been highly unnatural.

Liberal democracy is government that rests lightly upon people. It
has existed rarely, and only during the two centuries of rapid economic
growth in the West. It probably has been made possible by that growth,
by the belief that a rising tide raises all boats, a belief that dampens the
worst social conflicts. And viewed against the backdrop of history, our
experience of liberal democracy has been, as Saul Bellow says, "brief as
a bubble."

April 20 will be an important day in the history of popular government. What Carter says will be evidence of a free people's willingness
to be far-sighted, to be governors of rather than governed by appetites.
To be, in a word, mature.

[April 18, 1977]

Jiggling the Mobile, Gently

For weeks the conversation of the capital was even thicker than usual with jargon and strange numbers as lawmakers pondered the laws of thermodynamics. Then President Carter went to Capitol Hill, looked the nation in the eye, and said:

We must not *Balkanize* the nation, we must *retrofit* and *weatherize* with a *weatherization service,* using *geothermal* energy, *peak-load pricing, cogeneration, wellhead* and *sliding-scale taxes, intangible tax credits,* and uranium from *centrifuge technology* (better than *gaseous diffusion plants*), but without *fast-breeder reactors,* and avoiding *earthquake-fault zones.*

The nations' eyes glazed over. Never have so many people heard, in such a short span, so much unfamiliar stuff. Aristotle began his *Metaphysics* with the thought, "All men by nature desire to know," which may be true, but Carter's Aristotelian speechwriters got carried away. People want to know things, but not *everything.* What they want and need to know is how, and how much, they must pay.

Actually, no one knows. Modern society is like a Calder mobile: disturb it here and it jiggles over there, too. And because every facet of life has an energy component, energy policy disturbs everything at once. Energy policy illustrates this axiom: government cannot do one thing. That is, government cannot do *just* one thing. Every significant action has unanticipated consequences.

For example, if Congress imposes stiff taxes on the big cars the public prefers, it will not intend to disrupt an important redistribution of useful wealth toward low-income Americans. But it will.

Measured in terms of miles driven, American automobiles give more service than other nations' automobiles. But many first owners use only a fraction of the transportation potential of their cars before trading them. A car depreciates substantially the moment it leaves a showroom, and depreciates rapidly for a few years. So, second and third owners pay much less per mile (considering purchase and operating costs) than the first owners, who pay a considerable premium (with big cars, more than the highest contemplated excise tax) for the fleeting pleasure of newness.

But a tax on big cars will be passed along in their prices as used cars. And first owners will keep their cars longer if they must pay a stiff tax to replace them with cars of comparable size. This, too, will hurt low-income persons who, because of inadequate public transportation, must depend on the used-car market for cheap transportation. This does not mean a tax on inefficient cars is unjustifiable. It does demonstrate how hard it can be to conserve, simultaneously, energy and equity.

Carter's policy radiates American hardheadedness. "There's only one way to hold a district," said Boss Plunkitt, Tammany Hall philosopher. "You must study human nature and act accordin'." And that is how Americans govern. Self-interest, not love, makes the world go round, and Carter's energy program is a more or less carefully tossed salad of pecuniary incentives and disincentives. He says energy conservation is "a matter of patriotism," but his policy depends on monetary motives. Government always involves carrots and sticks, and Carter's program is tax breaks and taxes.

Mountainous Tip O'Neill, the speaker of the House, his marvelous face marvelously straight, intones: "I'd hate to see any logrolling done —you support our area, and we'll support your area." Logrolling? Perish the thought. But conceivably, political considerations may intrude. After receiving an advance briefing on the administration's policy, a congressman briefed the briefer:

"You're asking me to vote for things that will cost my constituents money and make life less convenient, and they won't see any benefit from it for the next five elections. And I'll tell you something else: if I do what you want, the last four of those elections, I'll be gone."

Perhaps he ought to just pull up his socks and walk the plank. What makes democratic politics different from most other professions is that, occasionally, the politician has a duty to risk his job by performing it conscientiously. But we need not wax philosophical. That congressman who says, so plaintively, "I'll be gone" does not face a choice between political duty and political death. He is exaggerating the severity of the sacrifices at issue.

Melodrama is a reflex in the nation that invented soap operas, so today the air is rent by excited cries about the Carter Terror. Surely the fear and trembling are not caused by Carter's suggestion that government subsidize (with tax credits) people who insulate their homes, or businesses that buy solar heaters. If he gets all his taxes at the wellhead and the gas pump, and he won't, in three years a gallon of gasoline may cost about ninety cents, slightly more than half of what Germans pay today.

Carter understands the carrot-and-stick approach to getting things done, and his basic energy policy is: talk loudly and carry a big carrot. Most of his sticks are small (imposing new building insulation standards in 1980 rather than 1981). Some are twigs (no more tax rebate of two cents on a gallon of motorboat fuel). Some would be employed delicately (price controls would be removed from gasoline this autumn, *after* the peak driving season). Others would be applied gradually (three stages for the wellhead tax, taking until 1980 to bring domestic crude oil to the world price).

Carter is right to proceed carefully: he does not want to jiggle the mobile violently. Carrots are proper instruments of government, and are friendlier than sticks. And loud talk is necessary to stir a comfortable nation. But Carter's immediate problem is the conservation of his political energy. He will run down his batteries if he continues issuing high-voltage declamations about energy "catastrophe," and couples them to an energy policy that is the moral equivalent of a *very limited* war.

He used television extravagantly to bring the energy argument to a simmer, but now he must pour his proposals into Congress, a shallow saucer where they will cool, and perhaps congeal. What Congress will do is uncertain, in part because public opinion is still unformed concerning the heart of the matter: is there an energy crisis? Is government *reliably* forecasting an alarming and unavoidable shortage that requires prompt action?

Last week, Carter put the country through an energy seminar. If his mild policy is adequate, then his hot rhetoric justifying it is excessive. That is one reason why many Americans, when told some of their habits and assumptions must change, still react with indignant disbelief, as Lord Palmerston did when his doctor told him he was dying. "Die, sir?" Palmerston snapped. "That's the last thing I shall do."

[May 2, 1977]

Lance, Carter, Babbitt and Gantry

It is said that when the flames reached Joan of Arc she gave an especially terrible scream, the cry of an innocent who had been sure that God protects his instruments and would save her from burning.

There is that piercing quality to White House cries about press coverage of Bert Lance's problems.

The rough justice is that Carter has been singed by fires he stoked. There is a hunger for righteousness in Carter's Washington that should gratify the man who advocated dismissing Clarence Kelley, the FBI director, because FBI carpenters made his valances.

Carter marched into unregenerate Washington under memorable banners proclaiming his nearly inexpressible love of virtue. But as De Gaulle warned, "We are too ready to accept the idea that men believe in their banners."

Americans believe and disbelieve simultaneously. They always are armed with the defensive cynicism of idealists who expect disappointment. American life always has been a blend of high purposes and high rollers. Americans know Moses was real, long ago and somewhere else, but Elmer Gantry was an American. The Lance affair has reminded Americans that Carter is not a Moses, entrusted by God with new standards of goodness. It also has made Carter seem a bit like Gantry, not about to live by the rules he preaches.

Of his fictional evangelist, Sinclair Lewis wrote dryly, "He said grace, at length." That is a capsule summary of Carter's campaign, the theme of which was love, with special reference to government as good, loving, etc., as thee and me. "He spoke very nicely of love," Lewis wrote about Gantry. "He said that Love was the Morning Star, the Evening Star, the Radiance upon the Quiet Tomb, the Inspirer equally of Patriots and Bank Presidents, and as for Music, what was it but the very voice of Love?" Until the case of his friend the bank president, Carter spoke nicely and too much about the need to be pure as driven hounds' teeth, or whatever.

Sinclair Lewis was of the "Mencken generation" of intellectuals, too fond of the passion of contempt, too comfortable despising the comfortableness of their countrymen. But Lewis was the authentic voice of a modern impulse: disdain for the business class. The importance of this disdain has grown with the growth of those classes of Americans (teachers, bureaucrats and others) who do things but do not make things. Now Lance is making matters worse, evoking echoes of another of Sinclair Lewis's characters:

"His name was George F. Babbitt. He was forty-six years old now, in April, 1920, and he made nothing in particular, neither butter nor shoes nor poetry. But he was nimble in the calling of selling houses for more than people could afford to pay."

Like Babbitt ("He was not fat but he was exceedingly well-fed"), Lance, forty-six, fits the stereotype of, well, Babbittry. Lance has less malice and more public-spiritedness than many who prosper in Washington. But his business comportment before he came to Washington has made him a symbol, useful to those who believe that American business is a low calling that rewards bounders who are more nimble than productive.

Lance is another chapter in a long story of missed opportunity for American business. The perceived failure of Great Society legislation, the obvious failure of Vietnam policy, and the moral failure of Watergate caused the public to become skeptical about the competence and motives of the public sector. These could have been years of renewed respect for the private sector. But the names that have dominated the news have been Penn Central, Gulf, Lockheed, other corporations in trouble, and now Lance.

Thoughtful businessmen quickly understood that Carter's continuing embrace of Lance injured businessmen generally because it implied that Lance's behavior was "normal" in business. This was felt especially by bankers, whose sense of their vocation as a public trust is keen, and compares favorably with that of, say, journalists and politicians.

Any town full of journalists and politicians will be fond of blood sports. Washington hasn't had a good kill for a while, and Lance is a large, slow stag. The chase has been invigorating, but blood dries quickly, and there will be other thirsts for judgment. Before the next, Carter's Washington might consider if its moralism is well focused.

Yes, it is better, other things being equal, to have government staffed by people who have impeccable pasts, rather than by people who, like Lance, do not. But a striking aspect of the Lance affair is this: the crystallizing consensus that he is unfit for government service is only loosely, and perfunctorily, concerned with how the defects of his private past are relevant to his current public duties.

Moral, and even aesthetic, judgments about individuals are becoming more important in Washington than political judgments about the wisdom of policies. Prevailing standards of "political morality" are not closely, or hence reasonably, related to the substance of politics. Washington has had so little success with policies in recent years that it wants to believe that policies are somehow not the point of government. Washington has jumped on and fallen off so many policies, and takes so little pride in much of what it does, that it wants to believe that the test of good government is the personal goodness of those in government, rather than how government handles issues. It generally is more satisfying, and less embarrassing, for Washington to ask: "Is this an admirable person?" than "Is this a just policy?"

Someday, discerning historians may marvel that while Lance was being judged unfit for public service because of what he once did with private power, those in public office who judged him were conducting public business as usual. That business included being honorable, in the local fashion, by delivering favors to those who paid for them.

Last year, maritime interests (unions, shipowners, shipyards) invested $1.1 million in contributions to Carter and many congressmen. Today Carter supports, and Congress probably will pass, a bill to require a rising portion of oil imports to be carried in U.S. ships at uncompetitive rates. Government studies say it would cost consumers at least $240 million a year, and would increase inflation and unemployment. President Ford's veto of a similar bill quickened the civic spirit of the maritime interests, as measured by their campaign gifts.

It is conceivable that, in God's eyes, what Lance did in private life with his checkbook, his bank's airplane and his depositors' trust and money was more reprehensible than what public officials are trying to do with public power and the public's money to thank maritime interests. But it seems that morally as well as architecturally Washington is, more and more, a city of pompous facades and a poor sense of proportion.

[September 19, 1977]

Civic Liturgies

President Carter is mightily, not to say ostentatiously, in favor of republican simplicity, and to that end a stern memo has circulated through the Interior Department: "President Carter has asked that the official presidential photograph be limited to those places where absolutely necessary." *Absolutely necessary?* The memo's list of such places indicates that this means photos should be "limited" to the usual places.

This exquisite memo, part of a pompous crusade against pomp, is an interesting counterpoint to events across the water. There Elizabeth II is celebrating her Silver Jubilee. So this is a fine time to sing the virtues of monarchy and the usefulness of public liturgies.

Many monarchies are among the wreckage of the century. The most venerable casualty was the House of Hapsburg whose Franz Josef was for sixty-eight years by the grace of God Emperor of Austria, Apostolic King of Hungary, King of Bohemia, King of Dalmatia, King of Croatia and Slavonia, King of Galicia and Lodomeria, Duke of the Bukovina, Duke of Upper and Lower Silesia and Margrave of Moravia. Austrian nobility was a bit backward on questions of equality. A Viennese dictum was, "The human race begins with barons." Still, Hungarians, Bohemians and others were better off under Franz Josef than under today's commissars.

Monarchy was an indispensable answer to the problems of sovereignty and succession during the evolution of the nation-state system. And one of the best recent events in Europe was the creation of a king, Juan Carlos of Spain. There are nine other monarchies in Western Europe. But monarchy is a neglected subject. Scholars seem to assume it is a relic without function. What notice monarchy receives is disparaging because monarchy offends the arid rationalism of intellectuals.

Monarchy, say its critics, is the importation of mystery into the life of a nation. Indeed it is, and that is its signal virtue. Shakespeare's Henry V lamented:

> And what have kings, that privates have not too,
> Save ceremony, save general ceremony?

And Elizabeth II, unlike Henry V, really does have little but ceremony. She is part of what Walter Bagehot called the "dignified" as distinct from the "efficient" aspect of the government process. But as Bagehot well knew, the "dignified" or ceremonial dimension of government is itself efficient for important purposes.

"There can be no society," said the sociologist Emile Durkheim, "which does not feel the need of upholding and reaffirming at regular intervals the collective sentiments and the collective ideas which make its unity and its personality." Monarchy exists for such reaffirmations. Stirred by the transformation of a young woman into a sovereign, sociologists Edward Shils and Michael Young in 1953 wrote a brilliant essay on the function of Britain's monarchy, and on the nature of man and society. The coronation, they said, was an act of "national communion," a "moral remaking" of the values that constitute Britain as a community.

The qualities that are distinctively human find expression in community, a group of persons united by shared values. But human beings are deeply ambivalent toward authority, be it of parents, state or church. So, Shils and Young argued, there is a recurring need to reaffirm moral rules in ceremonies that emphasize the relationship of the rules to those values and powers "which restrain men's egotism and which enable society to hold itself together." The coronation was an "intricate series of affirmations" of values necessary for a good and well-governed society.

"The most poignant moment in the ceremony," says Henry Fairlie, a gifted journalist who was in Westminster Abbey, "was when she had been stripped of the robes in which she had arrived at the Abbey, and stood alone among her peers and subjects in a white shift." Symbolically divested of authority, she promised to conform her will to laws and customs. She was presented with a Bible to be with her always, and enjoined to respect transcendental values.

Kneeling in submission before God's agent, the Archbishop of Canterbury, she was anointed, thereby placed in the tradition of the kings of Israel and England; she was given a sword, symbol of power, and an orb, symbol of her wide sphere of responsibility. Finally she was ornamented with the bracelets of sincerity and wisdom, wrapped in the robe royal and crowned. "With these dramatic actions," Shils and Young wrote, "she is transformed from a young woman into a vessel of the virtues which must flow through her into society."

To be sure, the real continuity of society derives from the commonplace activities of the humble rather than from the ceremonial activities of the majestic. But as Shils and Young said, the monarchy as "symbolic custodian of awful powers and beneficent moral standards is one weighty element" in Britain's consensus. Strict rationalists find it paradoxical that two strong British traits are common sense and a jolly passion for pomp and circumstance. But as Shils and Young noted, the coronation instilled "the common sentiment of the sacredness of communal life and institutions. . . .

"People became more aware of their dependence upon each other, and they sensed some connection between this and their relationship to the Queen. Thereby they became more sensitive to the values which bound them all together. Once there is a common vital object of attention, and a common sentiment about it, the feelings apt for the occasion spread by a kind of contagion."

Americans are uneasy about civic liturgies and respond most warmly to anti-majestic gestures, like Carter's walk down Pennsylvania Avenue. This Republic's annual "national communion" does not focus on a central personage. On the Fourth of July Americans parade, wave flags, chase the cat with sparklers. The nation's central personage, the President, is both head of government and chief of state. In him inhere the "dignified" and "efficient" aspects of government, the power and the glory. So, it is said, the trappings of glory can be incitements to abuse of power. (Some people seem to think silver trumpets and "Hail to the Chief" caused Watergate.) And the populist temper of the times rewards a President who, with a flourish, bans ruffles and flourishes.

But there is such a thing as being too proud of humility. Besides, as one of Her Majesty's subjects said of her coronation, "What people like is the sheer excess of it. We lead niggling enough lives these days. Something a bit lavish for a change is good for the soul." In a republic, too, statecraft is, to some extent, soulcraft.

[June 13, 1977]

"How Free She Runs"

The distinctive American aesthetic is republican simplicity. It offers the reasonableness, moderation and freedom from pretense of Jefferson's Monticello and his University of Virginia. It was a natural social complement to Jeffersonian government. There is republican simplicity in pre-Civil War domestic architecture from Marblehead, Massachusetts, to Savannah, as there was in other functional things, like ships.

The melancholy aspect of an otherwise exhilarating occasion is that Samuel Eliot Morison, historian and seaman, died a month before the "tall ships" appeared on the horizon, the apogee of the Bicentennial. Such an armada is sufficient to stir even a prosaic mind. Morison's was not, least of all when recalling American clipper ships:

"Their architects, like poets who transmute nature's message into song, obeyed what wind and wave had taught them, to create the noblest of all sailing vessels, and the most beautiful creations of man in America. With no extraneous ornament except a figurehead, a bit of carving and a few lines of gold leaf, their one purpose of speed over the great ocean routes was achieved by perfect balance of spars and sails to the curving lines of the smooth black hull; and this harmony of mass, form and color was practiced to the music of dancing waves and of brave winds whistling in the rigging. These were our Gothic cathedrals, our Parthenon; but monuments carved from snow. For a few brief years they flashed their splendor around the world, then disappeared with the finality of the wild pigeon."

They still are our symbols of gracefulness. The most elegant baseball player was called "the Yankee Clipper." When westward the course of empire took its way, the pioneers sailed the plains in "prairie schooners."

The clipper ships may have been poetry in motion, but they were ruled by the prose of life—money. Their special mission was to speed to the California gold fields, and they did not long survive the gold rush. Similarly, the great whaling ships were doomed in 1859: oil was discovered in Titusville, Pennsylvania. Steel and steam finally defeated all sails.

There is something compelling and appropriate about the recurring metaphor of sailing in our politics. Thus, Jefferson: "The tough sides of our Argosie [sic] have been thoroughly tried. Her strength has stood the waves into which she was steered to sink her. We shall put her on her Republican tack, and she will show by the beauty of her motion the skill of her builders."

When Longfellow wrote "Sail on, O Ship of State," he may have remembered that the first American "state" was a ship, the *Mayflower*. Our first political document, the Mayflower Compact, was named for the ship that carried the persons who would be bound by it. And Woodrow Wilson, in a 1912 campaign speech, used a sailing analogy to express his idea of prudent statecraft:

"We say of a boat skimming the water with light foot, 'How free she runs,' when we mean, how perfectly she is adjusted to the force of the wind, how perfectly she obeys the great breath out of the heavens that fills her sails.

"Throw her head up into the wind and see how she will halt and stagger, how every sheet will shiver and her whole frame be shaken, how instantly she is 'in irons,' in the expressive phrase of the sea. She is free only when you have let her fall off again and have recovered once more her nice adjustment to the forces she must obey and cannot defy."

The Founders' constitutional handiwork is frequently and reasonably described as the application to politics of Newtonian physics, the precise balancing of mutually checking forces. But it also is true that the Founders represented a seafaring nation that was mostly coast and ports—Boston, New York, Philadelphia, Baltimore, Charleston. The Constitution embodies a sailing people's sense of equilibrium, an instinct for balanced tension in the rigging. It involves the prudent accommodation of government to the turbulent nature of citizens, a "nice adjustment to the forces she must obey and cannot defy."

Appropriately, the most pithy foreign criticism of the Constitution was cast in nautical language, Lord Macaulay complaining that it is "all sail and no anchor." But, then, Columbia, the Gem of the Ocean, was made for movement.

[July 4, 1976]

This, Too, Shall Pass Away

Those of us from central Illinois manage, even in Bicentennial season, to keep the Founding Fathers in perspective. They were, of course, splendid fellows. But launching a republic is less difficult than perpetuating it. So before the fireworks mark our celebration of those who founded the Union, note these words from the man who preserved it: "It is said an eastern monarch once charged his wise men to invent him a sentence, to be ever in view, and which should be true and appropriate in all times and situations. They presented him the words: 'And this, too, shall pass away'."

"And yet," Lincoln said in 1859, "let us hope it is not *quite* true." We may hope to endure "by the best cultivation of the physical world, beneath and around us; and the intellectual and moral world within us."

The physical world was an understandable American preoccupation. In 1776 there were just 2.5 million free Americans, most living within twenty miles of tidewater. Their unexplored continent needed canals, bridges, railways, highways, Holiday Inns. Whew! But mighty chores are the mother of invention.

"Tom appeared on the sidewalk with a bucket of whitewash and a long-handled brush. He surveyed the fence, and all gladness left him and a deep melancholy settled down upon his spirit. Thirty yards of board fence nine feet high. Life to him seemed hollow, and existence but a burden."

But as is well known, Tom was tricky. Aunt Polly, a regular victim of his tricks, could not hide the admiration a red-blooded American feels for cunning in the service of self-interest: "But my goodness, he never plays them alike, two days, and how is a body to know what's coming?" Before Tom's buddies knew what hit them he had swapped the *right* to do the whitewashing, in exchange for such treasures as a one-eyed cat, and "a dead rat and a string to swing it with." His trickiness made him a hero to a nation whose most popular recent President was adored for his "Tom Sawyer grin" and was trickier than you would expect anyone called Ike to be.

Through two centuries of cultivating the physical world Americans have been prodigies of productivity. And they have been secretly tickled that "Yankee ingenuity" connotes a slightly disreputable knack for sharp practices. But for two centuries some Americans have worried that preoccupation with the physical world has been at the expense of "the intellectual and moral world within us." This is a political worry, because it concerns our fitness for republican government. As James Madison wrote in Federalist Paper No. 55:

"As there is a degree of depravity in mankind which requires a certain degree of circumspection and distrust: so there are other qualities in human nature, which justify a certain portion of esteem and confidence. Republican government presupposes the existence of these qualities in a higher degree than any other form."

But in Federalist Paper No. 51, the most important short essay on the American government and psyche, Madison, seeking to persuade a skeptical public to ratify the Constitution, emphasized how little the founders counted on the finer human qualities. Convinced that most people most of the time are self-seeking, the unsentimental founders devised a system on this Madisonian premise: "Ambition must be made to counteract ambition." In the American maelstrom of self-seeking, self-seekers would check each other. Put plainly by Madison: "This policy of supplying by opposite and rival interests, the defect of better motives, might be traced through the whole system of human affairs, private as well as public."

The rub is that government shapes as well as reflects national character. And Madison's government, with its candid conformity to mankind's least noble attributes, does not encourage those higher qualities which, Madison says, republican government presupposes.

In making government safe for a nation of adult Tom Sawyers, the government might make a nation of Tom Sawyers—clever cusses, but short on public-spiritedness and other forms of disinterestedness that occasionally come in handy even in a fat and sassy nation such as ours. This is one reason why many thoughtful Americans have worried that the Republic peaked a little early, and has been going downhill since Bunker Hill.

A nation blessed at birth with a shimmering golden age finds that the past becomes a reproach, especially if the founders fix upon posterity a reproving squint, as George Washington did: "The foundation of our empire was not laid in the gloomy age of ignorance and superstition, but an epoch when the rights of mankind were better understood and more clearly defined, than at any former period. . . . At this auspicious period, the United States came into existence as a nation, and if their citizens should not be completely free and happy, the fault will be entirely their own."

Seventy years later, in 1852, a gloomy Emerson was haunted by the portrait of Washington hanging in his dining room: ". . . I cannot keep my eyes off of it . . . the heavy, leaden eyes turn on you, as the eyes of an ox in a pasture, and the mouth has a gravity and depth of quiet, as if this man had absorbed all the serenity of America, and left none for his restless, rickety, hysterical countrymen."

Leaving aside the question of whether Washington's grave and quiet mouth expressed character or the tortures of wooden false teeth, we do know what Emerson meant. His malaise was as American as apple pie. In 1852 it already was true, as it still is true, that Americans are Jeffersonians in spirit but Hamiltonians in practice.

Jefferson, whose *Notes on Virginia* expresses a preference for a minimal state sufficient for a republic of agricultural yeomen, won our hearts. Hamilton, whose *Report on Manufactures* envisions a powerful central government sufficient for a great commercial nation, wrote our future.

This matters because Jefferson, the most optimistic of the founders, was optimistic only to the extent that he hoped the United States would not become anything like the urbanized industrialized nation it has become:

"I think our governments will remain virtuous for many centuries; as long as they remain chiefly agricultural; and this will be as long as there shall be vacant lands in any part of America. When they get piled upon one another in large cities, as in Europe, they will become corrupt as in Europe." He trusted "the people," but only as long as they tilled the soil, far from "pestilential" cities.

We have paid a price for our successful cultivation of the physical world. One must have a heart of stone to feel no pang of regret about the vanishing of Jefferson's Republic. "And *this*, too, shall pass away."

[June 28, 1976]

PART 6

Foreigners

The English Malady

LONDON, ENGLAND—Santayana understood: "What governs the Englishman is his inner atmosphere, the weather in his soul." Today, as usual, in spite of social tempests, that inner atmosphere is temperate. But it is tinged with an interesting nostalgia for the war years. The mind's mists have done their work, blurring memories of privation, fear and death. What remains is the feeling that 1939–1945 was a brief shining moment when purposefulness and social democracy existed in this class-riven society. Such feelings, accurate or not, are social facts.

In *The Girls of Slender Means,* novelist Muriel Spark depicts the scene at Buckingham Palace when Germany surrendered: "They became members of a wave of the sea, they surged and sang. . . . The huge organic murmur of the crowd. . . . It was something between a wedding and a funeral. . . . The next day everyone began to consider where they personally stood in the new order of things. Many citizens felt the urge, which some began to indulge, to insult each other, in order to prove something or to test their ground." A character wondered, "And now what will become of us without barbarians? Those people were some sort of a solution."

Socialists had a solution to the problem of what to do next. They came to power in 1945 determined to harness governments, so effective in war, to produce a just and modern nation. But what has been produced is evidence for Peter Drucker's axiom that government is effective only at waging war and inflating currency. In three decades, Britain has changed from the second richest nation in northwest Europe (behind Sweden) to the second poorest (ahead of Ireland). If relative rates of growth of the last twenty-five years continue for another twenty-five, Britain will be poorer than many Mediterranean and Asian nations.

Public spending consumes 60 percent of gross national product. Income taxes are 83 percent on earnings over $40,000, and the top rate on investment income is 98 percent. Predictably, low returns on investments of capital (or of intellectual and muscle power) have discouraged those investments. Thus *The Times* of London warns that "the present trend portends the progressive liquidation of manufacturing industry, and with it, most of the difference between the standard of living now and the standard of living 200 years ago, when the Industrial Revolution began." In the later nineteenth century, some English landowners granted leases for 999 years, serenely confident that currency values, like all Victorian values, were permanent. And between the Battle of Marston Moor (1644) and the Battle of the Marne (1914) the pound's purchasing power changed little. But since 1970, every extra 130 pounds in a worker's pay has produced a one-pound increase in real income. As *The Economist* says, "Every 130 steps forward brought 129 steps back."

What drives Americans frantic with incomprehension is the English placidity in the face of this. The English *must* be at least melancholy; we would be hysterical. But friendship should, indeed, be an education in complexity, so Americans should ponder a few things about this complicated island race. For example: our English friends lack two things we have in abundance, economic aggressiveness, and the space, both physical and "social space," in which to let such aggressiveness rip.

The English resist thinking economically. It was in England that economics was labeled "the dismal science." Economics is about the cost of our appetites. That is a tiresome concern, undignified, and unsuited for gentlemen. For centuries the governing classes have considered commerce ("trade," they call it) vulgar, and have considered economic complexities impenetrable. "Damned dots" is what Lord Randolph Churchill called decimal points when he was Chancellor of the Exchequer. In 1962, a year before he became prime minister, Lord Home cheerfully admitted: "When I have to read economic documents I have to have a box of matches and start moving them into position and illustrate the point to myself." Regarding the lower orders, it is untrue that England is, in Napoleon's disdainful description, *"une nation de boutiquiers."* Far from being natural shopkeepers, the English have underdeveloped commercial instincts. The aggressive gospel of getting on has not taken root in their souls.

In addition, as a perceptive American, Henry James, noted: "We seem loosely hung together at home as compared with the English, every man of whom is a tight fit in his place. It is not an inferential but a palpable fact that England is a crowded country." England is more densely settled than the Netherlands or Japan. And the sense of social denseness, of a person having a "tight fit in his place," shapes economic life. Thus, an English worker tends to regard a job as a form of property, a defensive sense of possessiveness deepened by decades of economic sluggishness. But this is not new. By the seventeenth century, more than 200 years before the Luddites smashed threatening machines, a sawmill had been outlawed to protect the sawyers' jobs. In the twentieth century, British socialism is devoted to "protecting jobs" and is more interested in redistributing the community's wealth than in expanding the wealth. This reflects an ancient habit of mind on this snug island.

So what ails England? Englishness does. Of course, socialism is a debilitating silliness. But the English "problem" is not just this or that government or party or class. England itself is, perish the thought, inefficient. A thousand years of history have produced a society deficient in those socially useful forms of controlled aggression (striving, coveting, risk-taking, competing) that bring a nation's economic life to a rolling boil. But it is idle to say that a nation's character is the nation's defect.

If the English problem is rooted in what we mean by Englishness, then it is not a "problem" to be solved, it is an intractable fact of life. It is worth recalling the words of Arthur Koestler, a continental who became an Englishman: "I have found the human climate of this country particularly congenial and soothing. . . . When all is said, its atmosphere still contains fewer germs of aggression and brutality per cubic foot in a crowded bus, pub or queue than in any other country in which I have lived." There again is the word Henry James stressed, "crowded." And inevitably, Santayana's word "atmosphere" recurs. What it means is "national character," something England has a lot of. England's character is placid, unaggressive, temperate, civil, and conservative even (perhaps especially) in its socialist manifestations. England's character is, undoubtedly, an economic handicap in many ways. If the modern world rewards, indeed *demands,* other character traits, that is too bad for England, but not just for England.

[February 23, 1976]

British Squabbling

LONDON, ENGLAND—In the nineteenth century British politics was, as Lord Salisbury said, "two sets of gentlemen squabbling for place." Today there are fewer gentlemen, and the following is a London newspaper's report of a recent squabble:

"Without once having to read from notes, Mr. Tom Swain—a Derbyshire miners' M.P. whose syntax is normally more muddled—yesterday exclaimed to a Tory: 'If you say that outside I'll punch your bloody head in.'

"It was the best, and most articulate, speech Mr. Swain had made in his seventeen years in the House. A moment later he illustrated his observation with a diagram: he seized the Conservative by the tie, but by that time other Members had understood the broad sweep of Mr. Swain's argument and had interposed themselves. . . ."

In spite of such ungentlemanly squabbling, what Churchill said after the war is approximately true today: "Four-fifths of each party in Britain agrees about four-fifths of the things that are to be done."

Four centuries ago, Elizabeth I affirmed the idea that relief of the poor is a state responsibility. Seven decades ago, Lloyd George began building a welfare state. And now in the twenty-fifth year of the reign of Elizabeth II that state is in crisis. But the consensus in its favor rests on a tradition long and durable. Since 1945, the Conservative party has accepted the premises of the welfare state. Like democratic parties everywhere, only more so, Britain's parties have treated electoral politics as an auction, bidding up benefits.

The welfare state, far from being regarded as a threat to personal liberty, or an irrational waste of resources, is still a source of consensus and stability. The costs—although currently more than the nation can afford to divert from investment—are not generally regarded as outweighing the benefits of economic and psychological security that the welfare state has meant for most citizens.

True, Britain's system of incentives is currently out of joint. It has been said that life should be like skiing—exhilarating, but also a bit scary. Britain's system of taxes and social services dampens the exhilaration of enterprise, and also dampens the fear of not being enterprising.

From the moderate left-center of the Labor party through the entire Conservative party there is agreement that it is time to cut the confiscatory marginal tax rates of 83 percent on earned income and 98 percent on investment earnings. Such rates are not revenue-raising measures, they are ideological gestures which express a passion for equality, and disapproval of investment earnings, which Socialists call "unearned income."

In addition, a man with two children must earn sixty-three pounds a week—about the national average—in order to be five pounds better off than he would be receiving unemployment compensation. It is among people who are earning at or below this level that resentment of the welfare state is growing.

But Britain's fundamental problem is not the generosity of its social programs. Pensions, sick pay and a number of other benefits are below the Common Market average, and far below Germany's benefits, which do not discourage industriousness.

Nor is Britain's fundamental problem that government controls a percentage of gross national product that is obviously intolerable. Computed in accordance with Common Market accounting procedures, the British government controls 46.3 percent of GNP, slightly less than the average in Common Market nations (46.5), and less than in Germany (48), Europe's economic pacesetter.

Britain's fundamental problem is productivity. Britain launched the Industrial Revolution, but it has never wholeheartedly come to terms with the values and demands of industrial civilization.

True, in the nineteenth century, especially, the British had exuberant entrepreneurs, such as the textile manufacturer who moved from the city to a country estate, and promptly cut a swath through the screen of trees so that when he sat in his drawing room he could see the smoke rising from the chimney of his mill. But the late D. W. Brogan, a historian interested in national character, captured the essence of Britain's problem (and perhaps our problem, too): Americans love machines, the British love dogs.

[December 23, 1976]

A Touch of Class

LONDON, ENGLAND—When members of Parliament were recently asked what thinkers influenced them, Conservatives cited a former member, the venerable Burke. Laborites cited Marx. Actually, it is as difficult to imagine the typical Labor MP slogging through Marx's *Herr Eugen Dühring's Revolution in Science* as it is to imagine the typical Tory savoring Burke's *On the Sublime and Beautiful.*

But the Labor party always has had a faint Marxist tinge, and in at least one sense it thinks of politics as Marx did, more in terms of collectives—classes—than individuals. The Labor party's strength is that of the working class, which has a confidence in the rightness of its self-interest—and an aggressiveness in pursuing it—that Britain's middle class now lacks.

The American middle class has the élan of a class to which virtually everyone claims to belong, and which believes it runs a successful nation. Britain's middle class feels vaguely responsible for Britain's failure to match the postwar performance of other middle-class nations.

The French middle class is tough, mean, grasping. It has been a provocation to Flaubert, Balzac and others who have made it the unlovely subject of some of the world's great literature. But it is a source of national vitality and prosperity. The British middle class seems disarmed by doubts, by the suspicion that the Socialist indictment of it may be right: that its values are vulgar, and an impediment to social justice.

Whether the Labor party owes more to the strength of the working class or the uncertainty of the middle class is unclear. But no other party in a democracy has had such a profound impact without ever winning 50 percent of the vote in a general election. Even in 1945, its first and greatest victory, Labor received only 48 percent. In the last election, in 1974, Labor won with just 38 percent of the vote, 28 percent of eligible voters.

Two things make Labor's future problematic. One is the softening of the class system; the other is the growing contradiction between Socialist inclinations and the nation's needs.

The declining importance of class is apparent in university admissions and business management. And Professor P. T. Bauer of Cambridge says this of politics:

"No one reading the standard pieces on class in Britain . . . would imagine that people like Disraeli, whose Jewish origin was announced in his very name, or Lloyd George, a very poor Welsh orphan brought up by an uncle who was a shoemaker, or Ramsay MacDonald, illegitimate son of a fisherwoman, could all have become prime minister. None of these men had been to university: Lloyd George and MacDonald had simple education only, and Disraeli attended a relatively unknown secondary school."

British socialism is a product of two of the nineteenth century's mightiest forces, the cooperative movement and religious dissent. Its raison d'être has been the promotion of equality. But today, from the moderate Labor left to the far Conservative right, there is agreement that industrial regeneration must take precedence over all other objectives.

The current level of public spending was justifiable, if at all, only on the assumption that there would be a rate of economic growth that has not been achieved. Now there is a broad consensus that Britain cannot maintain its current standard of living at current levels of productivity; and that one key to higher productivity is a higher level of investment; and that this can best be achieved by cutting public borrowing, which means cutting public spending.

So today a Socialist government is cutting public spending, not enough, but contrary to the lifetime passion of its members. It is doing so under pressure from a creditor, the International Monetary Fund, from which Britain is to receive $3.9 billion in loans. But there is something like relief that, at last, the nation has been forced in a direction it sooner or later had to move.

The Labor government never gave serious consideration to the alternative strategy proposed by its relatively few members whose socialism is a more hard-edged, continental sort. They would reject IMF conditions, and would adopt a "siege economy." This would be a policy of autarky, including import controls, and a state-controlled investment program involving nationalization of banks.

The Labor government's rejection of the far left is evidence that what Arthur Balfour said of British socialism seven decades ago is still true: "We always catch continental diseases . . . though we usually take them mildly."

[December 19, 1976]

The Chunnel

Like a thin shaft of Britain's watery winter sunshine comes welcome news that the "Chunnel" will not be built, at least not now.

The Chunnel was to be a thirty-two-mile railroad tunnel beneath the English Channel connecting Britain and France. After tunneling 350 yards—the French had advanced 450 yards the other way—Britain withdrew from the project.

Why would reasonable men want to go tunneling about under the Channel? Surely not just to cut in half the six-hour train trip (including ferry) between London and Paris. No, the tunneling beneath the Channel was for reasons at least partly metaphysical, like: "Because it is there."

That mountaineer who, when asked why he was climbing Everest, replied, "Because it is there," has a lot to answer for in the Hereafter. Notwithstanding the fact that his answer lacked a certain analytic rigor, the Everest Answer has become a standard "reason" for unreasonable exertions.

Another reason for the Chunnel was that a chunnel is technically feasible. Technology generates its own momentum. As J. Robert Oppenheimer said of the hydrogen bomb, "It was so technically sweet, we had to do it."

Britain's ostensible reason for canceling the Chunnel is that it would cost too much. The estimated cost had more than doubled to $4.7 billion in the first eighteen months of construction, and probably would have continued to soar until the Chunnel opened in the early 1980s.

All praise to the British government, which has unsheathed a novel principle that could revolutionize government in Britain and around the world: governments should not buy things they cannot afford. This principle is especially conspicuous in the hands of Britain's socialist government, which seems to derive its understanding of political economy from Rumpelstiltskin.

But absent convincing counterevidence, I choose to believe that the cancellation of the Chunnel is as well explained by poetry as by public finance. Remember Chesterton's splendid poem:

> Before the Roman came to Rye or out to Severn strode,
> The rolling English drunkard made the rolling English road.

Perhaps the crux of the Chunnel matter is in these Chesterton lines:

> I knew no harm of Bonaparte and plenty of the Squire,
> And for to fight the Frenchman I did not much desire;
> But I did bash their baggonets because they came array'd,
> To straighten out the crooked road an English drunkard made.

That is a sound not heard much—not heard enough—these days, the unapologetic growl of the British bulldog defending the nation's right to be its sometimes potty self. Chesterton's ample shade must be growling contentedly about the fate of the Chunnel.

At the heart of Britain's lack of enthusiasm for the Chunnel there is, I hope and suspect, a goodly measure of skepticism about the homogenizing forces loose in postwar Europe, forces that seem to disdain and threaten national differences.

The Chunnel was a symbol, and rejection of it is, for some Britons, an invigorating gesture. Enthusiasts defended the Chunnel as "physical recognition of our links with Europe," which was sufficient reason for at least as many Britons to resent the Chunnel.

Le Monde, the finest flower of Gallic journalism, grumpily editorializes that "Great Britain is an island and intends to remain an island." Trust the intellectual descendants of Descartes to alight upon the obvious with a sense of discovery.

Today Britons are hectored mercilessly, from within and without, about their duty to be less "insular" and to "get with it." The pronoun "it" usually refers to some lambent abstraction like the "New Europe." But many Britons fear that surfers on the Wave of the Future must be lightly clad, and hence must shed such encumbrances as national identity. This they are sensibly reluctant to do, and for that reason—whether reasonable or not—they were reluctant tunnelers.

This feeling will be a force of historic importance later this year when Britons vote on continued membership in the Common Market (nee "New Europe"). Meanwhile, Britain, "bound in with the triumphant sea," shall stay that way, secure in the knowledge that if God had wanted a Chunnel, He never would have created Horatio Hornblower.

[January 25, 1975]

North Sea Story

FORTIES FIELD, THE NORTH SEA—In 1776, that busy year, James Boswell, the biographer, visited Matthew Boulton, James Watt's partner in building steam engines. Boulton proclaimed to Boswell: "I sell here, sir, what all the world desires to have— power." Power is still what the world hungers for, more ravenously than ever. That is why as the helicopter churns through the fog, beating against a wall of wind, you suddenly see flames upon the sea. These are lanterns in a horizonless world, where gray sea meets gray sky; they are gas flares atop oil-pumping platforms, crouched like huge insects upon rolling water.

Of all seas bounded by civilization, the North Sea is the most devouring. It has been gnawing into land south and west since the Ice Age.

The Dutch have not yet reclaimed from it all the land that was arable in the time of Julius Caesar. Bits of England—Dunwich, an important East Anglian port in the Middle Ages, and Ravenspur in Yorkshire, where Bolingbroke landed in 1399 to overthrow Richard II—have disappeared into the sea. For 273 days of an average year North Sea weather is officially, and almost euphemistically, called "bad," and on forty-nine more days the weather is smilingly called "marginal." Few of the year's forty-three "fair" or "good" days occur in December. But in a December when the world has oil on its mind, a visit here does concentrate the mind on the power of oil.

At the dawn of the age of enterprise, between 1450 and 1600, when men sailed off the map in fragile boats, spice was what oil is today, a universal commodity the demand for which set men and nations in motion. Pepper, clove, nutmeg and other spices were used to flavor meat that often was on the edge of rottenness; they also were needed to make edible the ghastly fat that provided much of the common folk's calories before sugar—another cause of wars and dangerous expeditions—was added to Europe's diet.

Today, from the North Slope of Alaska to this dark patch of sea 110 miles off Scotland's coast, the world's unslakable thirst for oil has energized daring enterprise under wretched conditions. The suck of billions of carburetors has pulled steel and men off shore into deep, dangerous water. Billions of flashing spark plugs have ignited the attempt to produce a Texas on the floor of a mean sea.

In 1959 a large gas field was discovered under a beetroot field in Holland. Petroleum geologists, who are paid to notice things, noticed similarities between the geological formations beneath the beetroot fields and beneath a small gas field in Yorkshire, England. British Petroleum decided to find out what lay beneath the intervening sea.

On October 4, 1970, a BP drill punctured the sea floor. Three days later, a mile and a third beneath the floor, the bit hit oil-bearing sandstone. A coded message to BP's London headquarters produced a laconic announcement: "Indications of hydrocarbons have been found. . . ."

The black slime we call oil, which we consume so casually, was produced by nature accidentally and slowly. Complicated civilization is utterly dependent on the decayed remains of simple organisms. They were buried beneath the sediments of ages and transformed, under extreme pressure, into a substance that man can transform into something that makes a Buick go zoooooommmmmmmmmm.

Getting North Sea oil to the surface is an achievement comparable to the construction of the Panama Canal, or the Netherlands counterattacking the North Sea. The platforms operate where fifty-foot waves are common. They must be able to withstand ninety-foot waves, and 130-mph winds. They are nearly 700 feet from seabed to derricktop, weigh 34,000 tons, and support 17,000 tons of equipment. They are expensive. So is oil.

No one knows how much oil is beneath the water, and no one can know how much is in commercially exploitable fields. Much depends on what the oil producers' cartel does to the price of oil, and what inflation does to the steep and rocketing price of producing oil under North Sea conditions. Already the capital costs per barrel of developing a North Sea field are twenty-five times the cost of developing a Saudi Arabian field. British inflation could destroy the commercial prospects of the smaller, marginal fields that must be exploited if Britain is to become a substantial oil exporter in the 1980s.

If there are forty billion barrels beneath the North Sea (twenty-five billion British, fifteen billion Norwegian) this oil constitutes a tiny portion—less than 4 percent—of the world's known reserves. (Saudi Arabia has 25 percent.) But there is enough North Sea oil to make a decisive difference to Britain.

Norway, with a population of just four million, soon may replace Sweden as the world's richest industrial nation in per capita income. It already is ahead of Britain. For Britain, with a population of fifty-five million, the oil will be even more important. It should end the trade deficits that for thirty years have been an impediment to Britain's growth.

But oil could be bad for Britain, as silver was for seventeenth-century Spain. Silver looted from the New World financed the freezing of Spanish society. It enabled the aristocracy to resist a modernizing evolution of society. Today oil is pouring off tankers and through undersea pipelines into Britain at a moment when economic pressure may be about to force a modernizing transformation of industrial attitudes among Britain's workers and managers. At long last a sense of urgency has seized the British people; they are prepared to swallow stronger medicine than their government is prepared to administer. If the flow of oil drowns that sense of urgency, the British will have made a sow's ear out of a silk purse. Then the oil will indeed be a wasting asset, and not only in the sense that it cannot be replaced and probably will have passed peak production by the end of the next decade.

Of course some stern moralists will argue that Britain's North Sea oil is a windfall and hence it is *necessarily* bad for the soul. They will say that it is wealth bestowed capriciously by nature and, like gambling earnings and other results of luck, it is "ill-gotten gain." But that notion of "ill-gotten gain" is anachronistic in any nation that has a stock market. Besides, Britain's oil wealth was not *bestowed* by nature, it has been wrested from nature at her angriest.

So what if Britain "found" a Texas. The United States conquered its Texas. And speaking of luck, we cannot be absolutely sure that God was rewarding American virtue when he placed the energy equivalent of several Saudi Arabias on the United States in coal deposits beneath the American West. Finally, if food, the energy source for human motors, is the scarce universal commodity of the future, then no nation in the world, not even Saudi Arabia, has a natural blessing comparable to the topsoil of the American Midwest.

[December 27, 1976]

Highland Spirits

LONDON, ENGLAND—The Lent Term lecture list at Cambridge University cryptically announces: "Human Genetic Variation will take place on Tuesday." Alarming announcements are routine in Britain. The lecture list for last Easter Term announced: "Representation, Participation and Democracy is canceled."

Last summer, when Britain was in the throes of a nearly 30 percent inflation and bitter labor militancy, some sober persons believed that such a cancellation was probable. That cancellation did not happen, but now a headline here announces another social horror:

Danger of Scotch Shortage
in 1980s, Say Distillers

According to the Scotch Whisky Association, the "ineptitude and greed" of the United Kingdom government may produce a Scotch shortage. The combination of increased taxes and price controls has starved distillers of capital needed for expanding capacity.

World Scotch consumption over the next six years should increase to 190 million gallons a year. Allowing for 10 percent evaporation (a heartrending fact of life), the industry should have produced 210 million gallons this year. But it produced only 150 million, so in the early 1980s, there will be a shortage of mature whisky.

A distiller, voicing the grim Scottish suspicion of the government in London, predicts darkly that an "over-authoritarian" government, wanting to protect an important export, will ration home consumption of Scotch either by decree or steep excise taxes. If that happens, the United Kingdom may be canceled.

But, then, cancellation may happen anyway, if the large and growing number of Scottish nationalists have their way. They want to secede the union. The fuel for their resurgent nationalism is a fluid—not amber, black. It is North Sea oil, discovered beneath what suddenly are called "Scottish waters."

Denis Healey, now chancellor of the exchequer, called discovery of North Sea oil the first big piece of luck Britain has had in this century. When the first oil arrived by tanker in Scotland another minister called for "a day of national celebration." Ah, but in which nation? If the Scottish National party continues to grow as it has since the discovery of oil in 1970, the oil may be the proximate cause of the breakup of the United Kingdom, the emergence of Scotland as an independent nation.

Today the SNP is the fastest growing party in Europe. Its program, like its slogan ("Rich Scots or Poor Britons"), is the soul of simplicity: become independent and seize all the oil in its national waters. Although it has only eleven members of Parliament, a recent opinion poll gives it 37 percent support, making it the strongest party in Scotland, 7 percent ahead of Labor. If that holds up—and it has been growing—the SNP could win more than half of Scotland's seventy-one seats in Parliament.

Most would be won from the governing Labor party, which depends on forty-one Scottish seats. An SNP contingent of forty would hold the balance of power in Parliament, and would drive a hard bargain seeking independence.

Relaxing in a House of Commons bar, two SNP members explain why they are sure Scotland will have an easier time than, say, South Carolina had when it tried to leave a union. They express, in a single breath, a reason for seeking independence and a reason for thinking they can win it on the cheap: "Britain is not a nation you can be very proud of these days. The British are not—to put it mildly—at their peak. They are used to giving things away."

These two MPs feel like foreigners in London, and their words are heavy with ancient resentments. As Big Ben strikes 7 P.M. over the mother of parliaments, they sip that amber fluid. Warmed by it, and by thoughts of the black fluid, they vent their disdain for "Britain," meaning England. What delicious fun, these words rolled around on the tongue like Highland malt. They are reciprocating the disdain directed at Scotland for centuries by many Englishmen.

"Much may be made of a Scotchman, if he be *caught* young," said Doctor Johnson, looking down his upturned nose toward the northern border. "Sir, let me tell you," he said, "the noblest prospect which a Scotchman ever sees, is the high-road that leads him to England." Today for the Scottish nationalists the enticing prospect is an oily road away from England, a road so high they can at last look down on England.

[February 15, 1976]

Scotland: A Tartan Norway?

EDINBURGH, SCOTLAND—This is a city of histrionic geography. At its center an enormous eruption of rock is a pedestal for a dominating castle. The city is located on the Firth of Forth, where the ferocious North Sea slices into Scotland's narrow waist. In a heavy sea mist the city resembles the prow of a ship beating east.

For Scotland, eastward the course of empire takes its way, eastward into the North Sea, where fifty-foot waves and 70 mph winds are common. There you see the oil rigs crouched on the roiled sea like spindly legged waterbugs. And if you sail on east you reach Norway, which is an incitement to Scottish independence.

Because of oil, Norway (population four million, 1.2 million less than Scotland) soon will be the richest (per capita) industrial nation. It is an example not lost on Scotland, which has many links with Norway. The Shetland islands, an archipelago 130 miles north of the Scottish mainland, belonged to Norway until 1469. Scottish regiments were used (ill-used, actually) in the 1940 invasion of Norway. Edinburgh's Heriot-Watt University, with a strong curriculum in oil economics and technology, attracts many Norwegian students.

An independent Scotland, a "tartan Norway," would be larger in population than three Common Market members (Denmark, Iceland,

Luxembourg) and much richer. But there is a problem. Most of the oil is not under what would be independent Scotland's waters. Most of the oil would be Britain's if, as seems certain, the Shetlands refused to join Scotland in secession.

(The Shetlands are inhabited by 250,000 clever, industrious sheep that are busy producing sweaters, and by 19,000 humans who have the equivalent of a GNP of $13 million, most of it from scrumptious herring. To get a sense of scale, note that oil companies are spending $1 billion for terminal facilities that may make little Voe, a Shetland hamlet, into Europe's largest oil port, larger even than Rotterdam.)

But rather than appear rude, the British government talks only about "devolution," a ghastly name for a plan to create a Scottish assembly with power to do not much, and no power to do important things against Westminster's wishes. As an attempt to dish Scottish nationalists, it is inadequate. It is more apt to whet than slake the thirst for independence.

The resurgence of Scottish nationalism began in the 1960s, before the discovery of nearby oil. But separatist sentiment was checked by the fear that Scotland, heavily subsidized from London, could not go it alone. Today nationalists argue that Scotland cannot afford not to go it alone. Otherwise, oil revenues will flow through London, and Scotland will have missed its chance to escape the downtow of Britain's sinking economy.

Scotland was a separate kingdom until 1603. In 1707 the Scottish parliament was dissolved. It was a hard year, 1707, with snow in summer, crop failures, starvation, and a lot of bribery of Scottish politicians to grease the skids for absorbing Scotland into the United Kingdom.

In Tom Stoppard's play *Rosencrantz and Guildenstern Are Dead* a character wonders if England is not "just a conspiracy of cartographers." Scottish nationalists believe Great Britain is a conspiracy of *English* cartographers.

Certainly Britain remains splendidly unhomogenized. If you doubt the resilience, not to say chewiness, of Scotland's national identity, remember this axiom: cuisine is destiny. And spend a few unforgettable hours at table in one of my favorite European restaurants, Edinburgh's Cafe Royal.

Begin with Hotch Potch (mutton broth) or Partan Bree (a fish soup) or Cock-a-leekie (fowl simmered with leeks) or Cullen Skink (broth made with Finnan Haddock, fish smoked over peat fire). Next try pancakes Mary of Lorraine (diced chicken, sweatbreads, Drambuie liqueur, folded into a pancake with butter sauce). Or try Stoved Howtowdie wi' Drappit Eggs. (It is too complicated to explain.) Don't miss Rumblede-thumps (boiled potatoes and cabbages, that are, well, *thumped* together).

Always, there is Haggis, the pluck (including heart, lights and liver) of a sheep cooked together, then mixed with suet and oatmeal, stuffed into a sheep's paunch, boiled, and served with—sometimes drenched with—Scotch whisky. And it is not the watery blended stuff Scotland sells to America, but robust malt whisky. I hold this truth to be self-evident: with or without oil, people who pour malt whisky on oatmeal are people to be reckoned with.

[February 22, 1976]

Drunk on History

DUBLIN, IRELAND—During the 1932 election campaign, posters on the walls of this fair city proclaimed an arresting slogan: "Remember 1167." That is Irish politics—drunk on history.

It was about then (1170, actually) that Strongbow came to Ireland from England with fierce troops and the blessing of the English king, Henry II, to support one Irish faction against another. In 1932 the slogan meant: oppose candidates who are insufficiently anti-British.

The decisive event in modern Irish history was the potato blight and famine of the late 1840s. A million Irish died and a million more emigrated, starting a tidal pull that reduced Ireland's population from eight million to four million. The Irish character was transformed and tempered, like steel, in the furnace of the famine.

Just as Israel is a consequence of the holocaust, Irish independence is a consequence of the famine. By 1912 toughened Irish political agitation had Britain ready to grant Home Rule.

But the Protestant majority in the six northern counties denounced it as "Rome Rule" and prepared for military resistance to any rule from Dublin. The British army was politically unreliable, with army officers proclaiming their unwillingness to "coerce Ulster."

In 1914 the British government flinched from confrontation with Ulster's private armies. It amended the Home Rule bill to exclude Ulster, ostensibly for just six years. The lesson—violence works—was noted by extremists in the south.

On Easter, 1916, while 150,000 Irishmen—all volunteers—were in British uniforms, in Flanders, fighting for the king, about 2,000 rebels in Dublin rose against the British. Most of the Irish people were appalled. A Dublin crowd hissed at the rebels when they surrendered after a week of fighting.

But when the British executed the leaders, the rebels became martyrs, legitimizing the smoldering resentment of the oppression that had followed Strongbow across the Irish Sea. Between 1919 and 1921 the British fought a losing war of pacification against the guerrilla army of national liberation.

But the provisional Irish government had to accept a treaty that, in effect, excluded Ulster from Ireland. Irish resentment exploded into a savage two-year civil war, won by supporters of the treaty.

Today Dublin is one of the most agreeable cities in Europe. Belfast, the capital of Northern Ireland, is in the grip of a near-civil war fought by extremists who, drunk on history, "remember 1690." That year the Battle of the Boyne completed the Protestant conquest of Ireland.

Today the remnant of Protestant ascendancy is the Ulster enclave. There the brutish use of political and economic power by the Protestant majority has brought forth the brutishness in Catholic extremists.

As in 1912, Protestant power is backed by well-armed private armies. The Catholic extremists are the Provisional IRA (Irish Republican Army), a terrorist organization espousing infantile leftism and receiving substantial financial support from Irish-Americans.

More than 1,200 people have died in Ulster since 1966. That many might die in days if the British government withdraws the troops, which the Provisionals seek. They assume the Dublin government would intervene to protect the Catholic community, which is outnumbered two-to-one. But Dublin's army is not up to an urban guerrilla war against the armed Protestants. Thus British withdrawal might lead to a flight of Catholics to the border, with Dublin imposing a modest northward revision of the border.

The hand of history is heavy on Ireland. So heavy, in fact, that here in Dublin responsible people, even those who "remember 1167" and want a united Ireland, someday, want British troops to stay on Irish soil until some way is found to ameliorate the consequences of 1690. But a solution seems as many years away as the events that created the problem.

[June 13, 1975]

Ulster's Dark and Bloody Ground

BELFAST, NORTHERN IRELAND—The British official visiting Washington had said: visit Ulster and see that bombs don't explode constantly, and life is nearly normal. So it is 11 P.M., and the taxi has been selected randomly from traffic. I am standing in the drizzle while an enormous British soldier, his face blackened for night patrol, his rifle in one hand, my shaving kit in the other, searches in my suitcase for gelignite. This is part of normal life in Ulster on a weekend when two bombs will explode and five men will die.

Nearly 1,500 have died since 1968, and ten times that many have been maimed. Near center city words six feet high proclaim: "Don't let your child play with toy guns." A toy may draw fire from a nervous soldier who is at war in his own nation with a nearly invisible and often young enemy.

Such signs are aspects of war here. One warns against letter bombs: a letter that is lumpy, or "unexpected" should be shoved into a pan of water. Another sign says: "A firebomber could be anyone in the crowd"—not very helpful advice. Another says: "If you are suspicious call 652155." But if anyone can be a firebomber, everyone looks suspicious. This demoralizing social climate is the terrorists' triumph.

Since 1968 the provisional Irish Republican Army, a small band of gangsters and terrorists, has waged the most successful urban guerrilla operation since the Algerian war. It has carved out areas of Ulster—for example, the Bogside in Londonderry, and southern County Armagh—where the queen's law does not run. Neither police nor army can function there, and the IRA is the de facto civil power.

A senior official estimates that a mere twenty terrorists have made Armagh "bandit country." But 2,000 soldiers could ransack 2,000 houses without catching them. The same official estimates that in 1968 the provisional IRA consisted of fewer than ten men, and today there are fewer than 400 prepared to fire a gun or hurl a bomb. But there always are *enough* for IRA purposes. The purposes all relate to producing chaos, which is supposed to produce British withdrawal from this piece of Britain.

The Protestant majority has contributed its fair share to making Ulster a dark and bloody killing ground. A leading Protestant politician is the Reverend Ian Paisley, whose head swims with anti-Catholic rubbish, like the idea that the Common Market will adopt the Virgin Mary as its patron saint. Protestants can match Catholic viciousness. A Catholic mother of eight was murdered in front of her children. A Protestant woman suspected of informing police about Protestant terrorists was beaten to death by a mob of Protestant women.

In *Harry's Game,* a novel about Ulster, Gerald Seymour gives this authentic detail:

"The bottom eight feet of a wall at the end of the avenue had been whitewashed, the work of housewives late at night—so that at night, in the near darkness . . . a soldier's silhouette would stand out all the more clearly and give the boyos [sic] a better chance with a rifle. Most corners in the area had been given the same treatment, and the army had come out in force a week later and painted the whitened walls black. The women had then been out again, then again the army, before both sides called a mutual but unspoken truce. The wall was left filthy and disfigured from the daubings."

Perhaps the story of the Belfast walls is a metaphor for the best outcome that can be reasonably expected for Ulster. Perhaps in ten years or so, exhaustion, the great pacifier, will do the work of reason. Perhaps an unspoken truce will just happen, leaving Ulster filthy and disfigured, but mendable.

Meanwhile, what happens here involves the United States, in several ways. The current phase of "troubles" began eight years ago with a civil rights movement modeled, in part, on the U.S. movement. Today the violence is funded, in part, from the United States, where there are four times more "Irish" than there are in all of Ireland, North and South.

[February 5, 1976]

The State Within the State

BELFAST, NORTHERN IRELAND—Animosities 400 years old combined with modern grievances to produce a civil rights movement here in 1968. The televised example of the U.S. civil rights movement provided a strong draft for the fire.

The Ulster movement sang "We Shall Overcome" and chanted "One Man, One Vote." At the time, Catholics—one-third of the 1.5 million residents of Ulster—were substantially disadvantaged by property qualifications and gerrymandering in local elections. Such elections are crucial where, as in Ulster, local government allocates a large portion of scarce jobs and houses.

The spark that lit the dry kindling in Ulster in 1968 was the decision of a local council to rent a house to a nineteen-year-old unmarried Protestant girl. A Catholic politican, incensed that the girl was given preference over Catholic families, protested by occupying the house, and the movement was on the move.

Ulster government had changed not at all for nearly fifty years, but it changed a lot in less than two years. Pressure from London produced legislation meeting the movement's original demands, banning discrimination in jobs, housing, voting. But the bitterness between Protestants and Catholics in Ireland was nearly four centuries old when Ireland was partitioned in 1921. The violence begun in 1968 immediately took on a mad logic of its own, irrespective of this or that legislation.

Today, many of the killers in the provisional Irish Republican Army are seventeen-year-old boys who were ten in 1968. They have entered the terrorist trade as one enters any other trade. It can be a lucrative one. The IRA finances itself primarily with bank robberies, burglaries, and extortion in the Catholic community, as well as with contributions from Irish-Americans.

Although parking is severely restricted in downtown Belfast (too many cars have been used as bombs), and although shoppers are constantly bothered by searches, retailers had a good 1975. This was, in part, because of "hot" money that sticks to the fingers of those who say they steal or raise it for charitable purposes.

Agencies collecting for the IRA in the United States admit to having collected $1.2 million, which buys a lot of $280 Armalite "widow maker" rifles, a civilian version of the M–16. And the Defense Department believes that IRA sympathizers are responsible for a significant amount of the 6,900 weapons and 1.2 million rounds of ammunition stolen from bases between 1971 and 1974.

British officials diplomatically say they are sure most U.S. supporters of the IRA think that money contributions are going for charitable purposes. But the sulphurous propaganda churned out by the IRA's U.S. agents makes clear that most contributors must know that they are subsidizing murder.

To appreciate the forbearance the British government has shown in this matter, imagine the U.S. reaction if U.S. soldiers were employed putting down a U.S. rebellion financed with British contributions.

Imagine, too, the forthrightness with which U.S. politicians—say, Senator Henry Jackson, to take an example not quite at random—would denounce the collection in the United States of more than $1.2 million for the Palestine Liberation Organization, a terrorist organization no worse than the IRA.

The Social Democratic Labor party, which represents Ulster Catholics, has denounced the IRA. But Jackson, pandering to confused Irish-American voters in the Massachusetts primary, merely declares that as President he will encourage the "parties and governments involved" to meet in a "neutral place" for "fruitful discussions."

Jackson ignores the fact that only one government—the British government—has any rightful say in the governance of Ulster, which is as much a part of the United Kingdom as Hawaii is of the United States. Admittedly, the British government blurred this crucial fact in December 1974 when it adopted a de facto cease-fire with the IRA. By negotiating, however informally, a cease-fire in its own country, the British government treated the rebel organization as a state within the state, thereby encouraging extremists to doubt that Britain has the will to be sovereign here.

In fact, Britain has no choice. The Irish Republic would not accept Ulster as a gift. And Ulster can not be independent. What a South Carolinian once said of his secessionist state can also be said of Ulster: it is too small to be a nation and too large to be an insane asylum.

[February 8, 1976]

Too Few Factions

BELFAST, NORTHERN IRELAND—In an old Irish joke a traveler asks directions to a village and is told, "If I were you, I wouldn't start from here at all." No reasonable person is confident that, in the foreseeable future, Ulster can get from here, a slowly simmering chaos, to there, tranquility.

Ulster Catholics know that if the Provisional Irish Republican Army terrorists achieved their aim—withdrawal of British forces—that would be the signal for a civil war. Catholics, outnumbered two-to-one, would lose. This is a point too recondite for those clothheaded Americans (like Paul O'Dwyer, New York city council president) whose grasp of Ulster realities does not extend beyond the slogan "Brits out!"

The Provisional IRA is willing to gamble that the Irish Republic would intervene in a civil war after a British withdrawal. But the Republic's army is smaller than the Republic's telephone and telegraph staff, and would be no match for Ulster Protestants' paramilitary organizations in a guerrilla war.

British withdrawal would amount to dismemberment of the United Kingdom. It is one thing for the British to have withdrawn from colonial governance of distant peoples of different races and cultures. It would be something very different for the authority of the British government to be driven from a portion of the United Kingdom seventeen miles from Scotland.

Ulster would founder as an independent sovereignty outside the United Kingdom. Ulster's two mainstay industries, shipbuilding and aircraft, would collapse without the steady, heavy flow of subsidies from the government at Westminster. And because it is poorer, the Catholic community—even more than the Protestant community—needs the benefits of the British welfare state, benefits far more generous than those enjoyed south of the border, in the Republic.

Most Protestants do not believe what a few prominent Protestants continue to assert: that most Catholics desire prompt union of Ulster with the Republic. And certainly the Republic does not want such a union. The Republic's economy is weak and plagued by unemployment, and would be shattered by trying to integrate a war-torn and unsubsidized Ulster. Moreover, the government of the Republic knows that it would be the next target of a Provisional IRA that had forced Britain's withdrawal from Ulster.

True, the Social Democratic Labor party, which represents Ulster Catholics, formally favors the union of Ulster with the Republic. But that is of no more practical significance than the Republic's formal commitment to "eventual" union, a commitment far less important than the Republic's desire for the British army to stay in Ulster until Ulster has undergone some unforeseen transformation.

Ulster's tragedy is not that it has two factions, but that it has only two. The secret of political happiness under popular government is a multiplicity of factions. The more the merrier: a multiplicity of factions produces single-issue majorities that are fluid coalitions of minorities. Such transient majorities, unlike Ulster's monolithic Protestant majority, are unable to impose a stable, oppressive regime.

Protestant leaders cleave to an unsubtle and demonstrably unworkable democratic theory: the majority must get its way and the minority must knuckle under. That would be crude but workable if the minority consisted of 3 percent of the population. In Ulster the minority is 33 percent.

Two years ago Protestants used a three-week strike to destroy a compromise "power-sharing" plan devised in London. Later, IRA terrorists destroyed a similar plan that had won important Protestant support. On Friday several Protestant leaders supported power sharing. Over the weekend IRA terrorists attacked some Protestant social clubs, killing fathers and sons. On Monday, most of the support for power sharing evaporated.

Something like those power sharing plans—inclusion of a minority of Catholics in the Ulster executive—will be one price eventually paid in an attempt to staunch the community's bleeding. But it will not happen soon.

In 1868, Prime Minister Gladstone said: "My mission is to pacify Ireland." The problem was then more than 300 years old. And in 1921, Prime Minister Lloyd George "solved" the problem by isolating it in these six northern countries.

[February 12, 1976]

Zionism and Legitimacy

In the mid-seventeenth century, Oliver Cromwell contemplated exterminating the Irish and settling Ireland with Europe's persecuted Jews. Regarding the Irish, the plan appealed to Cromwell's passionate side. Regarding the Jews, the plan appealed to his common sense.

It acknowledged that the Jewish people were, indeed, a "people." Possessing a common past and culture, they lacked only land, which is not the essence of a nation. The Jews were a nation in need of a home.

More than 300 years later the Jewish people and the legitimacy of their nationalism are under attack in the organization misleadingly named the United Nations. A U.N. committee has voted 70–29 to declare that Zionism is a form of racism. This move was sponsored by Arab regimes and was supported primarily by dictatorial regimes—China and Chile, the Soviet Union and Spain.

I refer to regimes, not nations, because given the nature of the regimes of U.N. members, there are very few nations, meaning peoples, represented there. The vote censuring Zionism was a vote by numerous regimes, representing nothing but themselves, against a single nation, Israel.

The U.N. majority of dictatorial regimes is guilty of many things, but not of sincerity. Those regimes know that Zionism, far from being racism, is an especially defensible form of nationalism.

Zionism (the word was first used publicly in 1892) is the belief that the Jewish people, having come this far through a uniquely hazardous history, deserve a common future. And it is the belief that a national homeland is important to that future.

Zionism appeals to Jews who believe that, since emancipation in the nineteenth century, a Jewish state has been the only alternative to assimilation and loss of identity. Zionism also appeals to Jews who feel that they, like Italians and Germans and Americans and others, can more easily, more "naturally" achieve personal fulfillment in a nation state that embodies their common culture.

Zionism is supported by many non-Zionist Jews who are not anti-Zionist, but who believe that Jews everywhere will be more secure, culturally and physically, if a Jewish state exists as an embodiment of cultural values, and as a potential refuge.

To call Zionism racism is to assert that Jews are held together, where they are held together, by "racial affinity," whatever that means. And it is to assert that Israel is an expression of racial, as opposed to cultural—and especially religious—cohesion. In fact, Israel is a religious state in somewhat the same sense that Spain, after a series of concordats with Rome, is a Catholic state. Every nation's laws are to some extent authoritative expressions of values, and Israel's laws are anchored in a particular religious expression of values.

Zionism, like the Italian Risorgimento, like German nationalism, like a lot of other things, is a product of the French Revolution, which injected into European history the idea of a people attaining true fulfillment only through a revived nation. Zionism became, as it were, self-conscious, acquiring a name, a literature, and a leadership, at the end of the nineteenth century, in the heyday of philosophic nationalism.

Zionism became a fighting faith in response to resurgent anti-Semitism (especially the Dreyfus affair) that was one manifestation of militant nationalism. Nationalists attacked Jews as "a nation within the nation."

The original Zionists shared the great nineteenth century obsession with nationalism. And today Israel is, arguably, the clearest example of cultural affinities, as opposed to racial principles or coercion, as the basis of national organization.

Most U.N. members are police regimes. Many of these regimes rule over ersatz nations. Many use their energies to pound together human elements that lack cultural affinities. To such regimes Israel, a real nation, is either unintelligible or a reproach. Regimes resting on force are bound to find fault with the rich legitimizing sources of Israel's nationhood.

Israel became a nation after the United Nations was born. But in a sense Israel is one of the oldest nations (with Egypt and China) represented there. One hundred years hence, if historians bother to remember the United Nations at all, they may remember it as a mob of regimes representing force without legitimacy, all power and no authority, venting their rage against one of the few nations truly represented there.

[October 29, 1975]

Israel: State of Siege

JERUSALEM, ISRAEL—A small glass case in a museum here contains only a child's dusty, crushed shoe. Found at Treblinka death camp, the shoe reminds: no calamity is unthinkable. That is the premise of Israel's statecraft.

As prosecutor of Eichmann, Gideon Hausner was attorney general for six million victims. Today he ticks off statistics. One-third of all Jews died. Today there are twenty million more Japanese than before the war, fifteen million more Germans, but fewer Jews. Like most Israelis of governing age, Hausner is haunted by the fact that Goethe's nation became Göring's nation, and no other nation acted to save the Jews. When Coventry Cathedral was bombed, reprisal raids were visited upon Germany. When Jews begged that Auschwitz be bombed they were told that bombs were needed for "essential" operations.

Since the Holocaust there have been four wars (five, counting the 1969–70 "war of attrition") against Israel, the survivors' haven. The Arab warmakers have half a point when they say Israel is an "outpost of the West." They are wrong in suggesting that most Israelis are Westerners; most Israeli Jews are from families of immigrants from Arab nations. But Israel is an enclave of Western democratic values in a region where the only other nation even tolerant of such values (Lebanon) is bleeding to death. In Israel these values, like the orchards, are prospering against high odds.

The 3.4 million Israelis (less than three million Jews) pay the world's heaviest taxes. Taxes take 71 percent of national income. The inflation rate is 30 percent. No wonder emigration and immigration rates are converging. Israel's economic policy is, tersely put, "being comes before well-being." Thirty-five percent of GNP goes for defense. The 1973 war cost a year's GNP.

Statistically, but not psychologically, Israel is a "garrison state." Israelis insist that theirs is not a "real" army, that it is just society in khaki. Arabs would not agree. But the full-time professional army is small. There is universal conscription for both sexes and reserve duty through age fifty-five. The deputy governor of the Bank of Israel glances at his watch and excuses himself from the lunch table. He must, in the constantly heard phrase, "go to the army." He will be in uniform in an hour. When virtually everyone is in uniform, or soon will be again, there is no military caste insulated from democratic values.

Israel's values are being tested in its occupation of the Jordanian West Bank. Israel has made only three significant changes in local law. There is no longer capital punishment, or property qualifications for voting, and women can vote. West Bank residents elect local officials and publish anti-Israeli newspapers. In Arab countries only the latter is permitted.

Israel has been an occupying power for nearly a third of its twenty-eight years of national existence. It has had "Kent States." Violent demonstrations have panicked Israeli soldiers. There have been fatalities. But the occupation administrators think the demonstrations may be part of a new and yeasty political process that will produce an indigenous West Bank Arab leadership, independent of the Palestinian radicals in Beirut and willing to work and eventually negotiate with Israel. At least leaders of stone-throwing mobs cannot easily be dismissed as quislings.

A few Israelis hope that occupation will become annexation. They are not facing demographic facts. There are 700,000 Arabs on the West Bank, where the average age is under thirty and the birth rate is high. Annexation might eventually produce an Arab majority in expanded Israel. Such a state could not be both democratic and Jewish.

The occupied territories in Egypt, Jordan and Syria were conquered during the 1967 war. That war came at Israel from across the Sinai wastes, across the Jordan River, and down from Syria's Golan Heights. For nineteen years, from 1948, most Israelis were in range of random violence launched from territory now occupied, as any Israeli will tell you at the drop of a map.

Israelis are map junkies. They have maps on their walls, in their pockets, in their heads. They draw maps in the desert dust. In the vicinity of a map an Israeli's index finger is drawn to the spot where, prior to 1967, Israel's waist was ten miles across. There an Arab armored column could have sliced Israel in half in an hour. The combination of 1967 weapons and 1967 borders was intolerable. Israel's military leaders insist that more powerful 1976 weapons (especially anti-aircraft missiles) at the 1967 borders would be fatal.

The dialectic of changing Arab weaponry and changing Israeli perceptions of defensible borders complicates the pursuit of peace. But that is less important than the fact that key Israeli leaders are mildly optimistic about the possibility of pursuing peace. They have adopted a determinedly hopeful evaluation of the events set in train by the 1973 Yom Kippur war.

Syria and Egypt attacked on October 6. After several desperate days Israel turned the tide. Concerned for his Arab clients, Brezhnev summoned Kissinger. They promptly imposed a cease-fire in time to save an Egyptian army, and Egyptian President Sadat's skin.

The cease-fire turned a military shambles into a success, of sorts, for Egypt.

The Arabs launched their surprise attack with a tank advantage better than Montgomery's advantage over Rommel at El Alamein. But Egypt's meager prize, salvaged in Moscow, was a tiny strip of sand on the east side of the Suez Canal. That was enough. It removed the pressure on Sadat to do something violent to Israel.

Israel gave up some strategically significant Sinai terrain, and its only energy source, the Abu Rudeis oil fields that were providing 60 percent of its oil consumption. Israel got nothing tangible from Egypt, but got generous aid agreements from the United States. In effect, the United States bought some of the Sinai for Sadat.

Today President Ford is urging Israel to "dare the exchange of the tangible for the intangible." Israel just might dare because it thinks the interim agreement, an exercise in trading tangibles for intangibles, marginally improved Israel's long-term political hopes and decreased the short-term military threat. The Israelis believe it drove a wedge between Egypt and Syria. And they see the agreement as part of the process of weaning Egypt away from Soviet military aid to a U.S. diet that emphasizes economic aid. In their muted, guarded optimism Israeli leaders reason as follows:

As a nation's middle class grows, the nation is apt to become less bellicose. Egypt is a far cry from a bourgeois society, but development is an irresistible Egyptian aspiration.

Egypt is one of the world's poorest nations. It is the most urbanized Arab nation, and public services are groaning under the strain. Egypt has a GNP the size of Israel's but a population eleven times larger.

The tragedy of the Middle East is captured in this fact: Egypt's per capita income is $280 a year, less than the price of two tank shells. The hope for the Middle East may be in the movement of many hundreds of thousands of Egyptians into the Suez Canal cities Sadat is rebuilding. Sadat is putting people and precious resources smack in the probable path of any future war.

Obviously this Israeli optimism is fragile. It rests on one man. Regarding Sadat's putative new priorities, the evidence is meager. Besides, Sadat is mortal, and has a heart condition. Anyway, in this region regimes sometimes change with unseemly speed. And then there is Syria, another nettle to seize. As always, making peace will be like making whole eggs from an omelette.

At the moment, Israel is like the man who, having been convicted of a grave crime, was told by the king: "I intend to sentence you to death, but not for two years, and I will reconsider if by then you have taught my horse to talk." Later, to puzzled friends the man explained his optimism: "In these two years I may die a natural death. Or the king may die. Or the horse may talk."

But Arabian stallions don't talk, least of all to utter the words that would change everything: "Israel has a right to exist." So because a dialectic of ideas is impossible for now, Israel will try for a dialectic of actions. It will consider major concessions on all three fronts in exchange for progress toward a peace that is more than merely an absence of violence, a peace of normalized relations, involving freer movement of goods, ideas and people between nations.

In Israel's eyes, the principal obstacle to peace is the Arabs' position on Israel's right to exist, not Israel's position on a "national entity" for Palestinians. In 1948 Israel was created from one-sixth of 1 percent of lands inhabited by Arabs. Between 1948 and 1969 when Arabs spoke of "occupied territory" they meant Israel. No Arab nation has disavowed that notion.

Today most Palestinians live in either occupied or unoccupied Jordan, which is, historically and geographically and ethnically, part of Palestine.

The Israeli Government put the "Palestinian question" at the bottom, not the top, of a three-item Middle East agenda. A Palestinian entity must link the West and East banks, and must be negotiated between Israel and the regime in Amman, Jordan. That regime probably will be too timid to negotiate unless Cairo and Damascus have already come to terms with Israel.

A place to ponder Israel's situation, absent peace, is in the ruins of Belvoir, a twelfth-century Crusader castle.

Various Arab leaders have said that Israel is like the Crusades, a short-lived intrusion by people who never belonged in the region. Belvoir is on a promontory in lower Galilee. To the north, across the Sea of Galilee, are the Syrian hills from which so much trouble has come.

To the south runs the Jordan valley, across which tanks have driven toward Jerusalem. Belvoir is on a sparsely populated line that runs from Syria's Mount Hermon south to Sharm el Sheikh, where in 1967 Nasser blocked Israel's access to the Red Sea, thereby causing the war that produced the current borders.

A chain of Israeli settlements stretches along this line. These settlements are a terrible idea. They stimulate the fantasies of Israeli expansionists, including (but not confined to) those fanatics who find in the Bible divine injunctions to expand Israel. Fortunately, today's Israeli government will not let the settlement stand in the way of peace. But pending peace, a mixture of settlements and military posts along the occupation borders are Israel's way of providing the "defensive depth" it thinks it needs.

The "confrontation countries" surrounding Israel have standing armies that can be on the move six hours after receiving marching orders. Israel's reserve-based army requires thirty-six hours to mobilize. In a surprise attack, as in 1973, territory is time: "defensive depth" must provide the thirty-six hours that are Israel's life.

On the stony ground beneath the blazing Mediterranean sky, with one's back to the sea, life is real, life is earnest. Born from ashes and surrounded by armed hosts still vowing that to ashes it shall return, Israel understands what Stephen Crane meant:

> *A man said to the universe:*
> *"Sir, I exist!"*
> *"However," replied the universe,*
> *"The fact has not created in me*
> *A sense of obligation."*

The poem contains the premise of Israel's statecraft.

[June 14, 1976]

Israel and Munich

Various Jewish religious observances commemorate calamities or narrow deliverance from calamities, and the short history of the Jewish state is replete with such experiences. Today, friction between Israel and the Carter administration is building up a dangerous charge of static electricity. No Israeli government casually risks the U.S. government's displeasure: diminished support for Israel could lead to a calamity from which there would be no deliverance. But the contagious crossness between Washington and Jerusalem that originated in Washington is a compound of Washington impatience and Israeli anxiety. The anxiety is more reasonable than the impatience.

For a decade, since the Six Day War of 1967, U.S. policy has been that Israel should trade territory for peace. As President Ford put it, Israel should "dare the exchange of the tangible for the intangible." The secure are always exhorting Israel to be daring. Similarly, the governments of the world constantly insist that Israel be more forthcoming than those governments ever are.

Theodore Draper, scholar and journalist, notes that of all the millions of square miles of territory conquered in recent decades, only Israel's occupied territories are expected to be returned. Norman Podhoretz, editor of *Commentary* magazine, notes that of the thirty-five million refugees created since 1945, only the fraction of a million created by Israel's war of independence are expected to be repatriated.

Saul Bellow notes: "In this disorderly century refugees have fled from many countries. In India, in Africa, in Europe, millions of human beings have been put to flight, transported, enslaved, stampeded over borders, left to starve, but only the case of Palestinians is held to be permanently open. Where Israel is concerned, the world swells with moral consciousness. Moral judgment, a wraith in Europe, becomes a full-bodied giant when Israel and the Palestinians are mentioned. . . . What Switzerland is to winter holidays and the Dalmatian coast is to summer tourists, Israel and the Palestinians are to the West's need for justice—a sort of moral resort area."

Today the U.S. government is anxious to bestow upon Israel the honor of leading a life more daring than other nations choose to live. The United States became a mighty continental nation through conquest in the name of "manifest destiny." But the U.S. government is irritated because Israel is reluctant to commit itself, before negotiations, to return land it conquered from aggressors who still deny its right to exist on the coast of Palestine. U.S. security has always been a function of broad oceans and placid neighbors. But the U.S. government is irritated because Israel is wary of turning a geographical buffer (the occupied West Bank of Jordan) into a Palestinian "homeland" that probably would be dominated by the Palestine Liberation Organization, terrorists committed to the destruction of Israel.

The U.S. position is that Israel should withdraw to the 1967 borders (perhaps with slight revisions) and the Arab states should take "steps toward" normalization of relations with Israel. But even if Israel were to withdraw in exchange for full peace (recognition of its right to exist, plus free movement of people, ideas and commerce in the region), there still would be an inherent asymmetry of risk in a trade of the physical for the political. Arab political concessions could be repudiated overnight; Israel's physical concessions could not be reclaimed without war.

Nevertheless, Israel has accepted this asymmetrical policy. It has asked two things. One is that the United States not intrude itself so much that it spares Arab states the need to negotiate directly with Israel. The second is that the United States not propose a specific outcome (such as withdrawal in exchange for "steps toward" normalization). Israel thinks that if Arab states regard withdrawal as a given, they will have no incentive to give anything. After four wars, Israelis are unmoved by the idea that their security depends less on their toughness than on their malleability. And since the fourth war they are especially impatient with assurances that the "conscience of the West" will be their shield. In the October 1973 war Israel not only suffered debilitating losses comparable to Britain's in the First World War, relative to national strength. Israel also suffered an acute understanding of the "conscience of the West" under oil pressure. Israel was isolated.

Israelis are obsessively interested in U.S. diplomacy, and were fascinated by Jimmy Carter's May meeting in Geneva with Syrian President Assad. Carter praised Assad's helpfulness, constructive attitude and "intimate knowledge." That, Carter said, "has helped me a great deal to understand" the Mideast. Now, diplomacy always involves a lot of solemn nonsense, but Carter went a tad far. In recent years Assad has called Israel "a basic part of southern Syria," and his controlled press has asserted that Israel "shall be destroyed." Today the United States is pleased to regard Assad as a "moderate." Has Assad changed, or has the United States? Today Assad says "the Palestinian problem has two parts," the first concerning the West Bank and the Gaza Strip. "On this territory a Palestinian state might be established, as is now envisaged. This state could not accommodate all the Palestinians. This leads us to the second part of the Palestinian problem, namely the refugee problem. These refugees . . . have a right to return to the land from which they were driven in 1948."

The idea that Assad is a moderate, an idea enjoying currency in the

U.S. government, is part of a way of perceiving Israel, a way that reminds Podhoretz of autumn 1938 on the eve of Munich:

"As Czechoslovakia, a democratic country, was accused of mistreating the German minority in the Sudeten regions, so Israel, also a democratic country, is accused of mistreating the Arab minority within Israel itself and also, of course, in the occupied territories. As the creation of the Czechoslovak state after World War I was called a mistake by Hitler and Neville Chamberlain, so the creation of the Jewish state after World War II is called a crime by contemporary totalitarians and their appeasers. The insistence by the Czechs that surrendering the Sudeten regions to Hitler would leave Czechoslovakia hopelessly vulnerable to military assault was derided, especially on the Left, as a shortsighted reliance on the false security of territory and arms; so a similar insistence by the Israelis with regard to the occupied territories is treated today with lofty disdain by contemporary descendants of these believers in the irrelevance to a nation's security of territorial buffers and arms."

Made malleable by diplomatic pounding, Czechoslovakia, by spring 1939, had no shield except "the conscience of the West," and no deliverance.

[July 11, 1977]

Italian Baroque

ROME, ITALY—Seven centuries ago Dante, an alarmist, called Italy "a vessel without a pilot in a loud storm." But the "ship of state" is a metaphor, often pernicious, and especially inappropriate when applied to Italy.

Ships have unelected captains protected by the laws of mutiny from disobedient crews and passengers. Germans, who are vigorously obedient, can believe in a ship of state. Italians, who have an adversary relationship with their government, cannot.

Italy has had thirty-seven governments since fascism fell in 1943. But the fact that Italian governments are fragile does not mean Italian democracy is unstable, even though the Italians, ever operatic, call each government change a "crisis." Italy's central government is weak, but stable to a fault. Each "crisis" produces a Christian Democratic government run by men familiar for decades.

Democracy has had a hard time taking root in the three European peninsulas in the Mediterranean—the Iberian, Italian, Balkan. It has done best, and remarkably well, in Italy. Unlike Portugal, Spain and Greece, Italy is a relatively new nation. Unlike even the French, who were a people before they were a nation-state, Italy has become a nation-state before Italians have come to feel like a people.

In the nineteenth century Italy was stitched together from a lot of ancient principalities and city-states. Not surprisingly, many Italians think of themselves as Milanese, Romans, Neapolitans. Especially as regards tax collections, but in other matters, too, Italians tend to consider the national government in Rome a quasi-foreign entity, an occupying power.

Italy's greatest postwar leader, Alcide de Gasperi, was born near Trent, a subject of the Austrian emperor; was elected to parliament in Vienna at the time of the First World War; and spoke with what some considered a foreign accent. De Gasperi made, as perhaps only a partial foreigner could have done, the hard decisions that shaped postwar Italian politics. He expelled Communists from government, linked Italy with the Western alliance, and launched recovery.

Economic recovery—the "miracolo"—has enabled Italy to enjoy, or at least experience, thirty years of inelegant but uninterrupted democracy. It produced, with remarkable speed, the modernizing middle class that is a necessary, if not a sufficient, condition for democracy.

Since the war, six million Italians have moved from farms into cities. Two million more have left Italy to seek industrial jobs in northern Europe. As late as 1956 more Italians were employed in agriculture than industry. Other nations took fifty to 100 years to transform themselves with an industrial revolution. Italy telescoped that socially shattering experience into one generation. Spain has undergone a similar transformation, but used repression to cope with the social strains. Portugal and Greece have not been transformed.

Italy's economic problems have caused a worldwide migraine about Italy's impending "bankruptcy." But national "bankruptcy" is just another metaphor. No one comes in to repossess a nation's furniture. The strength of the Italian economy is in the 94 percent of Italian companies that have fewer than 100 employees, but which employ 60 percent of all industrial workers. These firms, unlike the large state-run or dominated firms, are relatively invulnerable to Italian government, and quick to respond to improved conditions.

Fifty years ago an Italian legislator had a bright idea for recapturing the greatness that was Rome's. Italy should sell her art treasures abroad, use the money to buy weapons, use the weapons to conquer the world and reclaim the art.

The idea was, like many of Italy's finest achievements, baroque. Unfortunately, baroque government has not been an ornament to modern Italy.

The French newspaper *Le Monde* has said, "Italy is the only country besides Tibet in which it is impossible to communicate through a postal service." That is an exaggeration. But many Christmas cards from the United States do arrive at Easter, and letters can take six months to get from one Italian city to another. Many Italians who live near Switzerland drive there to mail letters to Rome. When kidnappers cut off J. Paul Getty III's ear, they mailed it from Naples to Rome. It arrived twenty days later. Last year some enterprising mailmen sold 400,000 pounds of mail to a processing plant as scrap paper.

At least the constant strikes, like those that frequently shut Rome's airport until 2 P.M., run on schedule. And most political demonstrations allow participants to get home for lunch. This is the radicalism of the Austrian Communists who, storming City Hall in Vienna, obeyed the signs commanding them to stay off the grass.

For years people have been predicting that, any day now, Rome will suffer a terminal traffic jam. But still traffic whizzes through the piazzas with an unCatholic disregard for the preciousness of life. That is typical. Italy refuses to satisfy the non-Italian sense of order by surrendering to the disorder that would, indeed, be terminal to more orderly peoples. Italians, having experienced everything, are dismayed by nothing.

In 1943 A. L. Marion, who is my father-in-law and lives in West Hartford, Connecticut, visited Italy with the U.S. Army. On a ridge near Monte Casino he and his friends were firing their carbines in the direction of unseen German infantrymen who were on another ridge, firing their carbines in the direction of the unseen Americans.

While this drama of civilization was being played out on the ridges, Mr. Marion glanced down in the valley. There some peasant men were sitting under trees watching their wives hoe their gardens. They were not distracted by the arrival of World War II in their neighborhood.

The inhabitants of this peninsula have been more or less stoically enduring invasions and other inconveniences since visits by the Visigoths, Ostrogoths, Huns, Lombards, and Franks. If Americans had to endure for a day a fraction of the nuisances Italians routinely endure from their form of nongovernment, Americans would rebel—again.

But impatience is a function of youth. Americans are more patient today than they were when, just 200 years ago, they rebelled against a few featherweight nuisance taxes. Italians have passed beyond stoicism to wry indifference. Such wryness is a function of age.

In 1797, while negotiating the Treaty of Tolentino, Napoleon, another in an endless line of nuisances to Italy, asked a member of a great Roman family, Camillo Francesco Massimo, if it was true that he was descended from Fabius Maximus, opponent of Hannibal. Massimo, a proper Italian, knew an upstart when he saw one, and replied: "I could not prove it; the story has only been told in our family for twelve hundred years."

[June 4, 1975]

A People in Search of a State

ROME, ITALY—White tunic, white helmet, white gloves on mesmerizing Toscanini hands, the policeman at the center of the piazza directs traffic as though he is directing Verdi. He is the only graceful manifestation of the Italian state. But he is a fitting symbol of the state: he is irrelevant to the traffic that cascades heedlessly around him.

Since Mussolini, Italians have had governments galore, thirty-nine since fascism, but not a state capable of governing Italy's dynamism. And the growing strength of Italy's Communists is a product of Italian success, and failure.

At the end of the war, Italy was possessed by a primordial urge to transform itself from a predominantly rural Mediterranean society into an industrial society. Its economic "miracle" was fueled by the movement of large numbers of persons from low-productivity agricultural employment to high-productivity industrial employment. The astonishing dynamism of Italy's recovery from war and fascism has inflicted on the nation a vast population movement, from the land to the cities, from the stagnant south to the booming north.

Few things are more frightening to a peasant society, based on family solidarity, than the son who spurns life on the land for the promises of the city. Thus, in the Middle Ages, a father could lawfully break the leg of a son who proposed to leave the farm. Since 1945 the number of Italians living in cities has tripled.

The Italian Constitution begins with the firm declaration that Italy is "a democratic republic based on work." But the workers who poured into congested cities like Milan and Turin were ill-prepared for Italy's urbanization. And the ancient cities were not equipped to receive them.

As late as the beginning of this decade, more than one-sixth of the Italian work force was illiterate, and nearly two-thirds had not gone beyond elementary school. These people stepped from north-bound trains into cities that have an average of less than three square yards of green space per capita, compared with thirty square yards in London, twenty in Amsterdam, eighteen in New York.

Public services are things of shreds and patches. The judicial system requires a decade to resolve the average civil case. In a recent school year there was a 40 percent shortage of classrooms. Most hospitals and health insurance programs are deeply in debt.

Peter Nichols, a Briton by birth and a Roman by vocation, writes about Italy with the true friend's mixture of affection and dismay. He believes the condition of the state is both a cause and an effect of the role of the family as the strong atomic unit of Italian society: "Italy has not yet reached the point where it is safe to attempt to live one's life without the family close at hand. . . . The readiness of the family to help any of its members in trouble is one of the reasons why governments have done comparatively little governing and why there is less pressure than there should be on the politicians to introduce social reform."

After thirty years in power, the Christian Democratic party has no discernible idea but anti-communism, and no public purpose but the retention of power, which is not a public purpose at all. The reaction against the Christian Democrats is now almost revulsion, a reaction as much aesthetic as ideological. The party's primary efficiency is in dispensing ad hoc favors to compensate for the comprehensive failures of the government it ostensibly runs.

Luigi Barzini writes of the Italians' "absurd discrepancy between the quantity and dazzling array of the inhabitants' achievements through the centuries and the mediocre quality of their national history." One explanation for this is that Italy, although an ancient civilization, is a young nation. In slightly more than a century since unification it has not produced an Elizabeth I or a Bismarck, a powerful energizer of the state who left behind the ligaments of national authority.

Even Mussolini's fling at totalitarianism was (as a critic said) "a tyranny tempered by the complete disobedience of all laws." Whatever else the Communists offer, they offer something that, increasingly, Italians crave—a sense of a state.

[June 10, 1976]

Italy's "Ambivalent" Communists

ROME, ITALY—The possibility that next week's elections will give Communists a place in the government has loosed a torrent of talk about Italian Communists being "different." And the air is thick with sentimental nonsense about Italians being so democratic and undisciplined that they will frustrate any tyrant's designs. People are recalling Mussolini's wisecrack that governing Italy is possible but pointless.

But to those who say "it can't happen here" the answer is that it has happened here before. The first Italian experiment with liberal democracy collapsed more than a decade before the Weimar and Spanish republics collapsed.

True, there was in even the most sinister of modern Italian governments, Mussolini's, an aspect of opéra bouffe. When Hitler visited Rome in 1938 Mussolini ransacked his regiments for tall, blond blue-eyed Italians to line the parade route. And cardboard building facades, like film sets, were created on the assumption that Rome itself was not sufficiently grand to impress the frustrated architect from Berlin.

But Italian fascism is remembered with amused detachment only by persons who never lived under it. The fact that Italian fascism was "different," was less systematically vicious than another European movement of the same species, does not rehabilitate it in the judgment of history.

Today Italy's economy is desperately ill, and the Italian Communist party is the largest Communist party in the non-Communist world. A recent poll named the party's leader, Enrico Berlinguer, Italy's most popular and most trusted politician.

It is fair to note that the poll constituted faint praise from a nation whose politicians are a reminder that a disproportionate number of villains in Elizabethan dramas were Italians. And the depth of cynicism here can be gauged by the fact that 40 percent of Italians polled think a Communist government would not allow the electoral process to turn it out of power.

The Vatican is still resolutely anti-Communist. But, then, as recently as fifteen years ago confessional booths were adorned with lists of Communist and Communist-supporting organizations (including the largest trade union federation), and Catholics could be denied the sacrament until they repented any support of such organizations. Such opposition never halted the party's growth. And now the arrival of the party at the threshold of power has startled the Western world, which is like a homeowner who is surprised one morning by a glacier at his back porch. The glacier didn't come all the way overnight.

Today the question is not whether Communists will wreck the enterprise system. Relations between the Italian state and economy are so rococo that state control is hard to quantify. But through large holding companies, nationalized firms, and public credit sources the economy is thoroughly state-broken.

The real question is whether these "different" Communists respect fundamental freedoms. Local Communist administrations, as in Bologna, have been democratic. But that proves nothing. A local administration can hardly institute a "dictatorship of the proletariat" in the heart of a democratic nation. More revealing is the fact that when the party endorses a "democratic road to communism," it is endorsing a tactic it has no choice but to adopt. And it is speaking of democracy as a means to an end that has never been consistent with democracy.

Politicians define themselves at least as much through their voluntary gestures as through their response to necessities. Last year Berlinguer entertained representatives of East Germany's wall-building regime as his guests of honor at a Communist festival. And at last year's party conference, the representatives of Portugal's Stalinist Communist party received an ovation from the Italian Communists, who still proclaim "unbreakable ties of solidarity" with the Soviet Union.

The ties are indeed unbreakable, having survived the 1930s (the purge trials), 1948 (the coup in Prague and Moscow's order that Italian Communists oppose the Marshall Plan), 1953 (suppression of East Berlin), 1956 (Budapest), 1968 (Prague again).

The confusion of the West is apparent in mindless praise for Berlinguer as a man "ambivalent" about the Soviet Union and anxious to demonstrate "some independence" from that loathsome regime. His party still receives financial support from the Soviet Union, and he marches dutifully to Moscow for ritual minglings with the masters of the Gulag Archipelago and the other tyrants who rule in the name of communism. Italians should remember what the world reasonably and accurately assumed about Sir Oswald Mosley's values when, in the 1930s, the British fascist made similar pilgrimages to Berlin, and to Rome.

[June 13, 1976]

Castro's Revolution

Having subsidized the invasion of Fidel Castro's domain, and having planned with similar effect his assassination, the U.S. government may now be hatching another plot. It may dispatch Reggie Jackson and the other wage slaves of the New York Yankees to Havana to give the Communists a demoralizing demonstration of the superiority of capitalist motives.

Castro's durability next door to such hostility undermines the myth of American know-how, and suggests there is some advantage in being an enemy, rather than a client, of the United States. Remember, Castro arrived as a rookie in the big leagues of politics at approximately the same time as another hot prospect, Moise Tshombe (deceased), sovereign of Katanga (deceased).

But today there is in Washington an air of accumulating and irresistible conviction: it is time, sixteen years after the Bay of Pigs invasion, to let the dead past bury its dead, and to improve relations with Castro. That invasion was defended on the ground that Castro had "betrayed the revolution." This argument provoked columnist Murray Kempton to observe: "Ask no more why we do not give machine guns to people attempting to overthrow Trujillo; Trujillo never betrayed a revolution."

Castro is no less a dictator than was Trujillo. But Castro always has understood that Western society contains many leftists whose hatred of dictatorship is less constant than their hunger for political heroes. And he has mastered the symbols of heroism. The years have not dealt lightly with the hero's midriff, but at least it still is swathed in the green fatigues he wore in the heroic fight to replace the old dictator.

The fatigues are like Jimmy Carter's jeans and cardigans, a symbol of romantic origins. Wearing them conspicuously is the sort of gesture of humility that only the powerful can make; it is an assertion of the prerogative to disregard conventions. Castro's fatigues also are like the bib overalls favored by radical undergraduates who are sent from Scarsdale to expensive schools. They express reverence for the masses without expressing an aspiration to actually share their fate.

When Castro planted the red flag just 100 miles south of the pastel hotels along Miami Beach, he made Havana less colorful. To Graham Greene, who managed to make it the setting for a Cold War comedy *(Our Man in Havana)*, pre-Castro Havana was where "every vice was permissible and every trade possible." And the biggest vice, gambling, was potentially the biggest trade, even bigger than sugar.

But the famous casinos were early casualties of the revolution. In February 1960, in the fourteenth month of the Glorious Transition to Socialism, the last American gambler in Havana strode into the bar of the Riviera Hotel and exclaimed to some journalists who were lingering there to cover the revolution:

"Go to the casino. I just played a hand of blackjack. I had a jack and a seven; the dealer had three sixes. I looked at him and I said: 'Pay me; I got seventeen and you got fifteen.' He looked at the cards and he counted with his lips and he paid me. Sure, it'll be lonely here, but I'm living next to a casino where the dealer can't count."

Alas, soon the dealer couldn't deal because Castro closed the casinos, an act of historic folly which condemned Cuba to subsist on Communist agriculture. If détente with Cuba means restitution for American organizations whose economic interests were rudely jolted by the revolution, the Mafia will collect handsomely.

Casinos are where citizens of capitalistic societies rebel against the norms of capitalism: they abandon rational calculation and risk capital capriciously, on the roll of dice and the spin of wheels. If Castro were more clever, or more cynical, he would have kept the casinos and let Yankee tourists finance his regime. But Communist regimes generally are as Victorian as Karl Marx. Castro, like Marx and Lenin, is a child of middle-class parents. He probably believes, correctly, that gambling— the pursuit of unearned income—fosters frivolous attitudes toward work, accumulation and property.

The price in tyranny has been too high, but let us give the devil his due: Castro's revolution has prevented Havana from becoming a Las Vegas.

[March 10, 1977]

"American Century"

Halfway through the second half of what once was called "the American century" it begins to seem that centuries, too, do not last as long as they once did. But U.S. world preeminence came cheap, and began to pass, quickly, when the price of preeminence became apparent.

The United States became a great power through late participation in wars that cost other and earlier participants more dearly.

In World War I, the United States suffered 116,516 war-related deaths totaling one-tenth of 1 percent of its population. Great Britain lost one-thirtieth of 1 percent of its population between 7 A.M. and 7 P.M., July 1, 1916, at the Somme. There a four-month battle in the mud cost Britain 100,000 dead. In the four years of war Britain lost three-quarters of 1 percent of its population, France lost 5 percent.

In World War II, U.S. war-related deaths totaled one-quarter of 1 percent, and the United States suffered four civilian deaths from home-front bombing. (A Japanese balloon bomb launched from a submarine blew up an Oregon picnic.) The U.S.S.R. lost at least 8 percent of its population.

Korea severely strained the American public's tradition of deference toward the foreign policy elite, an identifiable, self-renewing group that served the executive branch, and that, not without reason, believed in itself and was believed in by the deferential public. The war undermined public support of President Truman, and made Secretary Dean Acheson, symbol of the traditional foreign policy elite, a subject of bitter controversy. The war was ended by a President whose single memorable election promise was "I will go to Korea," which meant: I will end the war.

An internationalist foreign policy has been possible only when Americans have subordinated their natural isolationism to their tradition of deference to the foreign policy elite. But Vietnam destroyed that tradition.

Vietnam was less a presidential war than a professors' war. It was too clever by half, with carefully calibrated violence—remember the "escalation ladder"?—sending "signals" to an uncomprehending enemy.

David Halberstam used the title of his best-seller about Vietnam—
The Best and the Brightest—to express the self-image of the professors,
soldier-scholars, systems analysts, and others who presided over the
escalation in Vietnam. More interesting than the Halberstam book is
the fact that the phrase "the best and the brightest" has entered the
nation's political lexicon as a piece of all-purpose sarcasm to express
disdain, not just for the "Kennedy-Johnson intellectuals" Halberstam
detests, but for elites in general. Today the very idea of an elite is
suspect.

I recently read a book on Art Deco in which the author denounced
"elitist furniture." I do not know what he meant, but I think that when
"elitist" becomes a trendy epithet, society has passed beyond disappointment with a particular elite, and is enraged at the general idea of
excellence.

In the United States, the general disdain for elites means the end
of deference toward an unnatural internationalism, and the rebirth of
the nation's natural isolationism. Human beings, in Tuscany or Tennessee, are "natural" isolationists. Sensibly, they do not want their treasure
or sons conscripted, and won't put up with it unless their government
is persuasive or coercive.

America is especially isolationist for historical and geographical
reasons. America tends to think it started fresh: it was immaculately
conceived, born without sin, and protected by God's oceans from unregenerate nations. This attitude—that America could dispense with the
world because the world had served its purpose, which was to be prologue to America—was good naturedly caricatured, but captured, too,
by a novelist from the American heartland:

"Main Street is the climax of civilization. That this Ford car might
stand in front of the Bon Ton Store, Hannibal invaded Rome and Eras-
mus wrote in Oxford cloisters. What Ole Jenson the grocer says to Ezra
Stowbody the banker is the new law for London, Prague, and the
unprofitable isles of the sea. . . ."

Sinclair Lewis published *Main Street* in 1920, the last sad year of
government by one of Princeton's "best and brightest," Professor Woodrow Wilson. That November the nation, weary of foreign entangle-
ments, elected Warren Harding, a choice accurately if unaesthetically
called a return to "normalcy."

[May 6, 1975]

Famous Victory

Last Sunday about 50,000 peace activists gathered for a last hurrah and hootenanny at the Sheep Meadow in Manhattan's Central Park. There, at the scene of so many peace rallies, they celebrated the peace that has come to Indochina.

It was like the good old days, with the folk singers and congressmen, and "the kids" who have kids now—kids clamoring for ice cream bars. Alas, one kid, little Peterkin, was not there to ask his question:

> "But what good came of it at last?"
> Quoth little Peterkin.
> "Why, that I cannot tell," said he;
> "But 'twas a famous victory."

'Twas a humdinger of a victory party at the Sheep Meadow but, according to the *New York Times,* it was tinged with melancholy:

"It was a joyous all-day carnival of songs and speeches in the perfect sunshine, hugging reunions of people who had last met at one demonstration or another. For some, there was an undercurrent of sadness, as if something more than the war—youth, perhaps—had ended too."

Ah, sweet bird of youth, flying away from the Sheep Meadow. But it was youth well spent, according to singer Peter Yarrow, who remembers singing during the moratorium in Washington in November 1969: "I remember the feeling then—that somehow by coming together we could make a life in which people would not kill or hurt each other any more."

Sing a little louder, Yarrow. Your message has not been received by the victors in Cambodia who are administering the peace you craved. In Cambodia today life is less than an all-day hugging carnival.

> With fire and sword the country 'round
> Was wasted far and wide
> And many a childing mother there
> And newborn baby died.
> But things like that, you know, must be
> At every famous victory.

In Cambodia the Communists, running true to form, are concentrating their fury on the ultimate enemy of any Communist regime, the people. The Communists have emptied the cities, driving upwards of four million people—young and old, childing mothers and newborn babies, the healthy, halt and lame—on a forced march to nowhere, deep into the countryside where food is scarce and shelter is scarcer still. Even hospitals have been emptied, operations interrupted at gunpoint, doctors and patients sent packing. The Communists call this the "purification" of Cambodia.

This forced march will leave a trail of corpses, and many more at its destination, wherever that is. But this is, according to the Communists, not an atrocity, it is a stern "necessity."

The Detroit *Free Press* contained a droll (I hope it was meant to be droll) sub-headline on events in Cambodia: "Reds Decree Rural Society." If one kind of society offends you, decree another. Communism, like its totalitarian sibling, fascism, is the culmination of a modern heresy: people are plastic, infinitely malleable under determined pounding. And society is a tinker toy, its shape being whatever the ruling class decrees.

To create a New (Soviet, Chinese, German, Cambodian) Man—and what totalitarian would aim lower?—you must shatter the old man, ripping him from the community that nourishes him. Send him on a forced march into a forbidding future. He may die. If he survives he will be deracinated, demoralized, pliant.

There is no atrocity so gross that American voices will not pipe up in defense of it. Today they say: it is "cultural arrogance" for Americans to call this forced march an atrocity, when it is just different people pursuing their "vision."

This is the mock cosmopolitanism of the morally obtuse. Such people say: only "ideologically blinkered" Americans mistake stern idealism for an atrocity just because it involves the slaughter of innocents. Such people will never face the fact that most atrocities, and all the large ones, from the Thirty Years War through Biafra, have been acts of idealism.

Of course, one must not discount sheer blood lust, and the joy of bullying. Totalitarian governments rest on dumb philosophy and are sustained by secret police. But they are a bully's delight.

Totalitarians have never been without apologists here, people who derive vicarious pleasure from watching—from a safe distance, of course; from the Sheep Meadow, with ice cream bars, if possible—other people ground up by stern "necessities." Apologists say that totalitarians only want totalitarianism for the sake of the revolution. The apologists, being backward, have got things backward.

[May 15, 1975]

Snubbing Solzhenitsyn

The U.S. government may have to expel Alexander Solzhenitsyn from the Republic, not only as a hands-across-the-barbed-wire gesture of solidarity with its détente partner, the Soviet government, but also to save the President and his attendants from nervous breakdowns.

This is not the first time Solzhenitsyn, winner of the Nobel Prize in literature, has taxed the nerves of the mighty. Last year Soviet Premier Leonid Brezhnev, having decided that he could not conveniently kill Solzhenitsyn and could not endure the sound of his voice, expelled him.

Solzhenitsyn became a nuisance to Gerald Ford when AFL-CIO President George Meany invited Solzhenitsyn to Washington to give a speech in which he reiterated his low opinion of détente, as the United States practices it. He believes this policy reduces the United States to craven, degrading reticence about slave labor, concentration camps, and other problems of human rights in the Soviet Union.

Solzhenitsyn is, of course, correct. The U.S. government thinks such reticence is "necessary" lest the Soviet government get angry and refuse to accept U.S. trade subsidies or engage in our memorable grain deals. But mere truthfulness does not redeem politically inconvenient speech, and Solzhenitsyn carries free speech to inconvenient conclusions.

His presence here posed a problem: should Ford meet with him? In coping with this problem the President contrived to confirm Solzhenitsyn's point while snubbing him for having made it.

Ford nervously diagnosed Solzhenitsyn's presence here as a foreign policy problem, and summoned advice from the National Security Council, which copes with such threats to the nation's security. He and aides brainstormed about how to justify snubbing the man who, outside U.S. and Soviet government circles, is recognized as one of the moral heroes of the twentieth century.

According to reports, several aides, showing a flair for baseness that would have stood them in good stead with the previous administration, questioned Solzhenitsyn's mental stability. The idea of American politicians rendering negative judgments about Solzhenitsyn's mental health has an antic charm, but such judgments were not publicly advanced to justify the snub, perhaps because they would not play in Peoria.

Other aides reportedly noted that during his visit to the United States, Solzhenitsyn is promoting the sale of his books. They said the President should not do anything that might even indirectly help a commercial promotion. The White House is selectively fastidious about such things.

A few days earlier Ford met with the Cotton Queen. A few days after that he summoned photographers to the White House lawn where he kicked a soccer ball with Brazilian star Pelé, for the benefit of the American entrepreneurs who are paying Pelé $4.5 million to help promote their soccer franchises.

Press Secretary Ron Nessen, keeper of the presidential image, explained that Ford could not see Solzhenitsyn because of a "crowded schedule." Nessen added: "For image reasons the President does like to have some substance in his meetings. It is not clear what he would gain by a meeting with Solzhenitsyn."

Nessen may have a point, but if so it reflects on Ford's ability to receive, rather than on Solzhenitsyn's ability to impart, wisdom. The President's image thus clarified, like butter, Nessen refrained from adding the salient point: Brezhnev frowns on Solzhenitsyn, but not on Pelé.

Obviously Ford decided that meeting Solzhenitsyn would be inconsistent with détente. Obviously Solzhenitsyn is correct: détente, as practiced by the United States, prevents even gestures of support for the cause of human rights in the Soviet Union. Certainly Solzhenitsyn was not surprised by Ford's snub. As he said in his Nobel Lecture:

"The spirit of Munich has by no means retreated into the past; it was not a brief episode. I even venture to say that the spirit of Munich is dominant in the twentieth century. The intimidated civilized world has found nothing to oppose the onslaught of a suddenly resurgent fang-baring barbarism, except concessions and smiles."

Détente has conferred upon Brezhnev veto power over the appointments calendar of the President of the United States. Perhaps Brezhnev, in the spirit of détente, would refrain from seeing people offensive to the U.S. government's moral sensibilities—if it had any.

[July 11, 1975]

Kissinger's Fatalism

When the professor came to Washington seven Januaries ago, he was just one planet in the solar system of presidential power. He became a sun, the brightest in the galaxy. But one senses that today he feels like a setting sun.

When he came to town he was a confident formalist. He believed, probably inordinately, in the importance of formal diplomacy. He believed—almost certainly more than he does today—in the efficacy of the formal arrangements wrought by "men of affairs." He believed, more than his last six years of experience encourage him to believe, in the ability of governing elites to channel and control the white-water river of events.

His early, undiluted formalism is expressed in his book about the diplomacy of Metternich and Castlereagh. The book is much better than its title: *A World Restored.* The title is, of course, preposterous. Diplomats, with their conferences and pieces of parchment, do not "restore" or even preserve "worlds." Certainly Metternich and Castlereagh, two of the best, did not.

While they were at the summit, restoring a "world" on parchment, something called the industrial revolution was picking up steam. Without even asking the diplomats for permission, it crumpled and blew away the diplomats' restored world, like so much parchment.

It is said that when the Russian ambassador died en route to a conference with Metternich, Metternich mused aloud: "I wonder why he did that?" He was being droll, but his remark reveals, while caricaturing, the formalist's mind: to understand events, understand the designs of statesmen.

Belatedly, and under the stern tutoring of events, reasonable formalists learn the extent to which politics, including diplomacy, is an epiphenomenon, an activity largely controlled by autonomous forces. Kissinger, the professor of statecraft, has probably learned slower than most formalists, because of his notorious disdain for rival arts, like economics.

Today there must be a jolting incongruity between that which Kissinger's intellect tells him is true, and that which, because of his vocation (statecraft), he must pretend is true. His daily experience must teach him how virtually impossible it is for governing elites to gain any purchase on the forces loose in the world. But he is paid to try.

His zest for the effort must be diminished by the fact that, increasingly, congressional restraints on his operating freedom, and press criticism, indicate that he has become the object of a vague and peculiar resentment in Washington. It is not correct to attribute this entirely to the fact (and it must be a fact; my hero, novelist Peter De Vries, says it is) that grudge is three-quarters of the burden of admiration. Rather, Kissinger's fall from grace represents nothing more profound than the fact, itself profoundly important, that Washington cannot admire anyone for more than a few years.

A few years is as long as it takes Washington to recognize that even the most tenacious, energetic and subtle men cannot alter the fact that today, as always, events are in the saddle, riding men and nations. Washington's most striking resoluteness is in refusing to face that fact. Washington is a town of formalists living by the belief that politics controls events.

The fact that the world has resisted Kissinger's increasingly frenetic attempts to order it has transformed Kissinger into a threat to Washington's sustaining myths. Kissinger, who of course has his faults, has become, through no fault of his own, a disconcerting paradigm of the limits of politics.

If his six years of service to his country have served to refute his own formalism, that speaks poorly of his original theory, not of him. If he is increasingly fatalistic, that is understandable: fatalism is the refuge of intelligent, disappointed formalists.

Anyway, fatalism has several virtues. It is broadly compatible with the basic facts of history. And it immunizes governing elites from delusions of grandeur, and the overreaching that such delusions stimulate.

It subtracts not a cubit from Kissinger's unique stature to note that, after six Kissinger years, the finest feature of U.S. foreign policy is something that is not happening. Kissinger knows what pleasure Sir Robert Walpole derived from telling Queen Caroline in 1734: "Madam, there are fifty thousand slain this year in Europe, and not one Englishman."

[January 11, 1975]

Kissinger: Strategic Pessimist, Tactical Optimist

We are sliding into one of those tedious Washington debates that begin wrong and go downhill from there. The debate is about Henry Kissinger. To what extent is the state of the world—from Lisbon to Saigon—his fault? It is another debate about a particular statesman, rather than the more troubling debate we should be having about the limits of democratic statecraft.

Kissinger's critics see him as the pilot in the following story:

A ship plying the coastal waters off Ireland picked up a pilot to guide the ship through the treacherously rocky waters. The ship's captain was appalled to learn that the pilot was drunk, but the pilot said: "Sir, I know every rock in these waters"—at which point there was the crash of hull hitting rock—"and, Glory be to God, there's one now."

Kissinger's critics tend to argue that if U.S. foreign policy is frustrated, some U.S. official must be to blame. This is unfair. But Kissinger is partly to blame for the unreasonable expectations that he has raised and cannot fulfill. His direct, personal involvement in the short-run tactics as well as the long-run strategy of foreign policy encourages people to think that he expects his statecraft to subdue events. For his own part, Kissinger feels like the sixteenth-century woman who was charged with witchcraft and was sentenced this way:

"The accused woman is to be thrown into the river—bound and gagged. If she sinks to the bottom and drowns, this will be proof of her innocence and she is to be given a proper burial; if she floats on the surface and breathes, this will be proof of her guilt, and she will be fetched immediately from the water and burned at the stake."

Kissinger has been criticized for an anti-institutional, over-personalized diplomatic style. And now that events beyond his control (beyond his congressionally diminished control) are unfolding unpleasantly, he is held personally responsible for them. In fact, Kissinger's problems today are a web of paradoxes.

Political forces have their own physics. Kissinger's vanity has provoked a matching force from those he considers his tormentors, the insurgents in Congress. What Napoleon said of the French Revolution is true of Congress' revolution against Kissinger's domination of foreign policymaking: "Vanity made the Revolution; liberty was only a pretext."

Kissinger does not have humility in the face of Congress because, increasingly, he has humility in the face of history. Congress believes that when its members say "aye" to (say) a "model cities" program, model cities should result. Kissinger lives day-by-day with an even more turbulent world than the one which frustrates Congress' will for "model cities."

Kissinger is a strategic pessimist and a tactical optimist. He knows that, strategically, time is not on the side of the bourgeois societies of the West. Totalitarian regimes, for all their stupidities, have one strength—staying power. Open consumer societies, devoted to the manufacture and gratification of appetites, have no appetite for the disciplines and deferred gratifications that protracted international competition entails.

But Kissinger, like a Confederate cavalry officer, believes that tactical daring in the short-run can partially compensate for the long-run weakness of a strategic position. This explains the fact that he is more ardent than discriminating in seeking agreements—pieces of paper.

The sobriety and pessimism of Kissinger's vision is, strictly speaking, un-American. It also is broadly correct: throughout history free societies have been short-lived rarities. Kissinger's view also is, literally, unspeakable. No official of a democratic government can express such skepticism about the long-run toughness and wisdom of his society.

The gathering strength of the totalitarian movements substantiates Kissinger's unspoken strategic pessimism, but seems to contradict the tactical optimism that is his only permitted public posture. This poses the ultimate paradox: the dangerousness of the world, from Lisbon to Saigon, may produce the sobriety and cohesion without which no democratic nation can have a purposeful foreign policy.

If you remember Lewis Carroll's poem "The Hunting of the Snark" you know that nervousness has its uses:

> But the valley grew narrow and narrower still,
> And the evening got darker and colder,
> Till (merely from nervousness, not from good will)
> They marched along shoulder to shoulder.

[April 17, 1975]

Kissinger's Dubious Monument: Détente

Stately, plump Henry Kissinger was lean and hungry in 1969. Since then he has prospered, eclipsing two Presidents. Never before has an appointed official given his name to an American era. Now the Kissinger Years are over, and it is time to sum up.

In 1969 the nation was weary of its postwar posture. Since then Kissinger has contrived to make retreat seem glamorous. His style, an elegant amorality, has reassured those who blame Vietnam on "moralistic" foreign policy.

Kissinger was fortunate that Richard Nixon was President. Like Kissinger, Nixon was instinctively secretive, and preferred to operate outside the leaky, creaky bureaucracy. Like Kissinger, Nixon believed in the supreme importance of personal relations in foreign relations. (Nixon to Elliot Richardson on the eve of the "Saturday Night Massacre": "Brezhnev would never understand it if I let Cox defy my instructions.") Like Kissinger, Nixon found economics tiresome. (Nixon: "I don't give a [expletive deleted] about the lira.")

Some say the U.S. approach to China ratified the end of the postwar era. Actually, it was as much Mao's approach to the United States. But in any case, historians who understand the sinews of national strength, and hence the determinants of international politics, are more apt to date the end of that era a month after Kissinger's secret flight to Peking. It was then that the dollar was floated, a concession to new economic realities.

The Nixon administration continued the Vietnam war through four divisive years because, it warned, the alternative was "unilateral withdrawal" at great cost to U.S. "honor." Then Kissinger negotiated, in effect, unilateral withdrawal. The "generation of peace" that began then has been punctuated by the destruction of South Vietnam, the most nearly successful attempt at the destruction of Israel, the achievement of Soviet war aims in Angola, the planting of Soviet bases in Somalia, the vigorous Soviet attempt to subvert Portugal, the Soviet naval buildup in the Mediterranean, and Soviet military production at what one Western expert calls a "near frantic rate . . . something akin to a war tempo."

It is unlikely that Middle East peace will result from those disengagement agreements Kissinger got by putting excruciating pressure on Israel, the victim and repeller of aggression. So détente must be Kissinger's monument, that by which he will be measured. And Soviet behavior is the measure of détente.

At first détente was said to be about "linkages." The Soviet Union would, like Gulliver, be restrained by tiny cords of political and economic agreements. But there was no chance that the Soviet Union soon would be linked to the West as tightly as, say, Germany was in 1939. And when "linkages" did not alter Soviet behavior, defenders of détente shifted ground. They said détente was a matter of the "basic principles" signed at the 1972 Moscow summit.

This "structure of peace" was built from platitudes like: the United States and U.S.S.R. disavow "efforts to obtain unilateral advantage at the expense of the other," and vow "to do everything in their power so that conflicts or situations will not arise which would serve to increase international tensions." Economically and politically, the "basic principles" have meant unreciprocated U.S. generosity and forbearance.

U.S. grain has rescued the Soviet Union from the dire consequences of agricultural problems that derive, in part, from policies of mass terror. The United States has acceded to Soviet demands for large, long-term subsidies. For their part, Soviet leaders exhorted the Arabs to continue their deadly oil embargo, which was the Yom Kippur war of aggression carried on by other means.

U.S. officials have expressed the hope that the Soviet Union will achieve a more "organic" relationship with the East European nations it tyrannizes. The Soviet Union has achieved the Helsinki agreement ratifying that tyranny, and has contemptuously disregarded the human-rights provisions that were supposed to redeem the agreement.

On June 25, 1973, Kissinger said, "It is safe to say that the Soviet Union and the U.S.A. agree on the evolution of the Middle East and how it should be resolved." Just 103 days later Soviet arms and incitements produced the fourth Arab-Israeli war. But Kissinger called Soviet behavior not irresponsible.

The latest rationalization for détente is that it is the only alternative to "a return to the cold war." But has Soviet behavior during détente been notably different from what it was during the cold war? Indeed, it has been in at least one particular. Only during détente has the Soviet Union dared anything as brazen as airlifting 12,000 troops from its Cuban client to combat in a proxy war 5,000 miles from Soviet territory.

Moving from one rationalization of détente to another, Kissinger resembles the Oxford don who delivered a deeply learned lecture on Plato and Saint Paul, and the next week announced: "Gentlemen, there is one detail you should remember when reviewing your notes of the last lecture. Whenever I said Plato I meant Saint Paul and whenever I said Saint Paul I meant Plato." When Kissinger first explained détente he said it would produce agreements. Now agreements are supposed to produce détente. Then détente was said to be possible because the Soviet Union had become a status quo power. Now détente is said to be necessary to moderate the Soviet Union's openly imperial phase.

This much *can* be said for détente. It certainly is not an excessively moralistic foreign policy. Kissinger and Brezhnev have denounced Congress for interfering in the "domestic affairs" of the Soviet Union. Their complaint is that Congress has made U.S. subsidies contingent on liberalization of Soviet emigration policies. To be sure, Congress has been seeking something virtually unprecedented, Soviet compliance with several international conventions it has ratified. But let the record show that Congress has rejected the détente doctrine that U.S. "interference" with Soviet tyranny must not go beyond the subsidizing of it.

In the name of détente the State Department urged President Ford to shun Solzhenitsyn lest Brezhnev be offended. Not even Watergate was as *fundamentally* degrading to the presidency as this act of deference to the master of the Gulag Archipelago. Leopold Labedz, editor of the invaluable journal *Survey,* is right to warn that détente has "contributed to a dangerous widening of the discrepancy between the American liberatarian tradition and American foreign policy." And unless the next administration does better, the story of U.S.-Soviet relations will continue to be, as Churchill said of appeasement in his day, the story of how the malice of the wicked was reinforced by the weakness of the virtuous.

[December 13, 1976]

PART 7

Personal

Half Promise, Half Threat

"Life," according to a fictional character who had just been run over by it, "is half promise and half threat. It is like Walter Cronkite giving us fair notice that he will be back with more news in a moment." Colonel Stoopnagel was half right to say "people have more fun than anybody," but if you are just seven pounds big and are blinking up at the lights in the Georgetown Hospital delivery room, life is no laughing matter.

Indeed, the meritocracy strikes early. Geoffrey Marion Will was about five seconds old when he took his first test. He earned a smashing ten—the highest possible—Apgar score, by which obstetricians grade a newborn infant on such vital signs as color, heartbeat, muscle tone. Pardon his "pinker-than-thou" attitude.

All men are born unto trouble, as the sparks fly upward. Geoffrey also was born into a desultory discussion of the energy crisis. It is 4:45 A.M., Geoffrey is in his second quarter-of-an-hour as an American citizen, his mother is a trifle tuckered, and both are listening to the obstetrician and the father muse about energy problems while the obstetrician repairs the limited damage Geoffrey's arrival did to Geoffrey's mother. Obviously having babies isn't what it used to be. For men, that is.

I grew up watching movies in which Randolph Scott waited in the next room of the cabin while plump midwives from neighboring ranches summoned pans of boiling water. And addled fathers pacing in waiting rooms were stock comic figures. But now, increasingly, fathers are going into delivery rooms where, with a good obstetrician presiding, a calm conducive to conversation prevails.

The experience should be part of every father's education. Fathers should understand the heroism of women giving birth, which is no less heroism because it is a natural process. Moreover, the sight of an infant at his or her most vulnerable moment is good for the soul.

I know (because Mrs. Will says so) that I have the temperament of a snapping turtle. But there is something about an infant that mellows a misanthrope, at least a little bit, briefly. And it causes this one to brood about the many of life's contingencies that can cause an agreeable infant, who gurgles and snoozes and spits up like a sensible infant, to become something disagreeable, like an adult.

Actually, there is little use worrying about it. The more we know about human beings the more we realize that we do not know much about them. Genes, not environmental factors, are in control. Hence, human beings are not as malleable or, fortunately, as manipulable as they were thought to be when people took seriously the more audacious pretensions of the behavioral "sciences."

So I will follow my father's example of harboring minimal hopes. He once said he didn't care what I did with my life, but he hoped I would never disgrace him by becoming a college cheerleader. (He never thought to fear having a columnist in the family.) I will not hope that Geoffrey attains the highest position our society can bestow (shortstop for the Chicago Cubs). I will hope to be as good a parent as Scott-King (one of Evelyn Waugh's few endearing characters) was a tutor. The headmaster of the school where Scott-King taught said to him:

"Parents are not interested in producing the 'complete man' any more. They want to qualify their boys for jobs in the modern world. You can hardly blame them, can you?"

"Oh yes," Scott-King replies, "I can and do." He added: "I think it would be very wicked indeed to do anything to fit a boy for the modern world."

Geoffrey was not born in the best of times, but he was born in the best of places, the United States. No threat in the modern world outweighs the promise of that blessing.

[January 10, 1974]

Gorilla in the Kitchen

The adult consensus is that two-year-olds are among the thorns and thistles of life because they lack the more subtle social graces. But exposure to two-year-olds teaches much about Life, and Our Troubled Times.

I work and eat lunch at home, often with Geoffrey Will, two, whose luncheon tastes run to onion soup and chocolate donuts. On a recent day, the intellectual conversation was waxing when suddenly Geoffrey stared across the kitchen like Cortez first seeing the Pacific. "Gorilla!" he shouted.

Such a shout might cause you, gentle reader, to drop your chocolate donut. But I already knew that the Will house, every nook and cranny, is infested with imaginary beasties—mostly porcupines and gorillas, an occasional yak, and a profusion of "monsters."

Usually Geoffrey's sangfroid makes the world seem agreeable, even when the north wind blows. But periodically he gives rein to his powerful imagination, and his blood turns to flame. Youth is resilient, and he quickly recovers from these storms of emotion. But such storms, when gorillas sometimes appear in the kitchen, are one facet of the "terrible twos"—an age when there is much turbulence in tiny vessels.

The monster a child sees under the sofa, like the monster he wants to hear about in scary fairy tales, often is (in the words of Bruno Bettelheim) "the monster a child knows best and is most concerned with: the monster he feels or fears himself to be." Bettelheim defends and recommends fantastic and scary fairy tales, such as Grimms', in which evil is as omnipresent as virtue. Fairy tales, such as those from the brothers Grimm, need to be defended in "progressive" opinion.

"Fairy tales," Bettelheim notes, "underwent severe criticism when the new discoveries of psychoanalysis and child psychology revealed just how violent, anxious, destructive, and even sadistic, a child's imagination is." The tales were blamed as an external cause of the inner turmoils of children. Nothing dies harder than the myth that society is the source of all human imperfections and anxieties.

"There is," Bettleheim says, "a widespread disinclination to let children know that the source of much that goes wrong in life is due to our own natures—the propensity of all men to act aggressively, asocially, selfishly, out of anger and anxiety. Instead, we want our children to believe that all men are inherently good." Only the theoretically inclined can believe anything so contrary to intuition and experience. Children know better.

Once—only once—I advised Geoffrey that there was no porcupine in the pantry. Geoffrey pulled himself up to his full height (thirty-two inches), fixed me with the stony stare of an earl about to address an earthworm, and then just shrugged in weary disgust. He will not tutor anyone who is too dim to realize that the monsters children see at the head of the stairs and the foot of the bed often are projections of the passions children are afraid they cannot control.

Bettelheim says frightening fairy tales, full of evil and strife and fear and trembling, are good for the half-formed soul. They enable children to ponder their real fears about a really difficult world. Bettelheim thinks that people raised on goo like *The Little Engine That Could* may believe that all difficulties really will yield to anyone who chants "I think I can, I think I can, I think I can. . . ."

That book is a child's version of *The Power of Positive Thinking.* It does nothing to help a child clarify his emotions as experience begins to teach him that earnestness and good intentions are unavailing against many of life's difficulties. But the story has been an American favorite for years. No wonder there are so many Democrats.

I have been slow to understand that the contrariness of the "terrible twos" is the bloody-mindedness of little people trying to get a grip on their partially formed selves. I used to think that a two-year-old's father needs only what a Washington columnist needs, the ability to look perfectly grave no matter what nonsense is being spoken to him. But I no longer think that what two-year-olds say is nonsense. When Geoffrey shouts "Gorilla!" I know that, in a sense, there really is a gorilla in the room.

[October 17, 1976]

The Chicago Cubs, Overdue

A reader demands to know how I contracted the infectious conservatism for which he plans to horsewhip me. So if you have tears, gentle reader, prepare to shed them now as I reveal how my gloomy temperament received its conservative warp from early and prolonged exposure to the Chicago Cubs.

The differences between conservatives and liberals are as much a matter of temperament as ideas. Liberals are temperamentally inclined to see the world as a harmonious carnival of sweetness and light, where goodwill prevails, good intentions are rewarded, the race is to the swift, and a benevolent Nature arranges a favorable balance of pleasure over pain. Conservatives (and Cub fans) know better.

Conservatives know the world is a dark and forbidding place where most new knowledge is false, most improvements are for the worse, the battle is not to the strong, nor riches to men of understanding, and an unscrupulous Providence consigns innocents to suffering. I learned this early.

Out in central Illinois, where men are men and I am native, in

1948, at age seven, I made a mad, fateful blunder. I fell ankle over elbows in love with the Cubs. Barely advanced beyond the bib-and-cradle stage, I plighted my troth to a baseball team destined to dash the cup of life's joy from my lips.

Spring, earth's renewal, a season of hope for the rest of mankind, became for me an experience comparable to being slapped around the mouth with a damp carp. Summer was like being bashed across the bridge of the nose with a crowbar—ninety times. My youth was like a long rainy Monday in Bayonne, New Jersey.

Each year the Cubs charged onto the field to challenge anew the theory that there are limits to the changes one can ring on pure incompetence. By mid-April, when other kids' teams were girding for Homeric battles at the top of the league, my heroes had wilted like salted slugs and begun their gadarene descent to the bottom. By September they had set a mark for ineptness at which others—but not next year's Cubs—would shoot in vain.

Every litter must have its runt, but my Cubs were almost all runts. Topps baseball bubblegum cards always struggled to say something nice about each player. All they could say about the Cubs' infielder Eddie Miksis was that in 1951 he was tenth in the league in stolen bases, with eleven.

Like the boy who stood on the burning deck whence all but he had fled, I was loyal. And the downward trajectory of my life was set. An eight-year-old could not face these fires without being singed, unless he had the crust of an armadillo, and how many eight-year-olds do?

Of the sixteen teams that existed in 1949, all have since won league championships—all but the Cubs. And which of the old National League teams was first to finish in tenth place behind even the expansion teams? Don't ask. Since 1948 the Cubs have played more than 6,000 hours of losing baseball. My cruel addiction continued. In 1964 I chose to do three years of graduate study at Princeton because Princeton is midway between Philadelphia and New York—two National League cities. All I remember about my wedding day in 1967 is that the Cubs dropped a doubleheader.

Only a team named after baby bears would have a shortstop named Smalley—a righthanded hitter, if that is the word for a man who in his best year (1953) hit .249. From Roy Smalley I learned the truth about the word "overdue." A portrait of this columnist as a tad would show him with an ear pressed against a radio, listening to an announcer say: "The Cubs have the bases loaded. If Smalley gets on, the tying run will be on deck. And Smalley is overdue for a hit."

It was the most consoling word in the language, "overdue." It meant: in the long run, everything is going to be all right. No one is really a .222 hitter. We are all good hitters, all winners. It is just that some of us are, well, "overdue" for a hit, or whatever.

Unfortunately, my father is a righthanded logician who knows more than it is nice to know about the theory of probability. With a lot of help from Smalley, he convinced me that Smalley was not "overdue." Stan Musial batting .249 was overdue for a hot streak. Smalley batting .249 was doing his best.

Smalley retired after eleven seasons with a lifetime average of .227. He was still overdue.

Now once again my trained senses tell me: spring is near. For most of the world hope, given up for dead, stirs in its winding linen. But I, like Figaro, laugh that I may not weep. Baseball season approaches. The weeds are about to reclaim the trellis of my life. For most fans, the saddest words of tongue or pen are: "Wait 'til next year." For us Cub fans, the saddest words are: "This is next year."

The heart has its reasons that the mind cannot refute, so I say:

Do not go gently into this season, Cub fans; rage, rage against the blasting of our hopes. Had I but world enough, and time, this slowness, Cubs, would be no crime. But I am almost halfway through my allotted three-score-and-ten and you, sirs, are overdue.

[March 20, 1974]

Cubs on the Banister of Eternity

The letter was suitably terse: "We regret to inform you that you have been nominated for membership in the Emil Verban Memorial Society."

Verban was the Chicago Cubs' second baseman, 1948–1950. Then he was sent to the Boston Braves and retired after four games to rest on his laurels. His most notable achievement was a remarkable ratio of home runs (one) to times at bat (2,911). Today he is a patron saint of Cub fans because he symbolizes mediocrity under pressure.

The sight of the first crocus, the song of spring's first robin, are harbingers of the lighter and brighter side of life. But for Cub fans, they are omens of hideous clarity—signs that the happy stagnation of winter is over and the season of suffering is beginning.

Spring is the winter of the Cub fan's soul. By June his lifeless eyes will resemble oysters on half shells; his complexion will be the color of Cream of Wheat. He must have Spartan stoicism and the skin of a rhino: the Cubs have lost 2,654 games since they last won a pennant, in 1945, when all but the halt and lame were fighting fascism.

Undaunted, I recently resolved to buy one share of Cub stock so that on my income tax form I could list my occupation as "baseball owner." While visiting Omaha, I confided this desire to Warren Buffett, who is a St. Louis Cardinal fan but not otherwise sinister. In a blaze of bonhomie uncharacteristic of Cardinal fans, he wrote to a friend in Des Moines, the estimable Joseph Rosenfield:

"Dear Joe: George Will, an otherwise quite competent young man, shares your irrationality regarding the Cubs. I mentioned to George that in this case it is possible to put your money where your aberrations are, and that it would be possible for him to buy stock in the club. Considering the fact that you and Phil Wrigley are the two largest shareholders, I don't see how he can help but upgrade the present membership. Why don't you write George directly and explain to him precisely how to get aboard this gravy train."

Cardinal fans are by nature sarcastic. Cub fans are of sweet disposition. Rosenfield wrote to me offering help and counsel:

"If I were you, I would not consult Warren about the stock as he will give you a long, learned treatise on price-earnings ratios, return on capital, and a bunch of other hogwash which has no place in a transaction between two true sportsmen."

Well put, don't you think? And the sainted Rosenfield continued:

"Before buying the stock, I would like to give you a word of caution based on personal experience. Some twenty-five years ago I was in a broker's office in Chicago and the trader, who knew of my interest in the Cubs, told me ten shares were available, which I bought as a kind of lark. However, I soon discovered that this was like taking your first shot of heroin and I could not wait to increase my holdings. Every time stock appeared. . . . I bought it and I kept pounding the brokers to find more.

"At the end of two years I had accumulated 274 shares and was otherwise bankrupt. After two unsuccessful holdup attempts in order to get additional buying power, I managed to kick the habit through sheer iron will and the help of three nationally known psychiatrists.

"However, the Cubs have given me something to live for. Some time ago Harry Reasoner was telling on the news about an old friend of his who had set as his goal to live until the Washington subway was completed. Being of a somewhat advanced age myself, I have determined to live until the Cubs win the National League pennant."

In the deepest sediment of my soul, I know that the Cubs have been good for me, too. They have taught me the first rule of reasonable living: discern the unalterable and submit to it without tears. The Cubs, although deeply painful to contemplate, symbolize Man's Fate. All of us are, like the Cubs, feathers on the breeze of Life, lonely sliders down the banister of Eternity.

[March 13, 1977]

"Have You Ever Known a Yankee Fan with Real Character?"

At 7:17 A.M. (the moment is forever fixed in my memory) the drowsy stillness of the Will house was broken only by the voice of Ray Gandolf, the reporter of sports for the CBS morning news, giving baseball scores: ". . . the mighty Cubs beat . . ."

Mighty Chicago Cubs? I have waited decades to hear a sportscaster say something like that.

Unpleasant people say the Cubs are like Hilaire Belloc's water beetle:

> He flabbergasts the Human Race
> By gliding on the water's face
> With ease, celerity and grace;
> *But if he ever stopped to think*
> *Of how he did it, he would sink.*

Stuff and nonsense. The Cubs' sudden ascent to greatness is the work of Providence. In the fullness of time it has come to pass, as the prophets prophesied: the meek, who have eaten the bread of affliction, are inheriting the earth.

The romantic saga of the rampant Cubs is a nice counterpoint to unromantic baseball news, such as the trade of Tom Seaver from the New York Mets to the Cincinnati Reds. The Seaver trade, and the restless mobility of "free agent" superstars, strains fan loyalty. Baseball is a business but such unsentimental capitalism is bad business. Baseball capitalism that respects only market forces is profoundly destructive because it dissolves the glue of sentiment that binds fans to teams. Besides, as Jacques Barzun says, baseball is Greek because it is based on "rivalries of city-states." Athens would not have traded Pericles for Sparta's whole infield.

Scholars concede but cannot explain the amazing chemistry of Cub fans' loyalty. But their unique steadfastness through thin and thin has something to do with the team's Franciscan simplicity.

The Cubs play on real grass, under real sunlight. Their scoreboard does not explode and they do not wear gaudy uniforms like those that have the Pittsburgh Pirates looking like the softball team from Ralph's Bar and Grill.

Iron has entered into the soul of this generation of Cub fans. World War II made the National League safe for the Cubs (they won in 1945, conscription having taken the able-bodied opposition), but since then, rooting for the Cubs has been the moral equivalent of war: hell. I became a Cub fan in Champaign, Illinois, in my seventh year, 1948, the year the Cubs management ran newspaper ads apologizing for the team. Thereafter, my youth was spent devising theories to take the sting out of summer.

One theory was that a .217 hitter was not a .217 hitter, he was a .295 hitter who really was, as broadcasters say, "overdue" for a hit. Never mind that most .217 hitters retired "overdue." Another theory was that each team would score only a certain number of runs each season, so when the Cubs lost twenty-two to nothing the winner squandered twenty-one of its allotted runs.

Baseball always was a sober experience for me. My friends played on Little League teams like the Piggly Wiggly Pirates and their colors were peppy red or green or blue. We on the Mittendorf Funeral Home Panthers wore black. Many children who had been trusted playmates revealed shocking flimsiness of character: they sank to rooting for the St. Louis Cardinals. I regarded these opportunists with lofty disdain, as De Gaulle regarded Vichyites. But now I know they were more to be pitied than censured, because rooting for a successful team is ruinous.

Have you ever known a Yankee fan with real character? Twenty years ago rooting for the Yankees was like rooting for IBM. It was for icky children who liked violin lessons and dreamed of being secretary of the treasury. What do Yankee fans know of the short and simple annals of the poor? Of lives of quiet desperation?

The most that can be said for most team loyalties is that they are poor preparation for here or the Hereafter. But rooting for the old, unregenerate Cubs was a complete moral education.

From 1946 through 1966 they finished seventh seven times and eighth six times. In 1962, the first year it was possible to finish ninth, the Cubs did, and the Mets had to extend themselves to a record 120 losses to wrest tenth from the Cubs. In 1966 the Cubs became the first non-expansion team to finish tenth.

In those days, Cardinal fans reading *Who's Who in Baseball* found glowing descriptions of Stan Musial and Marty Marion. Cub fans read about their Lenny Merullo: "He is always on the verge of being ousted from his job because of his frequent erraticness—but he probably will be around this season." Bill Nicholson: "He has been in the clutches of a prolonged batting slump for three seasons." Paul Minner: "His wins were meager, but his stamina tremendous." Dutch McCall: He suffered from "disheartening support—and lack of endurance." Roy Smalley: "His errors at short are many, but he keeps trying." And Ralph Hamner: "The tall stringy hurler abounds in tough luck."

The Cub teams of this era taught their fans an invaluable lesson about the inevitable triumph of ineptitude over sincerity. From these Cubs we learned life's bitterest truth: the race *is* to the swift. Cub fans have seen a relief pitcher stride to the mound, promptly injure himself by falling off the mound and leave without having thrown a pitch. Today, a quarter of a century later, I am an embattled parent, convinced that babies are born plump as peaches because they are packed full of dubious ideas. It is never too soon for them to learn that the world is not their oyster, and *nothing* teaches that lesson quicker than loyalty to a train wreck like the 1948 Cubs.

Still, one summer of happiness can't do irreparable harm to me or the two rising Cub fans at the Will house. Yes, the paths of glory lead but to the grave, but so do all other paths. Yes, we bring nothing into the world and can take nothing out, but Ray Gandolf's sweet reference to "the mighty Cubs" is one thing no one can take away from us.

[June 27, 1977]

Walking

Consider the pleasures and virtues of walking. It is the wine of life, good for body and soul, as many notable souls have known.

Henry Adams described his boyhood home at Quincy as "but two hours walk from Beacon Hill." Young Sam Johnson frequently walked the muddy road between Lichfield and Birmingham and back, thirty-two miles in a day. Immanuel Kant's daily walk through Königsberg was a categorical imperative, and so regular the local burghers set their clocks by him. Wordsworth's perambulations through the Lake District made literary history. Lord Macaulay plowed through London crowds with his nose in a book.

Jonathan Swift frequently walked thirty miles, recording his thoughts and observations in what we know as the *Journal to Stella*. Bertrand Russell would walk forty miles, rendering himself too weary to talk, an agreeable result.

Charles Dickens was a promiscuous walker through London by day and through the night "to still his beating mind." If you had ten children conceived in ten years you, too, would understand the charm of solitary walks even through streets that were neither safe nor sanitary.

Dickens' novels are a pedestrian's novels, capturing the texture of life seen at a walker's pace. Today a person's sense of the social material around him fades a few steps from the front steps, where he enters a vehicle.

I think these writers were walkers because they could compose while walking. The typewriter has, I suspect, altered habits of composition. Today's writers need to see their words materialize immediately and clearly before them. But Gibbon composed whole paragraphs—and what paragraphs—before writing a word.

Walking is the most civilized and civilizing exercise because it is the one most conducive to thinking. Only walkers can take in the country or city at appropriate speed, immersed in particularities.

Few are blessed with the piercing vision of William Blake, who could

... see a World in a Grain of Sand,
And a Heaven in a Wild Flower

But from a speeding car we cannot see even the grains or flowers, or gaze "down into a little ditch beneath a gray hedge," as C. S. Lewis could do when walking, and have a "sense of the mysteries at our feet where homeliness and magic embrace one another."

"I consider it among my blessings," Lewis said, "that my father had no car. . . . The deadly power of rushing about wherever I pleased had not been given to me. . . . I had not been allowed to deflower the very idea of distance."

For years Lewis, scholar and author, made daily walks, and an annual walking tour through rural reaches of Britain and Ireland, "to take in what is there and give no thought to what might have been there or what is somewhere else." His writings about his walks are lyrical:

". . . we struck inland again over the moor in one of those golden evening lights that pours a dreamlike *mildness* over the world. Light seemed to be liquid that you could drink . . . We had done well over twenty miles and felt immortal."

Now that people routinely cross continents between breakfast and lunch it is all the more exhilarating to cross a small valley afoot between sunrise and sunset. Lewis, as was his wont, not only thought but thought theologically in his exhilaration:

"But for our body one whole realm of God's glory—all that we receive through the senses—would go unpraised. For the beasts can't appreciate it and the angels are, I suppose, pure intelligence. They understand colours and tastes better than our greatest scientists; but have they retinas or palates? I fancy the 'beauties of nature' are a secret God has shared with us alone. That may be one of the reasons why we were made—and why the resurrection of the body is an important doctrine."

That, of course, is a matter of opinion, but it is a walker's opinion. Other exercises do not provoke or even permit interesting thoughts. Imagine what the world would have lost if Kant had been a jogger, or Dickens had taken up tennis.

[August 13, 1977]

Living with a Crick in the Neck

On a hill overlooking the bay, under the sort of slightly overcast sky that people here consider inclemency, San Diego University recently conferred upon me the commencement speaker's privilege of dispensing gratuitous advice to a captive audience. I was tempted to lecture sternly about shoes.

Academic gowns focus attention on the shoes that protrude beneath, and I tremble for my country when I see sneakers and semi-platform shoes worn to commencements by the rising generation. Fortunately, the other end of the student matters more. And this small university is properly proportioned to function, so students here have their minds well tended.

Having turned thirty-six, I am entitled to lecture the young: I am more than halfway through my allotted three-score-and-ten years and am on life's downward slope. It is traditional to pour upon graduates syrupy talk about the future belonging to them. But it is better to be able to offer congratulations because the past "belongs" to them.

A teacher of mine was fond of saying to me, "I want to make philosophy genius-proof, not fool-proof." It was painfully obvious to both of us that I would be no obstacle to the genius-proofing of anything. But his point was this: the world has suffered much from the bright ideas of clever people who are so uninformed about the past that they do not know that they are addressing old, wrong questions in old, wrong ways.

Understandably, undergraduates have employment on their minds. But they should be protected from the acquisition of merely "useful" knowledge. Commencement would be a melancholy ceremony if those graduating had devoted four years to looking ahead nervously to the next four decades of necessitous employment in the workaday world. Rather, education should be primarily an innoculation against the disease of our time, which is disdain for times past.

Emerson lamented that college education often was a ship "made of rotten timber, of rotten, honeycombed, traditional timber without so much as an inch of new plank in the hull." But education should produce people laden with traditional timber, not brimming with new ideas. After all, most new knowledge is false.

As Princeton's president, Woodrow Wilson wrote often and well about the university as society's "seat of vital memory," an "organ of recollection" for the transmission of the best traditions. He regarded education as a conserving enterprise, a way of making young people artificially "old" by steeping them in seasoned ideas.

"We seek to set them securely forward at the point at which the mind of the race has definitely arrived, and save them the trouble of attempting the journey over again," Wilson said. "We are in danger of losing our identity and becoming infantile in every generation. . . . We stand dismayed to find ourselves growing no older, always as young as the information of our most numerous voters. . . . The past is discredited among them, because they played no part in choosing it."

Some people think that such reverence for tradition violates the spirit of democracy, which they think involves nothing but deference to opinions of the most numerous voters. G. K. Chesterton knew better:

". . . tradition is only democracy extended through time. . . . Tradition may be defined as an extension of the franchise. Tradition means giving votes to the most obscure of all classes, our ancestors. It is the democracy of the dead. Tradition refuses to submit to the small and arrogant oligarchy of those who merely happen to be walking about. All democrats object to men being disqualified by the accident of birth; tradition objects to their being disqualified by the accident of death. Democracy tells us not to neglect a good man's opinion, even if he is our father."

The Class of '77 should live with a crick in its neck, figuratively speaking, a crick in its neck from looking backward. That may not be a heroic posture, but it is prudent. And it is the duty of conservatives like me to affirm homey virtues like prudence at movable feasts like commencements. So wherever they are heading, I wish graduates Godspeed, or such speed as is prudent while looking backward.

[May 26, 1977]

Frederick L. Will

Frederick L. Will, professor of philosophy and scourge of the elusive small-mouth bass, is ending thirty-nine years of service to the University of Illinois. His retirement is a loss to the unformed minds of

Illinois youth, a menace to bass throughout the upper Midwest, and an occasion for testimonials, including mine.

Fred grew up in many small Pennsylvania and Maryland towns where his father, a minister, looked after Lutheran souls. The Reverend Will struggled to reconcile the Lutheran doctrines of grace and free will, a struggle that stimulated his son's inclination to philosophy rather than piety. Fred has indulged that inclination at Illinois since 1938, with working sojourns at Cornell and Oxford.

The University of Illinois opened with three professors (not quite enough) and fifty students (about right) in 1868, six years after Congress passed the Land Grant College Act. One of the most fruitful laws in U.S. history, it endowed colleges in every state. It is primarily responsible for the broad dispersal of the nation's cultural resources, and for democratizing access to higher education. Seventy-six years after that act, Fred came here to introduce young Illinoisans to Immanuel Kant.

I was generally oblivious of, and impervious to, the presence of a gifted tutor at breakfast, lunch and dinner in the Will house. Being now a father myself, confronted at meals by two unruly male descendants, I well understand Fred's decision to cast his pearls before pupils more promising than I. He contented himself with the hope that I might absorb by osmosis something of his learning.

But there is no moral power like that of quiet example, and none more vivid to me than my father's. He had labored for nearly a decade completing a manuscript which skillfully reached conclusions broadly supportive of a longstanding consensus among philosophers about his special interest, the problem of induction.

Then one day while standing at a blackboard, there suddenly came to his mind an episode from a Thackeray novel which, when he later reflected about it, suggested that he and the conventional philosophic wisdom in the two centuries since David Hume were decisively mistaken about induction. So he set aside the manuscript that was the fruit of his career until then, and began again.

Leaving aside his exemplary conduct, I remember my father making only one overt attempt to instill sound principles in me.

In the mid-1950s, the Ford Motor Company manufactured a model called the "Crown Victoria." It came in color combinations best ignored (the chic combination was pink and charcoal). It had embryonic tail fins, and its crowning glory was a broad swath of chrome across the roof.

Fred, like all sensitive Americans, winced when he first saw one and murmured, with rare sarcasm, that it should be called a "Mozart Wagon."

"Why?" I inquired, thirsty for knowledge. "Because," he explained dryly, paraphrasing Shaw, "Mozart demonstrated that one could be a genius without being ostentatious." I admired the homily, but was dismayed that Fred might still hope that his son's potential justified a warning against allowing genius to become ostentatious.

Students of modern thought know that to have had an unhappy childhood is a prerequisite for subsequent intellectual seriousness and (what is proof of seriousness) angry rebellion against the awfulness of one's parents. So I have ransacked my memory for childhood traumas, and searched my soul for scar tissue.

All I recall is a mild dread that society might learn that Fred cuts shredded wheat biscuits into quarters, the stuff of four breakfasts, or that he might tell his favorite joke in public. It is the one about a man at a banquet who was asked why he was rubbing mayonnaise into his hair. He exclaimed: "Good God! I thought it was spinach."

The children of the happy union between Fred and Louise Will have gone astray. The daughter married an anthropologist and, worse yet, the son sank to journalism. But although the modern assumption is otherwise, I insist that the shortcomings of the son are not the father's fault.

This son has traveled far and lived twice eighteen years, but he has never dined with a more learned man, or better companion, than the one with whom he dined during his first eighteen years.

[April 3, 1977]

Index

Index